T0342310

TOPOLOGICAL INSULATORS AND TOPOLOGICAL SUPERCONDUCTORS

TOPOLOGICAL INSULATORS AND TOPOLOGICAL SUPERCONDUCTORS

B. ANDREI BERNEVIG

with Taylor L. Hughes

PRINCETON UNIVERSITY PRESS
Princeton and Oxford

Published by Princeton University Press, 41 William Street, Princeton,
New Jersey 08540
In the United Kingdom: Princeton University Press, 6 Oxford Street, Woodstock,
Oxfordshire OX20 1TW

press.princeton.edu

Library of Congress Cataloging-in-Publication Data

Bernevig, B. Andrei, 1978–
Topological insulators and topological superconductors / B. Andrei Bernevig
with Taylor L. Hughes.
pages cm

Summary: "This graduate-level textbook is the first pedagogical synthesis of the field of topological insulators and superconductors, one of the most exciting areas of research in condensed matter physics. Presenting the latest developments, while providing all the calculations necessary for a self-contained and complete description of the discipline, it is ideal for graduate students and researchers preparing to work in this area, and it will be an essential reference both within and outside the classroom. The book begins with simple concepts such as Berry phases, Dirac fermions, Hall conductance and its link to topology, and the Hofstadter problem of lattice electrons in a magnetic field. It moves on to explain topological phases of matter such as Chern insulators, two- and three-dimensional topological insulators, and Majorana p-wave wires. Additionally, the book covers zero modes on vortices in topological superconductors, time-reversal topological superconductors, and topological responses/field theory and topological indices. The book also analyzes recent topics in condensed matter theory and concludes by surveying active subfields of research such as insulators with point-group symmetries and the stability of topological semimetals. Problems at the end of each chapter offer opportunities to test knowledge and engage with frontier research issues. Topological Insulators and Topological Superconductors will provide graduate students and researchers with the physical understanding and mathematical tools needed to embark on research in this rapidly evolving field"– Provided by publisher.
ISBN-13: 978-0-691-15175-5 (hardback)
ISBN-10: 0-691-15175-X (cloth)
1. Energy-band theory of solids. 2. Superconductivity.
3. Solid state physics–Mathematics. 4. Superconductors–Mathematics.
I. Hughes, Taylor L., 1981– II. Title.
QC611.95.B465 2013
530.4′1–dc23

2012035384

British Library Cataloging-in-Publication Data is available

This book has been composed in ITC Stone

Typeset by S R Nova Pvt Ltd, Bangalore, India

10 9 8 7 6 5 4 3 2 1

Contents

TOPOLOGICAL INSULATORS AND TOPOLOGICAL SUPERCONDUCTORS

1

Introduction

Condensed matter physics deals with the behavior of particles at finite density and at low temperatures, where, depending on factors such as applied pressure, doping, spin of the particles, etc., matter can reorganize itself in different phases. Classically, we learn of phases such as liquid, crystal, or vapor, but the quantum world holds many more fascinating mysteries. Quantum theory predicts the existence of myriad states of matter, such as superconductors, charge density waves, Bose-Einstein condensates, ferromagnets and antiferromagnets, and many others. In all these states, the underlying principle for characterizing the state is that of symmetry breaking. An order parameter with observable consequences acquires an expectation finite value in the state of matter studied and differentiates it from others. For example, a ferromagnet has an overall magnetization, which breaks the rotational symmetry of the spin and picks a particular direction of the north pole. An antiferromagnet does not have an overall magnetization but develops a staggered-order parameter at some finite wavevector. Landau's theory of phase transitions provides the phenomenological footing on which symmetry-broken states can be explained. In this theory, a high-temperature symmetric phase experiences a phase transition to a low-temperature, less-symmetric, symmetry-broken state. The theory has very general premises, such as the possibility of expansion of the free energy in the order parameter: when close to a phase transition, this is likely to be possible because the order parameter is small. The transition can be first, second, or higher order, depending on the vanishing of the coefficient of the second, third, or higher coefficient of the expansion of the free energy in the order parameter. However, the major limitation of Landau's theory of phase transitions is that it is related to a *local* order parameter. In the past decade it has become clear that a series of phases of matter with so-called topological order do not have a local order parameter. For some (most) of them, a (highly) nonlocal order parameter can be defined, but it is unclear how a Landau-like theory of this order parameter can be developed. Among these phases, which boast excitations with fractional statistics, the experimentally established ones are the fractional quantum Hall (FQH) states. It has been shown that FQH phases have a nonlocal order parameter, which corresponds to annihilating an electron at a position and, crucially, unwinding a number of fluxes (for an Abelian state). The flux unwinding is highly nonlocal, and it is not clear if a true Ginzburg-Landau theory can be written for this order parameter. States with nonlocal order parameters could be called topologically ordered; more generally, however, the definition of a topological phase is a phase of matter whose low-energy field theory is a topological field theory. Recently, topological phases have been pursued because of their potential practical applications: it has been proposed that a topological quantum computer could employ the 2-D quasi-particles of non-Abelian FQH states (called non-Abelian anyons), whose world lines cross over one another to form braids [1]. These braids form the logic gates that make up the computer. A major advantage of a topological quantum computer over one using trapped quantum particles is that the former encodes information nonlocally and hence is less susceptible to local decoherence processes [1, 2]. Although the concept of topologically ordered phases has been around for more than a decade, most examples involved complicated states of matter, such as the FQH, quantum double models, doubled Chern-Simons theories, and others. The topologically ordered states that have been experimentally realized and theoretically investigated up to now have involved strong electron-electron interactions.

States of matter formed out of free fermions have been deemed to be topologically "trivial" in the past, mostly because the Hamiltonian spectrum is exactly solvable. However, a subset of such states, which we call *topological insulators*, have interesting properties, such as gapless edge states, despite being made out of noninteracting fermions. The paradigm example of such a state is the integer quantum Hall effect, which requires an applied magnetic field on a semiconductor sample. More recently, examples of topological phases that do not require external magnetic fields have been proposed, the first being Haldane's Chern insulator model [3]. Although this state has not been experimentally realized, a time-reversal-invariant (TR-invariant) version has been proposed and discovered [4, 5, 6, 7, 8, 9, 10]. The term *topological* (attached to phases or insulators) implies the existence of a bulk invariant (usually an integer or a rational number or set of numbers) that differentiates between phases of matter having the same symmetry. In the field of topological insulators, topological is usually associated with the existence of gapless edge modes when a system is spatially cut in two, but this is not the generic situation; topological phases or even insulators can exist without the presence of gapless edge modes.

Despite the fact that the field of band theory has been around since the foundations of quantum mechanics almost a century ago, it was realized only in 2006 that there exists an enhanced band theory, called *topological band theory* which takes into account concepts such as Chern numbers and Berry phases. It is quite remarkable and exciting that the theory we have used to understand insulators and semiconductors for almost a century is incomplete and that the underpinnings of the new theory, which includes topological effects, is being worked out during our lifetimes. In this book, we provide a foundation for understanding and beginning research in the field of topological insulators. We aim to provide the reader with both physical understanding and the mathematical tools to undertake high-level research in this emerging field. Recent work in the theory of insulators [4, 5, 6] showed that an important consideration is not only which symmetries the state breaks, but which symmetries must be preserved to ensure the stability of the state. A series of symmetries, the most important of which are time-reversal and charge conjugation, can be used to classify insulating states of matter. The trivial insulator is differentiated from the nontrivial, topological insulator, which (in most, but not all, cases) exhibits gapless surface or boundary gapless states protected from opening a gap.

A periodic table classifying the topological insulators and superconductors has been created. The table organizes the possible topological states according to their space-time dimension and the symmetries that must remain protected: TR, charge-conjugation, and/or chiral symmetries [11, 12, 13]. The most interesting entries in this table, from a practical standpoint, are the 2-D and 3-D TR-invariant topological insulators, which have already been found in nature [7, 8, 9, 10]. These are insulating states classified by a Z_2 invariant that requires an unbroken TR symmetry to be stable. There are several different methods to calculate the Z_2 invariant [5, 9, 13, 14, 15, 16, 17, 18], and a nontrivial value for this quantity implies the existence of an odd number of gapless Dirac fermion boundary states as well as a nonzero [13] magnetoelectric polarization in three dimensions [13, 19]. The current classification of the topological insulators covers only the TR, charge-conjugation, or chiral symmetries and does not exhaust the number of all possible topological insulators. In principle, for every discrete symmetry, there must exist topological insulating phases with distinct physical properties and a topological number that classifies these phases and distinguishes them from the "trivial" ones. This is the point of view taken in this book.

So far, in our discussion we have used the term *topological* cavalierly. In this book, we also hope to clarify what makes an insulator topological. We should start by first defining a trivial insulator: this is the insulator that, upon slowly turning off the hopping elements and the hybridization between orbitals on different sites, flows adiabatically into the atomic limit.

In most of the existent literature on noninteracting topological insulators, it is implicitly assumed that nontrivial topology implies the presence of gapless edge states in the energy spectrum of a system with boundaries. However, it is well known from the literature on topological phases that such systems can theoretically exist without exhibiting gapless edge modes [20]. Hence, the edge modes cannot be the only diagnostic of a topological phase; consequently, the energy spectrum alone, with or without boundaries, is insufficient to determine the full topological character of a state of matter. In the bulk of an insulator, it is a known fact that the topological structure is encoded in the eigenstates rather than in the energy spectrum. As such, we can expect that entanglement—which depends only on the eigenstates—can provide additional information about the topological nature of the system. However, we know that topological entanglement entropy (or the subleading part of the entanglement entropy) [21, 22], the preferred quantity used to characterize topologically ordered phases, does not provide a unique classification and, moreover, vanishes for any noninteracting topological insulator, be it time-reversal breaking Chern insulators or TR-invariant topological insulators. However, careful studies of the full entanglement spectrum [23] are useful in characterizing these states [23, 24, 25, 26, 27, 28, 29, 30, 31, 32, 33, 34].

This book covers most introductory concepts in topological band theory. It is aimed at the beginning graduate student who wants to enter quickly the research in this field. The philosophy behind the writing of this book is to mix physical insight with the rigorous computational details that would otherwise be time consuming to derive from scratch. The book starts with a review of the most important concept—that of Berry phases in chapter 2.

In chapter 3, we show how measurable quantities, such as Hall conductance, are related, through linear response, to Berry curvature and Chern numbers on both periodic and disordered systems, with and without applied magnetic field. We introduce the flat-band limit of an insulator, which simplifies tremendously the calculation of topological quantities. We present the Chern number as an obstruction in the full Brillouin zone (BZ) and as an integral of the Berry curvature over the full BZ of an insulator.

In chapter 4, we introduce the concept of TR symmetry for both spinless and spinful particles and show the vanishing of the Chern number in the presence of TR symmetry. We show that time reversal is an antiunitary operator and that its presence requires, for half-integer spin particles, the existence of doubly-degenerate states—a theorem known as Kramers' theorem. We then focus on the case of Bloch Hamiltonians in translationally invariant crystals and derive the conditions under which a Bloch Hamiltonian is TR invariant.

In chapter 5, we introduce and analyze the problem of 2-D lattice electrons in a magnetic field, with emphasis on the Chern numbers of the bands. We analyze the magnetic translation group on the lattice and explain the magnetic unit cell, showing that even though the magnetic field is uniform, translational invariance with the original unit cell is broken. We then obtain the spectrum of the 2-D nearest-neighbor-hopping Hamiltonian with rational flux p/q per unit cell, where both p and q are integers. We present the spectrum of the Harper equation and the Hofstadter butterfly, analyze its symmetries, and show the presence of an index theorem that guarantees the existence of Dirac fermions. To obtain the Chern number of the bands, we perform the explicit calculation of the Hall conductance, starting from the anisotropic 1-D limit, and analyze the gap openings upon reducing the amount of anisotropy to make the system 2-dimensional. We then prove the equivalence of this method with the Diophantine equation method, which allows for the determination of the Hall conductance through simple methods. All these were bulk calculations, but in chapter 6 we begin to investigate the intimate relationship between Hall conductance and edge modes. We present the bulk-edge correspondence, Laughlin's gauge argument, and the transfer matrix method for calculating the edge modes and the bulk bands of a system with open boundary conditions in an applied magnetic field.

In chapter 7, we start the discussion of graphene, a single layer of carbon atoms in a hexagonal lattice, which has created a tremendous amount of research in the past few years. We analyze the symmetries of the graphene lattice and show that the local stability of the Dirac nodes present in this material is guaranteed by TR and inversion symmetries. We then show that these Dirac nodes are also influenced by the C_3 symmetry present in the hexagonal lattice. This symmetry forces the Dirac nodes to stay at particular points in the BZ and makes them globally stable. We then move to analyzing graphene in an open-boundary geometry and show that the system has edge modes, even though it is not a bulk insulator but a semimetal. We show the different possibilities for the evolution of these edge modes when the system opens a gap and argue for the existence of topological insulators on simple, physical grounds, purely on the basis of the existence of these edge modes.

In chapter 8, we start the presentation of the simplest topological insulator, the Chern insulator, in two spatial dimensions. This insulator has a phase that exhibits a nonzero Hall conductance, so it breaks TR symmetry without having a net magnetic flux per unit cell (quantum Hall with zero applied field). We derive the physics of this insulator by looking at the physics of the closing and reopening of the Dirac fermion gap at points in the BZ. We analyze the behavior of the Berry potential on the lattice and point out, through an explicit calculation, the absence of a smooth gauge in cases where the Chern number does not vanish. We present Haldane's model of a Chern insulator on the graphene lattice and analyze its properties. We then end the chapter with an analysis of the physical properties of a Chern insulator in a magnetic field and of the edge modes that appear in a mass domain wall of the Dirac equation.

In chapter 9 we begin the subject of TR topological insulators by presenting the Kane and Mele model—the first model exhibiting such phase in two space dimensions. We analyze its symmetries and show that, in its simplest form, it is equivalent to two copies of Haldane's Chern insulator, for spin $\uparrow$ and spin $\downarrow$ with opposite Chern numbers. We then couple the two spins and show that the topological phase survives perturbations. We physically argue that there are two types of TR-invariant topological insulators, differentiated by the number of pairs of edge modes that they consider. If the number of pairs of edge states is odd, the topological phase is protected against small perturbations, whereas if the number of pairs is even, the insulator is not protected against perturbations and is adiabatically connected to a trivial one. We end the chapter by presenting the theoretical model and experimental predictions for mercury telluride (HgTe) quantum wells, which is the first experimentally observed topological insulator.

Chapter 10 is devoted to the introduction of Z_2 invariants for TR-invariant topological insulators. These invariants are used to characterize whether a TR-insulator is a nontrivial topological or a trivial one. There exist several equivalent formulations of these invariants in two dimensions. We adopt a chronological viewpoint and first present the initial invariants before moving to the modern description of the Z_2 invariant through sewing matrices. We introduce and expand on the modern theory of charge polarization (used to define the Chern number) before presenting its generalization to the theory of TR polarization used to define the Z_2 invariant. We show the intimate link between the TR polarization and the presence of pairs of edge modes and physically define the 2-D topological insulator even in the presence of interactions. We show that a trivial generalization of the Z_2 invariant from two to three dimensions allows us to define a 3-D topological insulator class of materials.

In chapter 11, we show how to understand topological insulators through simple band-crossing arguments in different dimensions. We supplement the Wigner–von Neumann classification of degeneracies with requirements such as TR and inversion symmetry and show how a trivial insulator can be transformed into a nontrivial one through a series of gap-closing-and-reopening transitions.

In chapter 12, we analyze insulators with inversion symmetry, with and without time reversal. When both inversion and TR symmetry are present, we give a simple expression of the Z_2 invariant, which can and has been used extensively to distinguish nontrivial topological insulators.

In chapter 13, we start analyzing the field theory of topological insulators by the procedure of dimensional reduction. We show that integer quantum Hall effects (Chern insulators) exist in any even space dimension, derive their response to magnetic and electric fields, and show that they are classified by an integer called nth Chern number, where n is an integer. We relate their response to electromagnetic fields with the existence of gapless surface states.

In chapter 14, we show how, by dimensional reduction, the 3-D TR-invariant topological insulator can be obtained from a four-space-dimensional topological insulator. We obtain the field theory of three-space-dimensional topological insulators and show that they can be described by a magnetoelectric polarization—the generalization to 3 space dimensions of the 1-D charge polarization. We analyze the magnetoelectric polarization and show that in the presence of TR symmetry (or inversion), it can be quantized to take two values.

In chapter 15, we discuss several experimental predictions of the physics of the topological insulators.

In chapter 16, we introduce the concept of topological superconductors via standard models of spinless fermions in 1 and 2 dimensions with p-wave superconductivity. These superconductors exhibit topological states that are stable even without conserving any special symmetries. These superconductors have unique properties, one example being the non-Abelian statistics of vortices in the 2-D chiral topological superconductor.

In chapter 17 we move on to study topological superconductors, which require time-reversal symmetry to be stable. These states are similar in nature to the 2-D quantum-spin Hall effect and the 3-D strong topological insulator discussed in the earlier chapters, but with some notable differences. Perhaps the most interesting distinction is that the 3-D superconductor states are classified by an integer instead of a Z_2 quantity, as in the insulator case.

Finally, in chapter 18, we capitalize on all the previous discussions to show how hybrid topological insulator and superconductor states can be created from time-reversal invariant topological insulators in proximity to magnets, superconductors, or both. Such composite structures lead to experimentally viable proposals to create exotic phenomena, such as Majorana fermion-bound states (the simplest non-Abelian anyon).

The first 15 chapters of this book have evolved from the notes prepared for Physics 536, the advanced graduate condensed matter course at Princeton University taught by B. A. Bernevig in 2010 and 2011. B. A. Bernevig would like to thank the students in this class who have submitted corrections of formulas and typos in the body of the manuscript. Chapters 16–18 on topological superconductors have been prepared by T. L. Hughes.

2

Berry Phase

The Berry phase [35] is the most important concept in topological band theory. However, its discovery was not specifically related to Bloch electrons but rather to the general idea that quantum adiabatic transport of particles in slowly varying fields (be they electric, magnetic, or strain) could in principle modify the wavefunction by terms other than just the dynamical phase. Joshua Zak then realized that Berry phases and, more generally, the issue of adiabatic transport should be present in Bloch-periodic systems, where the parameters (Bloch momenta) are varied in closed loops (bands or Fermi surfaces) by applying electric fields. In this chapter, we derive the Berry phase for a particle obeying Hamiltonian evolution under a set of slowly varying parameters, show how to compute the Berry phase without gauge smoothing, and then present the simple solution of a particle in a varying magnetic field. In subsequent chapters we use this formalism as the basis for defining a series of topological invariants such as Chern numbers, and Z_2 invariants and for the analysis of topological phase transitions in insulators.

2.1 General Formalism

We consider a physical system with a general time-varying Hamiltonian $H(\mathbf{R})$ that depends on time through several parameters (such as magnetic field, electric field, flux, and strain) labeled by a vector $\mathbf{R} = (R_1, R_2, R_3, \ldots.)$ where $R_i = R_i(t)$. We are interested in the adiabatic evolution of the system—the evolution of the system as the parameters $\mathbf{R}(t)$ are varied very slowly (compared to other energy scales—gaps—in the problem) along a path $\mathcal{C}$ in the parameter space. For our purposes, $\mathcal{C}$ can now be any path, closed or open. We introduce an instantaneous orthonormal basis of the instantaneous eigenstates $|n(\mathbf{R})\rangle$ of $H(\mathbf{R})$ at each point $\mathbf{R}$ obtained by diagonalizing $H(\mathbf{R})$ at each value $\mathbf{R}$:

$$H(\mathbf{R})|n(\mathbf{R})\rangle = E_n(\mathbf{R})|n(\mathbf{R})\rangle. \tag{2.1}$$

Equation (2.1) determines the basis function $\left|n(\mathbf{R})\right\rangle$ up to a phase. There is still a gauge freedom of an arbitrary phase (which in the case of degenerate states can be a matrix). This phase can be $\mathbf{R}$-dependent ($\mathbf{R}$ is not itself the position of the particle, so $H(\mathbf{R})$ commutes with an $\mathbf{R}$-dependent phase factor). We can make a phase choice (pick a gauge) to remove the arbitrariness. A natural choice could be to require that the phase of each basis function $|n(\mathbf{R})\rangle$ be smooth and single valued along the path $\mathcal{C}$ in the parameter space. As a side note, it turns out that sometimes such a smooth and single-valued choice is not possible: if a system has a nonzero Hall conductance, the occupied bands cannot have a single-valued smooth phase across the BZ. In those cases, we divide the path into several segments, which overlap slightly with each other and in which the phase can be defined smoothly. Smooth, single-valued gauges can always be found piecewise in finite neighborhoods of the parameter space. We will come back to this later.

We want to analyze the phase of the wavefunction of a system prepared in an initial pure state $\left|n(\mathbf{R}(0))\right\rangle$ as we slowly move $\mathbf{R}(t)$ along the path $\mathcal{C}$. The preparation in an initial eigenstate is crucial. Per the adiabatic theorem, a system starting in an eigenstate $|n(\mathbf{R}(0))\rangle$ will evolve with $H(\mathbf{R})$ and hence stay as an instantaneous eigenstate of the Hamiltonian $H(\mathbf{R}(t))$ throughout the process. But what is the phase? The phase is the only degree of freedom we have, and during the adiabatic evolution of the system with the Hamiltonian H, the phase $\theta(t)$ of the state $|\psi(t)\rangle = e^{-i\theta(t)}|n(\mathbf{R}(t))\rangle$ does not need to be zero. In fact, we know it cannot be zero because it must at least contain the dynamical factor related to the energy of the eigenstate. The surprise is that it contains more than that. The time evolution of the system is given by

$$H(\mathbf{R}(t))|\psi(t)\rangle = i\hbar\frac{d}{dt}|\psi(t)\rangle, \tag{2.2}$$

which translates into the differential equation

$$E_n(\mathbf{R}(t))|n(\mathbf{R}(t))\rangle = \hbar\left(\frac{d}{dt}\theta(t)\right)|n(\mathbf{R}(t)) + i\hbar\frac{d}{dt}|n(\mathbf{R}(t))\rangle. \tag{2.3}$$

Taking the scalar product with $\langle n(\mathbf{R}(t))|$ and assuming the state is normalized, $\langle n(\mathbf{R}(t))|n(\mathbf{R}(t))\rangle = 1$, we get

$$E_n(\mathbf{R}(t)) - i\hbar\langle n(\mathbf{R}(t))|\frac{d}{dt}|n(\mathbf{R}(t))\rangle = \hbar\left(\frac{d}{dt}\theta(t)\right). \tag{2.4}$$

The solution for the phase $\theta(t)$ is, hence,

$$\theta(t) = \frac{1}{\hbar}\int_0^t E_n(\mathbf{R}(t'))dt' - i\int_0^t \langle n(\mathbf{R}(t'))|\frac{d}{dt'}|n(\mathbf{R}(t'))\rangle dt'. \tag{2.5}$$

The first part of the phase is the conventional dynamical phase. The negative of the second part is called the Berry phase. If we write

$$|\psi(t)\rangle = \exp\left(\frac{1}{\hbar}\int_0^t E_n(\mathbf{R}(t'))dt'\right)\exp(i r_n)|n(\mathbf{R}(t))\rangle, \tag{2.6}$$

then the Berry phase is γ_n

$$\gamma_n = i\int_0^t \langle n(\mathbf{R}(t'))|\frac{d}{dt'}|n(\mathbf{R}(t'))\rangle dt'. \tag{2.7}$$

The Berry phase comes from the fact that the states at t and $t + dt$ are not identical. Time can been removed explicitly from the equation—the only thing needed being the dependence of the eigenstates on the parameters R_i, which are implicitly time dependent.

$$\gamma_n = i\int_0^{t_{\text{end cycle}}} \langle n(\mathbf{R}(t'))|\nabla_{\mathbf{R}}|n(\mathbf{R}(t'))\rangle\frac{d\mathbf{R}}{dt'}dt' = i\int_C \langle n(\mathbf{R})|\nabla_{\mathbf{R}}|n(\mathbf{R})\rangle d\mathbf{R}. \tag{2.8}$$

By analogy with electron transport in an electromagnetic field, in the last equation we can now define a vector function called Berry connection, or Berry vector potential:

$$\mathbf{A}_n(\mathbf{R}) = i\langle n(\mathbf{R})|\frac{\partial}{\partial\mathbf{R}}|n(\mathbf{R})\rangle, \quad \gamma_n = \int_C d\mathbf{R}\cdot\mathbf{A}_n(\mathbf{R}). \tag{2.9}$$

We notice that the Berry phase γ_n is real (it's not Berry decay), despite the fact that it has an imaginary sign in front of it because $\langle n(\mathbf{R})|\nabla_{\mathbf{R}}|n(\mathbf{R})\rangle$ is itself imaginary: $\langle n(\mathbf{R})|n(\mathbf{R})\rangle = 1 \to \langle n(\mathbf{R})|\nabla_{\mathbf{R}}|n(\mathbf{R})\rangle = -\langle n(\mathbf{R})|\nabla_{\mathbf{R}}|n(\mathbf{R})\rangle^*$ Hence the Berry phase can also be written as

$$\gamma_n = -Im \int_C \langle n(\mathbf{R})|\nabla_{\mathbf{R}}|n(\mathbf{R})\rangle d\mathbf{R}. \tag{2.10}$$

The Berry vector potential $\mathbf{A}_n(\mathbf{R})$ is obviously gauge dependent. Under a gauge transformation $|n(\mathbf{R})\rangle \to e^{i\zeta(\mathbf{R})}|n(\mathbf{R})\rangle$, with $\zeta(\mathbf{R})$ a smooth, single-valued function, the Berry potential transforms in the usual way:

$$\mathbf{A}_n(\mathbf{R}) \to \mathbf{A}_n(\mathbf{R}) - \frac{\partial}{\partial \mathbf{R}}\zeta(\mathbf{R}). \tag{2.11}$$

As such, the Berry phase γ_n will be changed by $-\int_C \frac{\partial}{\partial \mathbf{R}}\zeta(\mathbf{R}) \cdot d\mathbf{R} = \zeta(\mathbf{R}(0)) - \zeta(\mathbf{R}(T))$ where T is the (long) time after which the path C has been completed. Before Berry, many people had concluded that, by a smart choice of the gauge factor $\zeta(\mathbf{R})$, the phase γ_n can be canceled and, hence, is not of much relevance. This is wrong. We can consider closed paths C in parameter space for which, after a long time T (period), we return to the original parameters: $\mathbf{R}(0) = \mathbf{R}(T)$. For such paths, the fact that we chose our eigenstates basis to be single-valued means that when we return to the original parameter configuration, the basis state must be the same: $|n(\mathbf{R}(T))\rangle = |n(\mathbf{R}(0))\rangle$. Gauge transformations must maintain this property, so $e^{i\zeta(\mathbf{R}(0))}|n(\mathbf{R}(0))\rangle = e^{i\zeta(\mathbf{R}(T))}|n(\mathbf{R}(T))\rangle = e^{i\zeta(\mathbf{R}(T))}|n(\mathbf{R}(0))\rangle$ and hence $\zeta(\mathbf{R}(T)) - \zeta(\mathbf{R}(0))) = 2\pi m$, with m an integer. As such, under a closed path, the Berry phase cannot be canceled unless it is an integer itself.

In the remainder of this chapter, we will concern ourselves only with closed paths C and will consider the parameter space to be three-dimensional (R_1, R_2, R_3), which is good enough in condensed - matter physics. For a closed path, the Berry phase is a gauge-invariant quantity independent of the specific form of how $\mathbf{R}$ varies in time. Because C is a closed path, application of Stokes theorem gives

$$\begin{aligned}\gamma_n &= -\text{Im} \int d\mathbf{S} \cdot (\nabla \times \langle n(\mathbf{R})|\nabla|n(\mathbf{R})\rangle) = -\text{Im} \int dS_i \epsilon_{ijk} \nabla_j \langle n(\mathbf{R})|\nabla_k|n(\mathbf{R})\rangle) \\ &= -\text{Im} \int d\mathbf{S} \cdot (\langle \nabla n(\mathbf{R})| \times |\nabla n(\mathbf{R})\rangle).\end{aligned} \tag{2.12}$$

where $\langle \nabla n(\mathbf{R})| \times |\nabla n(\mathbf{R})\rangle$ is the Berry curvature (to be precise, $F_{jk} = \langle \nabla_j n(\mathbf{R})|\nabla_k n(\mathbf{R})\rangle - (j \leftrightarrow k)$ is the Berry curvature, obtainable from the preceding vector by contraction with an ϵ_{ijk} symbol). We can think of F_{jk} as a magnetic field in parameter space (the curl of the Berry vector potential).

2.2 Gauge-Independent Computation of the Berry Phase

We want to gain further insight into the Berry phase; specifically; we want to know how to compute it numerically. As it stands, numerical computation is nontrivial: the Berry phase thus far contains *derivatives* of the eigenstates. In a numerical diagonalization of the Hamiltonian to obtain an eigenstate basis at each R, the diagonalization procedure will output states with wildly different phase factors, thereby preventing the taking of derivatives. We must gauge-smoothen first, but this is a nontrivial procedure. We here provide a formula for the Berry phase that is manifestly gauge independent. Thus, we massage our formula by introducing a complete set of eigenstates $\sum_m |m\rangle\langle m| = 1$ (at each R):

$$\begin{aligned}&\epsilon_{ijk}\langle\nabla_j n(\mathbf{R})|\nabla_k|n(\mathbf{R})\rangle\rangle\\ &= \epsilon_{ijk}\langle\nabla_j n(\mathbf{R})|n\rangle\langle n|\nabla_k n(\mathbf{R})\rangle\rangle + \sum_{m\neq n}\epsilon_{ijk}\langle\nabla_j n(\mathbf{R})|m\rangle\langle m|\nabla_k n(\mathbf{R})\rangle\rangle.\end{aligned}$$

We can drop the first term because both $\langle\nabla_j n(\mathbf{R})|n\rangle$ and $\langle n|\nabla_k n(\mathbf{R})\rangle$ are imaginary and hence their product is real, giving no contribution to the imaginary value that is used to compute the Berry phase. Hence, we have

$$\gamma_n = -\mathrm{Im}\int\int dS_i \sum_{m\neq n}\epsilon_{ijk}\langle\nabla_j n(\mathbf{R})|m\rangle\langle m|\nabla_k n(\mathbf{R})\rangle\rangle.$$

We massage this equation further to remove the derivatives on the eigenstates:

$$E_n\langle m|\nabla n\rangle = \langle m|\nabla(Hn)\rangle = \langle m|(\nabla H)n\rangle + E_m\langle m|\nabla n\rangle, \tag{2.13}$$

where we have used $\langle m|n\rangle = 0$ for $m \neq n$. Hence,

$$\langle m|\nabla n\rangle = \frac{\langle m|(\nabla H)|n\rangle}{E_n - E_m}, \tag{2.14}$$

and a similar expression for $\langle\nabla n|m\rangle$ gives

$$\begin{aligned}\gamma_n &= -\int\int_C d\mathbf{S}\cdot\mathbf{V}_n\\ &= -\int\int_C d\mathbf{S}\cdot\mathrm{Im}\sum_{m\neq n}\frac{\langle n(\mathbf{R})|(\nabla_{\mathbf{R}}H(\mathbf{R}))|m(\mathbf{R})\rangle\times\langle m(\mathbf{R})|(\nabla_{\mathbf{R}}H(\mathbf{R}))|n(\mathbf{R})\rangle}{(E_m(\mathbf{R}) - E_n(\mathbf{R}))^2},\end{aligned} \tag{2.15}$$

where the component $V_{ni} = \epsilon_{ijk}F_{jk}$. Formula (2.15) is manifestly gauge independent. It has the advantage over other gauge-independent formulas such as the Berry curvature in that it does not depend explicitly on the phases of $|n\rangle$, so it can be evaluated under any gauge choice (in other words, for any eigenstate that the computer outputs) as the derivatives have been moved from the wavefunction to the Hamiltonian. When using formula (2.15), it is no longer necessary to pick $|n\rangle$, $|m\rangle$ to be smooth and single valued.

Before we leave this section, it is useful to make several remarks. Equation (2.12) includes only the eigenstate $|n\rangle$ and its derivatives, but Equation (2.15) shows that the Berry curvature can be thought of as the result of the interaction with the level $|n\rangle$ of the other levels $|m\rangle$ that have been projected out by the adiabatic interaction. If we sum over the Berry phase of all energy levels, we get 0, showing that the sum of all filled bands can have only zero Berry phase. The current formalism is valid for the case where the level E_n is singly degenerate. For degenerate energy levels, the Berry vector potential becomes a matrix of dimension equal to the degeneracy of the levels—it becomes *non-Abelian*. We will analyze this situation at later stages.

2.3 Degeneracies and Level Crossing

One of the most important applications of the Berry phase is the classification of degeneracies. This will be one of the main ingredients of band crossings in topological band theory presented in chapters 8–16. If the denominator of Equation (2.15) is close to zero, we now show that this level degeneracy point corresponds to a monopole in the parameter space. Because the denominator of Equation (2.15) is made of two-level terms (only two energies enter, that of level m and that of level n), the generic degeneracy point is at the intersection of two levels as we vary $\mathbf{R}$. For this reason, two-level systems will appear repeatedly in this book.

Consider two states $\pm$, of energies $E_+(\mathbf{R}) \geq E_-(\mathbf{R})$ with a degeneracy $E_+(\mathbf{R}^*) = E_-(\mathbf{R}^*)$. For $\mathbf{R}$ close to the degeneracy point $\mathbf{R}^*$, the Hamiltonian can be expanded in $\mathbf{R} - \mathbf{R}^*$ as $H(\mathbf{R}) \approx H(\mathbf{R}^*) + (\mathbf{R} - \mathbf{R}^*) \cdot \nabla H(\mathbf{R}^*)$, thereby giving the following for the Berry curvature:

$$\mathbf{V}_+((\mathbf{R})) = \text{Im}\frac{\langle +(\mathbf{R})|(\nabla H(\mathbf{R}^*))| - (\mathbf{R})\rangle \times \langle -(\mathbf{R})|(\nabla H(\mathbf{R}^*))| + (\mathbf{R})\rangle}{(E_+(\mathbf{R}) - E_-(\mathbf{R}))^2}. \tag{2.16}$$

We can immediately see that, because of the cross product, $V_-(\mathbf{R}) = -V_+(\mathbf{R})$, and hence $\gamma_-(C) = -\gamma_+(C)$ as obviously required by our previous condition, the sum of Berry curvature over all levels must vanish.

The generic form of the Hamiltonian of any two level system is

$$H = \epsilon(\mathbf{R}) I_{2\times 2} + \mathbf{d}(\mathbf{R}) \cdot \boldsymbol{\sigma} \tag{2.17}$$

where σ^i are the Pauli matrices and $\mathbf{d}$ is a 3-D vector that depends on the coordinates $\mathbf{R}$. This Hamiltonian describes many interesting systems in condensed matter such as graphene, spin-orbit-coupled systems, Bogoliubov quasiparticles, a spin-$\frac{1}{2}$ electron in a magnetic field, and many others. The energy levels are $E_\pm = \epsilon(\mathbf{R}) \pm \sqrt{\mathbf{d} \cdot \mathbf{d}}$. Notice that the constant term $\epsilon(\mathbf{R})$ is just an additive term, which can be neglected because it does not modify the eigenstates and also cancels in the subtraction of levels E_+ and E_-. Let us now determine the Berry phase upon varying the parameters $\mathbf{R}$ by the two formulas expressed in equation (2.12) and (2.15).

2.3.1 Two-Level System Using the Berry Curvature

We can employ spherical coordinates and parametrize the vector $\mathbf{d}(\mathbf{R}) = |d|(\sin(\theta)\cos(\phi), \sin(\theta)\sin(\phi), \cos(\theta))$. The two eigenstates with energies $\pm\frac{1}{2}|d|$ are

$$|-\mathbf{R}\rangle = \begin{pmatrix} \sin\left(\frac{\theta}{2}\right) e^{-i\phi} \\ -\cos\left(\frac{\theta}{2}\right) \end{pmatrix}, \quad |+\mathbf{R}\rangle = \begin{pmatrix} \cos\left(\frac{\theta}{2}\right) e^{-i\phi} \\ \sin\left(\frac{\theta}{2}\right) \end{pmatrix}, \tag{2.18}$$

which are orthogonal and normalized. For now we do not consider the Berry vector potential $A_{|d|}$. The two leftover Berry vector potential s, A_θ and A_ϕ and the Berry curvature, $F_{\theta\phi}$, are given by

$$A_\theta = i\langle -\mathbf{R}|\partial_\theta| - \mathbf{R}\rangle = 0, \qquad A_\phi = i\langle -\mathbf{R}|\partial_\phi| - \mathbf{R}\rangle = \sin\left(\frac{\theta}{2}\right)^2,$$

$$F_{\theta\phi} = \partial_\theta A_\phi - \partial_\phi A_\theta = \frac{\sin(\theta)}{2}. \tag{2.19}$$

To explicitly see the gauge invariance of the Berry curvature, we now notice that the eigenstates we have chosen previously have one point where they are not well defined. This will be extremely important in the Chern insulator problem: if we are able to find a gauge in which all wavefunctions are well defined, then the system cannot have nonzero Hall conductance. The wavefunction $|-, \mathbf{R}\rangle$ is not well defined if the system can reach, in its adiabatic evolution, the south pole $\theta = \pi$. At $\theta = \pi$ the first component of the $|-\mathbf{R}\rangle$ is $e^{-i\phi}$, despite the fact that ϕ cannot be defined at this point as both x and y-coordinates are zero. We can then choose another gauge by letting $|-\mathbf{R}\rangle \to e^{i\phi}|-\mathbf{R}\rangle$, and in this gauge we have

$$|-\mathbf{R}\rangle = \begin{pmatrix} \sin\left(\frac{\theta}{2}\right) \\ -\cos\left(\frac{\theta}{2}\right) e^{i\phi} \end{pmatrix}, \tag{2.20}$$

which is single valued everywhere except at the north pole $\theta = 0$. In the case of a model of Chern insulator, we will see that we cannot pick a gauge that is everywhere well defined—the sphere will be fully covered. This cryptic statement will become obvious later. In this new gauge we have

$$A_\theta = i\langle -\mathbf{R}|\partial_\theta| - \mathbf{R}\rangle = 0, \quad A_\phi = i\langle -\mathbf{R}|\partial_\phi| - \mathbf{R}\rangle = -\cos\left(\frac{\theta}{2}\right)^2,$$

$$F_{\theta\phi} = \partial_\theta A_\phi - \partial_\phi A_\theta = \frac{\sin(\theta)}{2}, \tag{2.21}$$

clearly showing the obvious fact that the Berry curvature is gauge independent. For a $\mathbf{d}(\mathbf{R})$ that is dependent on a set of parameters $\mathbf{R}$, we have

$$F_{R_i,R_j} = F_{\theta\phi}\frac{\partial(\theta,\phi)}{\partial(R_i,R_j)} = \frac{\sin(\theta)}{2}\frac{\partial(\theta,\phi)}{\partial(R_i,R_j)} = -\frac{1}{2}\frac{\partial(\cos(\theta),\phi)}{\partial(R_i,R_j)} = \frac{1}{2}\frac{\partial(\phi,\cos(\theta))}{\partial(R_i,R_j)}, \tag{2.22}$$

where $\frac{\partial(\theta,\phi)}{\partial(R_i,R_j)}$ is the Jacobian of the transformation from $\mathbf{R}$ to (θ, ϕ). To gain some physical intuition, let us work a simple specific example and assume that $d(\mathbf{R}) = \mathbf{R}$; we immediately find the following for the Berry field strength $V_{-i} = \epsilon_{ijk}F_{R_j,R_k}$:

$$\mathbf{V}_- = \frac{1}{2}\frac{\mathbf{d}}{d^3}. \tag{2.23}$$

This is the field generated by a monopole (in $\mathbf{R}$ parameter space) of strength $1/2$ (for the $|-\mathbf{R}\rangle$ band) at the degeneracy point $h = 0$. For the $|+\mathbf{R}\rangle$ band, the monopole has opposite strength, $-\frac{1}{2}$. Thus, we found that degeneracy points in parameter space act as sources and drains of the Berry curvature. If we integrate the Berry curvature over a sphere containing the monopole, we get 2π, and hence the Berry curvature integrated over a closed manifold is equal to 2π times the net number of monopoles contained inside. (If we have another degeneracy point, it would be 2.) The Berry curvature integrated over a 2-D surface is an integer (in units of 2π). This integer is called the Chern number.

2.3.2 Two-Level System Using the Hamiltonian Approach

Richard Feynman once famously said that in order to really understand something, we must be able to derive it in different ways. Following this reasoning, we now want to obtain the Berry phase by using the manifestly gauge-invariant approach. Without loss of generality, subtract the $\epsilon(\mathbf{R})$ from the Hamiltonian to give the generic Hamiltonian, $H(\mathbf{R}) = \mathbf{d}(R)\cdot\boldsymbol{\sigma}$, and assume the degeneracy point to be at $\mathbf{R}^* = 0$. This is just a harmless translation in parameter space. By performing an extra rotation, if needed, close to the transition point $d(\mathbf{R}) = 0$, we can take $\mathbf{d}(\mathbf{R}) = \mathbf{R}$, and the Hamiltonian will take the form $H(\mathbf{R}) = x\sigma_x + y\sigma_y + z\sigma_z$, with $\mathbf{R} = (x, y, z)$. The eigenvalues are $E_\pm = \pm|R|$, and the degeneracy is an isolated point at which all three parameters (x, y, z) vanish. This is a restatement of an old theorem by von Neumann, Wigner and which states that it is necessary to vary three parameters in order to make a degeneracy occur accidentally and not on account of a symmetry. In a 2-D Bz, where two of these parameters are the lattice momenta varying across the BZ, we need to vary only one parameter to make a degeneracy occur. In three dimensions, where we have three momenta varying by themselves in the BZ, Dirac points are stable. We then have $\nabla H = \boldsymbol{\sigma}$, and we use Equation (2.15) to determine the Berry phase. There is an easy and physically insightful method for computing the matrix elements needed. We rotate the axes so that the z-axis points along $\mathbf{R}$. In this case, we have the following matrix actions for the two eigenstates $|\pm\rangle$:

$$\sigma_z|\pm\rangle = \pm|\pm\rangle, \quad \sigma_x|\pm\rangle = |\mp\rangle, \quad \sigma_y|\pm\rangle = \pm i|\mp\rangle. \tag{2.24}$$

With this basis choice, we immediately see that $V_{+x} = V_{+y} = 0$, because they are proportional to matrix elements such as $\langle -|\sigma_z|+\rangle = 0$. The only nonzero matrix element is

$$V_{+z} = \mathrm{Im}\frac{\langle +|+\rangle\langle -|i|-\rangle - \langle +|-i|+\rangle\langle -|-\rangle}{4R^2} = \mathrm{Im}\frac{2i}{4R^2} = \frac{1}{2R^2}. \tag{2.25}$$

We remember we rotated the system so that $\mathbf{R}$ lies on the z-axis, and hence rotational invariance implies

$$\mathbf{V}_+(\mathbf{R}) = \frac{\mathbf{R}}{2R^3}. \tag{2.26}$$

This then gives, for the Berry phase,

$$\gamma_+(C) = -\int\int_c dS \cdot \frac{\mathbf{R}}{2R^3}, \quad \exp(i\gamma_\pm(C)) = \exp(\mp\frac{1}{2}i\Omega(C)), \tag{2.27}$$

which is the same formula we obtained earlier from the eigenstate formalism and is the flux through an area bounded by the curve C of a monopole with strength $-\frac{1}{2}$ located at the degeneracy (in this section, we computed the Berry phase for $|+\mathbf{R}\rangle$, as opposed to $|-\mathbf{R}\rangle$ in the previous section—this is the reason why the monopole strength has opposite sign). $\Omega(C)$ is the solid angle that C subtends at the degeneracy point. We can immediately see that the phase factor is independent of the choice of surface spanning C because Ω can change only through 4π multiplets.

We now illustrate the preceding derivation with the case of a Dirac-Weyl fermion. Dirac fermions will repeatedly appear in this book as we try to understand issues such as gauges, Chern numbers, Z_2 numbers, etc. These fermions have a Hamiltonian, which, in three-space dimensions, reads $k_i\sigma_i$, where k is Bz momentum—similar in spirit to the Hamiltonians analyzed in this chapter. We ask what happens to the wavefunction of a Dirac fermion as it is transported around a path C in momentum space. It will acquire a phase, the Berry phase. We can obtain its value by a simple application of the results learned so far. We now assume that the path C is confined to the plane $Z = 0$. This is the case for a gapless Dirac fermion in which x, y become the k_x, k_y momenta. $Z = 0$ is the necessary and sufficient condition for a Dirac fermion to be gapless in 2 dimensions and it corresponds to taking the one parameter left out of the Wigner–vonNeumann classification (besides the k_x, k_y momenta) and tuning it to zero. In this case, the solid angle subtended by the curve in closed space is 2π if the curve encircles the degeneracy and zero if it does not, with the Berry phase $\exp(i\gamma_{,\pm}(C)) = -1, +1$

in the two cases, respectively. Notice that the choice of contour is arbitrary: we do not need to pick a contour of equal energy (i.e., $x^2 + y^2$ constant), but we can pick any contour as long as it is closed. We have just proved that the Berry phase of the eigenstates of a gapless Dirac fermion in 2 dimensions have a Berry phase equal to π upon going around the Fermi surface. We will come back to this later.

2.4 Spin in a Magnetic Field

One of the most important applications of the Berry phase is in investigating the behavior of electrons with spin in a magnetic field. This example has deep consequences because it allows us to understand why half-integer spin particles behave differently than integer spin particles. In fact, the two-level system analyzed in the previous subsection is identical, physically, to a spin-$\frac{1}{2}$ particle—a spin-$\frac{1}{2}$ spin particle hence must acquire minus a $(-)$ sign upon a full rotation in the 2-D plane. In this section, we analyze the behavior of a spin-S particle in a magnetic field. The discussion in this section parallels the proof in Berry's paper. Consider the Hamiltonian of a spin-S particle in a magnetic field:

$$H(B) = \mathbf{B} \cdot \mathbf{S}; \quad E_n(B) = Bn, \quad n = -S, -S+1, \ldots, S-1, S. \tag{2.28}$$

At zero magnetic field there is a $2S + 1$ degeneracy in the spectrum. We now want to calculate the Berry phase when the system is prepared (started) in an eigenstate $|n, S(\mathbf{B})\rangle$ and time-evolves under the Hamiltonian in equation (2.28) as $\mathbf{B}$ is rotated slowly around a loop C. We wish to compute the Berry field strength:

$$\mathbf{V}_n(\mathbf{B}) = \mathrm{Im} \sum_{m \neq n} \frac{\langle n, S(\mathbf{B})|\mathbf{S}|m, S(\mathbf{B})\rangle \times \langle m, S(\mathbf{B})|\mathbf{S}|n, S(\mathbf{B})\rangle}{B^2(m-n)^2}. \tag{2.29}$$

As before, we choose the (instantaneous) z - axis parallel to B to obtain

$$\begin{aligned} S_z|n, S\rangle &= n|n, S\rangle, \\ S^+|n, S\rangle &= \sqrt{s(s+1) - n(n+1)}|n+1, S\rangle, \\ S^-|n, S\rangle &= \sqrt{s(s+1) - n(n-1)}|n-1, S\rangle. \end{aligned} \tag{2.30}$$

In this basis, we obtain for the field strength $V_{n,x} = V_{n,y} = 0$ and $V_{n,z} = n/B^2$. Restoring rotational invariance, we then find

$$\mathbf{V}_n(\mathbf{B}) = n\frac{\mathbf{B}}{B^3}, \quad \gamma_n(C) = -\int\int_C d\mathbf{S} \cdot n\frac{\mathbf{B}}{B^3}. \tag{2.31}$$

We see that the Berry phase is the flux through the area bounded by C of a monopole of strength $-n$ located at the origin of the degeneracy. The Berry phase is equal to n times the solid angle that the closed contour C subtends at $\mathbf{B} = 0$. It is insensitive to the $(2S + 1)$-degeneracy multiplicity at $B = 0$, but it is sensitive to the degree of coupling between the eigenstates of energy nB and $-nB$. For half-integer spin fermions ($n = (2m+1)/2$), a whole turn of $\mathbf{B}$ (i.e., a rotation through 2π, in a plane) gives $\Omega = 2\pi$, which in turn gives $\exp(i\gamma_n(C)) = -1$. Hence, for half-integer spin fermions, the sign change of spinors from a 2π rotation and the sign change of wavefunctions around a degeneracy point of a two-level system have identical origin.

2.5 Can the Berry Phase Be Measured?

No physical property is experimentally interesting if it cannot be measured. Fortunately, the Berry phase has important measurable consequences—the whole subject of this book is based, at some level, on consequences of the Berry phase. However, it is very easy to propose a simple experiment that does not involve any electrons on the lattice. Imagine the following experiment proposed in Berry's paper [35]. Split a beam of particles, all prepared in a definite spin state n in two paths. On one path, $\mathbf{B}$ is constant, whereas on the other path $\mathbf{B}$ is constant in magnitude, but its direction slowly varies around a closed path C subtending a solid angle γ_C. After passing through this field configuration, the two beams are combined at detector. The dynamical phase factor is identical between the two beams because the energy $E_n(B)$ depends only on the magnitude of $\mathbf{B}$ which is the same. However, the beam that has undergone the $\mathbf{B}$ change acquires a Berry phase. The intensity of the diffraction patterns will be

$$|1 + e^{i\gamma_C}|^2 = 4\cos^2(\gamma_C/2). \tag{2.32}$$

The intensity variation can be measured as the magnetic field is slowly varied to undergo the path γ_C.

2.6 Problems

1. *Non-Abelian Berry Transport:* Extend the derivation of the Berry curvature to the adiabatic transport of a degenerate multiplet of states separated by a gap from the excited states—you will have to take into consideration that under time evolution, the state can rotate within the degenerate manifold. Obtain the non-Abelian Berry potential matrix $\gamma_{mn}(t) = i\int_0^t \langle m(\mathbf{R}(t'))|\frac{d}{dt'}|n(\mathbf{R}(t'))\rangle$.

2. *Non-Abelian Berry Curvature:* Find the expression for the Berry curvature in the case of a non-Abelian Berry potential. Show that it must involve a matrix product term,$\gamma_{mn}\gamma_{np}$.

3. *$U(2)$ Berry Curvature of a Four-Band System:* Show that, in general, a four-band crossing $H = d_a(k)\Gamma_a$ described by Clifford-algebra matrices Γ_a, $a = 1, \ldots, 5$, $\{\Gamma_a, \Gamma_b\} = 2\delta_{ab}$ (such as the one that can occur in systems with both TR and inversion symmetry, as we will learn later in the book) can be described by a $U(2)$ monopole that gives rise to a non-Abelian Berry curvature in the same way that in the two-band crossing the Berry field strength is that of a $U(1)$ monopole. (See arXiv:cond-mat/0310005 as a good introduction and for help with details.)

3

Hall Conductance and Chern Numbers

Although interesting in its own right, the Berry potential and the Berry curvature became of widespread use in condensed matter physics because of their far-ranging consequences in the physics of electrons on a lattice. The manner in which the Berry phase and the Berry curvature enter in solid-state physics is strongly dependent on the dimensionality of the system and on whether the system is a metal or an insulator. In general, the Berry phase is the integral of the Berry potential over a closed curve (a 1-D manifold). As such, it is relevant for filled bands (insulators) in one spatial dimension (wires) — for which the closed curve could be the Bloch momentum $-\pi \leq k_x < \pi(\equiv -\pi \bmod 2\pi)$. Other 1-D closed manifolds (curve) that we can envision in crystals are the (1-D) Fermi surfaces of 2-D metals. The Berry phase is relevant in these cases too. As a rule of thumb, the Berry phase is relevant for 1-D insulators or 1-D Fermi surfaces of 2-D metals.

The Berry field strength is a 2-form, which implies it should be relevant in the case of surfaces. Its integral over a surface (2-D manifold) is relevant in two cases. First it is relevant in filled bands in two dimensions, for which the surface could be the full 2-D BZ. In this case, the integral of the Berry curvature over the full (filled) Brillouin zone is related to the Hall conductance of the insulator (as we will show in this chapter) and is identical to the Chern number of the filled band. Second, the Berry curvature is relevant for Fermi surfaces in 3-D metals. These are 2-D manifolds and are also characterized by a Chern number (integral of the Berry curvature over the 2-D Fermi surface).

In this chapter, we show that the Hall conductance of a 2-D system is related to the integral of the Berry curvature over the filled portion of the bands in a crystal. We provide a self-contained analysis, first deriving the current operators and then analyzing the linear response of a system to the application of electric and magnetic fields, obtaining the current-current correlation function and computing it for insulators through a trick called band flattening. We explicitly compute this and show that the Hall conductance equals the integral of the Berry curvature of the filled bands over the full BZ. In short, the purpose of this long section will be to prove that the Hall conductivity of a band insulator, when the Fermi level is in the gap, is the integral of the Berry curvature over the BZ:

$$\sigma_{xy} = \frac{e^2}{h}\frac{1}{2\pi}\int\int d_{kx}d_{ky}F_{xy}(k),$$

$$F_{xy}(k) = \frac{\partial A_y(k)}{\partial k_x} - \frac{\partial A_x(k)}{\partial k_y}, \qquad A_i = -i\sum_{\alpha\in\text{filled bands}}\langle\alpha\mathbf{k}|\frac{\partial}{\partial k_i}|\alpha\mathbf{k}\rangle. \tag{3.1}$$

3.1 Current Operators

In this section, and for a large part of this book, we will be working with one-body, noninteracting Hamiltonians. We aim to find the response of these Hamiltonians to applied fields. We pick a generic Hamiltonian

$$H = \sum_{i,j,\alpha,\beta} c_{i\alpha}^{\dagger}(h_{ij}^{\alpha\beta} - \mu\delta_{ij}\delta_{\alpha\beta})c_{j\beta} \tag{3.2}$$

with i, j lattice sites (on an arbitrary dimensional lattice) and α, β orbital indices (which, as of now, can include spins, orbitals, and any other quantum number). The hoppings between sites i and j of obital α to orbital β, respectively, are $h_{ij}^{\alpha\beta}$. Assume there is a total of m orbitals. Models with a unit cell containing two sites, such as graphene, can also be written in this form because we can consider the site index as an extra "orbital" index. We assume translational symmetry, which means $h_{ij}^{\alpha\beta} = h_{i-j}^{\alpha\beta}$, and we Fourier-transform:

$$c_j^\alpha = \frac{1}{\sqrt{N}} \sum_k e^{ikj} c_k^\alpha, \qquad H = \sum_k c_{k\alpha}^\dagger h_k^{\alpha\beta} c_{k\beta}, \tag{3.3}$$

where N is the total number of lattice sites. Here, k is the momentum quantum number, which is vector $\mathbf{k}$, depending on the space dimensionality. The Fourier-transformed Hamiltonian just shown includes the chemical potential μ in the matrix h (we need to remember that we are now measuring everything from chemical potential zero). We are now ready to obtain the current operators through two different methods, one of which we will use in linear response and one of which we will use in the Hofstadter problem of electrons in a magnetic field on a lattice.

3.1.1 Current Operators from the Continuity Equation

The current satisfies a continuity equation:

$$\dot{\rho}(x) + \nabla \cdot \mathbf{J}(x) = 0, \qquad \rightarrow \dot{\rho}_q - i\mathbf{q} \cdot \mathbf{J}_q = 0, \tag{3.4}$$

where $A_j = (1/\sqrt{N}) \sum_k e^{ikj} A_k$ is the Fourier transform of an operator A on the lattice and N is the number of lattice sites. The density operator on site is $\rho_i = c_i^\dagger c_i$ (orbital index summation is assumed), whose Fourier transform becomes

$$\rho(\mathbf{q}) = \frac{1}{\sqrt{N}} \sum_k c_{\mathbf{k}+\mathbf{q}}^\dagger c_{\mathbf{k}}, \tag{3.5}$$

where $c_j^\dagger = (1/\sqrt{N}) \sum_k e^{-ikj} c_k^\dagger$. The continuity equation then gives

$$-i\mathbf{q} \cdot \mathbf{J}_\mathbf{q} = -\dot{\rho}_q = i[\rho, H] = i\frac{1}{\sqrt{N}} \sum_{p,k} h_p [c_{\mathbf{k}+\mathbf{q}}^\dagger c_\mathbf{k}, c_p^\dagger c_p], \tag{3.6}$$

which after some operator gymnastics becomes

$$\mathbf{q} \cdot \mathbf{J}_\mathbf{q} = \frac{1}{\sqrt{N}} \sum_{p,k} (h_{\mathbf{k}+\mathbf{q}} - h_\mathbf{k}) c_{\mathbf{k}+\mathbf{q}}^\dagger c_\mathbf{k} \tag{3.7}$$

We now want to particularize to the small q limit — we are interested in low-energy, long-wavelength fields and will not analyze fields that vary on atomic length-scales. If the field variation is larger than several lattice spacings, then this approximation is good. We could make the approximation $h_{k+q} - k_k \approx \frac{\partial h_k}{\partial \mathbf{k}} \cdot \mathbf{q}$, but this would be valid only to first order in $\mathbf{q}$. We can make a better approximation, valid to second order in $\mathbf{q}$, by shifting $\mathbf{k} \rightarrow \mathbf{k} - q/2$:

$$\begin{aligned} \mathbf{q} \cdot \mathbf{J}_\mathbf{q} &= \frac{1}{\sqrt{N}} \sum_{p,k} (h_{\mathbf{k}+q/2} - h_{\mathbf{k}-q/2}) c_{\mathbf{k}+q/2}^\dagger c_{\mathbf{k}-q/2} \\ &= \frac{1}{\sqrt{N}} \sum_{p,k} \left(\frac{\partial h_k}{\partial \mathbf{k}} \cdot \mathbf{q} \right) c_{\mathbf{k}+\frac{\mathbf{q}}{2}}^\dagger c_{\mathbf{k}-\frac{\mathbf{q}}{2}} + \mathcal{O}(q^2) \end{aligned} \tag{3.8}$$

The linear term q is important to get right in some cases, and hence the shift performed is very important. The current operator at small q is, hence,

$$\mathbf{J_q} = \frac{1}{\sqrt{N}} \sum_{p,k} c^\dagger_{\mathbf{k}+q/2} \frac{\partial h_k}{\partial \mathbf{k}} c_{\mathbf{k}-q/2}. \tag{3.9}$$

3.1.2 Current Operators from Peierls Substitution

We can also obtain the current operator directly from the tight-binding Hamiltonian under the influence of a vector potential of an external field. The derivation presented here is informative because we use it later in the book to analyze electrons on a lattice in a magnetic field. The hoppings $h_{ij}^{\alpha\beta}$ in the Hamiltonian without the electromagnetic field (with i, j lattice sites and α, β orbital indices) come from evaluating overlap integrals between the atomic orbitals of the on-site atoms involved. For each such orbital, when we have magnetic field, the momentum operator in the Hamiltonian (before taking the tight-binding approximation) changes p to $p - eA$. We are poised to ask the question of how to introduce the minimal coupling of the vector potential in the lattice Hamiltonian. One proposal (used extensively, but not the only one) is the so-called Peierls substitution: it corresponds to changing every hopping matrix element $h_{ij}^{\alpha\beta}$ to $h_{ij}^{\alpha\beta} e^{i \int_j^i \mathbf{A}(l)\cdot d\mathbf{l}}$, where $\mathbf{A}$ is the vector potential of the electromagnetic field, which can depend on the position l. This substitution generates a phase equal to 2π (which amounts to no change in the Hamiltonian) when we surround a plaquette pierced by a unit flux. Because a unit flux can be removed by performing a gauge transformation, the Peierls phase gives the right result. There is now a clear ambiguity: the phase obtained between two sites i and j depends on the path taken between them. We solve it by always taking the *shortest* path — a straight line connecting two lattice sites. We then use an approximation that is valid if the vector potential does not vary wildly over the integration path, which is of the order of a few lattice constants (we assume h_{ij} contains only a finite number of tight-binding matrix elements):

$$\int_R^{R'} \mathbf{A}(s, t) \cdot d\mathbf{s} \approx (\mathbf{R}' - \mathbf{R}) \cdot \frac{1}{2}(\mathbf{A}(\mathbf{R}', t) + \mathbf{A}(\mathbf{R}, t)) \approx (\mathbf{R}' - \mathbf{R}) \cdot \frac{1}{2}\mathbf{A}\left(\frac{\mathbf{R}' + \mathbf{R}}{2}, t\right) \tag{3.9a}$$

We want to expand the Hamiltonian to first order in the vector potential because we are dealing with linear response. To get the usual diamagnetic term, we would need to expand the Hamiltonian to second order. However, we do not analyze this term here because it is diagonal in the spatial indices and we are merely interested in the off-diagonal (Hall) response. A Taylor expansion gives

$$H = \sum_{i,j,\alpha,\beta} c^\dagger_{i\alpha} h_{ij}^{\alpha\beta} e^{i \int_j^i \mathbf{A}(l)\cdot d\mathbf{l}} c_{j\beta} \approx \sum_{i,j,\alpha,\beta} c^\dagger_{i\alpha} h_{ij}^{\alpha\beta} \left(1 + i \int_j^i \mathbf{A}(l) \cdot d\mathbf{l}\right) c_{j\beta} = H_0 + H_{\text{ext}},$$

where H_{ext} is the Hamiltonian change due to the applied field. Let $r = i - j$ and remember that our system exhibits (in the absence of the electromagnetic field) translational invariance, so $h_{ij} = h_{i-j} = h_r$; this gives

$$\begin{aligned} H_{\text{ext}} &= \sum_{k_1,k_2,\alpha,\beta} c^\dagger_{k_1,\alpha} c_{k_2,\beta} \frac{1}{N} \sum_{r,j} e^{i(\mathbf{k}_2 - \mathbf{k}_1)\cdot \mathbf{j} - i\mathbf{k}_1 \cdot \mathbf{r}} h_r^{\alpha\beta} i \int_j^{j+r} \mathbf{A}(l) \cdot d\mathbf{l} \\ &\approx \sum_{k_1,k_2,\alpha,\beta} c^\dagger_{k_1,\alpha} c_{k_2,\beta} \frac{1}{N} \sum_{r,j} e^{i(\mathbf{k}_2 - \mathbf{k}_1)\cdot \mathbf{j} - i\mathbf{k}_1 \cdot \mathbf{r}} h_r^{\alpha\beta} i \mathbf{r} \cdot \mathbf{A}(\mathbf{j} + \frac{\mathbf{r}}{2}). \end{aligned} \tag{3.10}$$

We now take the Fourier transform $A_j = (1/\sqrt{N}) \sum_q e^{-i\mathbf{q}\cdot\mathbf{j}} A_q$ and rewrite the Hamiltonian as

$$H_{\text{ext}} = \frac{1}{\sqrt{N}} \sum_{k_1,k_2,q,\alpha,\beta} c^{\dagger}_{k_1,\alpha} c_{k_2,\beta} \delta_{\mathbf{k}_2-\mathbf{k}_1-\mathbf{q}} \sum_r e^{-i(\mathbf{k}_1+\frac{\mathbf{q}}{2})\cdot\mathbf{r}} h^{\alpha\beta}_r i\mathbf{r}\cdot\mathbf{A}_\mathbf{q}. \tag{3.11}$$

Shift the momentum $k_1 + q/2 = k$ and use

$$h^{\alpha\beta}_k = \frac{1}{\sqrt{N}} \sum_r e^{i\mathbf{k}\cdot\mathbf{r}} h^{\alpha\beta}_r \rightarrow \frac{\partial h^{\alpha\beta}_k}{\partial \mathbf{k}} = \frac{1}{\sqrt{N}} \sum_r i\mathbf{r} e^{i\mathbf{k}\cdot\mathbf{r}} h^{\alpha\beta}_r \tag{3.12}$$

to obtain the final form of the external perturbation Hamiltonian of the applied field:

$$H_{\text{ext}} = \sum_{k,q,\alpha,\beta} c^{\dagger}_{k+q/2,\alpha} c_{k-q/2,\beta} \frac{\partial h^{\alpha\beta}_k}{\partial \mathbf{k}} \cdot \mathbf{A}_{-q} = \sum_q \mathbf{j}_q \cdot \mathbf{A}_{-q}. \tag{3.13}$$

We see current operator is the same as before.

3.2 Linear Response to an Applied External Electric Field

Having found the current operator, we now try to understand the influence that an applied electric field has on a system. We assume a small applied field so that linear response theory is enough. However, as we will see later, the Hall conductance is fixed by general arguments having to do with gauge invariance, which are much stronger than the perturbative linear response calculations used here. The generic Hamiltonian for a many-particle system in an external field $A(x,t)$ is

$$\begin{aligned} H &= \sum_i \frac{1}{2}(p_i - eA(x_i,t))^2 + \sum_{i<j} V_{ij} \\ &= \frac{1}{2}\sum_i p_i^2 + \sum_{i<j} V_{ij} - e\sum_i \frac{1}{2}(p_i A(x_i,t) + A(x_i,t)p_i) + \frac{1}{2}\sum_i e^2 A_i^2, \end{aligned} \tag{3.14}$$

where V_{ij} is a two-body interaction potential (for example Coulomb interaction). As in the previous two subsections, if we define the total current operator as

$$J(x,t) = \frac{1}{2}\sum_i \left[(p_i - eA(x,t))\delta(x-x_i) + \delta(x-x_i)(p_i - eA(x,t))\right], \tag{3.15}$$

we immediately see that the Hamiltonian can be written as

$$H = H_0 - e\int d^3x \int (\delta\mathbf{A}(x,t))\cdot\mathbf{J}, \tag{3.16}$$

with H_0 is the Hamiltonian unperturbed by the electric or magnetic field. The second integral over the variation of the vector potential, $\delta\mathbf{A}(x,t)$, is needed because $J = \delta H/\delta A$; hence the coefficient $\frac{1}{2}$ in the last term of the Hamiltonian, $\frac{1}{2}\sum_i e^2 A_i^2$, becomes linear when differentiated to obtain the current. The total current is made of paramagnetic and diamagnetic contributions:

$$J(x,t) = j(x) - neA(x,t), \tag{3.17}$$

where $n = \sum_i \delta(x - x_i)$ is the uniform density of electrons, which we assume to be uniform to maintain translational invariance. To first order in $A(x, t)$, the Hamiltonian then becomes

$$H = H_0 - e \int d^3x \mathbf{A}(x, t) \cdot \mathbf{j}(x) = H_0 + H_{\text{ext}}. \tag{3.18}$$

We would now like to measure the response to the field $\mathbf{A}$ of the electric current in the system described by H_0. The easiest consistent way to do this is by linear response. Suppose that initially we are in some eigenstate $|E_N\rangle$ of the full Hamiltonian, H_0. As time evolves, the state will change according to both the original H_0 as well as to the new Hamiltonian of the applied field, H_{ext}:

$$i\frac{\partial |E_N(t)\rangle}{\partial t} = H|E_N(t)\rangle. \tag{3.19}$$

Without loss of generality, we choose to write the eigenstate at time t as a matrix evolution of the state at $t = 0$, or $|E_N(t)\rangle = \exp(-iH_0t)U_{\text{ext}}(t)|E_N\rangle$, and obtain

$$i\frac{\partial U_{\text{ext}}}{\partial t}|E_N\rangle = \exp(iH_0t)H_{\text{ext}}\exp(-iH_0t)U_{\text{ext}}|E_N\rangle = H_{\text{ext}}(t)U_{\text{ext}}|E_N\rangle \tag{3.20}$$

We find that U_{ext} satisfies the integral equation

$$U_{\text{ext}}(t) = 1 - i\int_0^t H_{\text{ext}}(t')U_{\text{ext}}(t')\,dt', \tag{3.21}$$

where $H_{\text{ext}}(t) = e^{iH_0t}H_{\text{ext}}e^{-iH_0t}$. To first order in H_{ext}, which is a small perturbation, we have

$$U_{\text{ext}}(t) = 1 - i\int_0^t H_{\text{ext}}(t')\,dt'. \tag{3.22}$$

What we want to obtain is the expectation value in the evolved ground state of the current density at time t:

$$\langle E_N(t)|\mathbf{J}(x, t)|E_N(t)\rangle = \langle E_N(t)|\mathbf{j}(x)|E_N(t)\rangle - ne\mathbf{A}(x, t). \tag{3.23}$$

Now we can substitute for the time-evolved ground state (remembering $|E_N\rangle = |E_N(0)\rangle$) and obtain, to lowest order, terms in H_{ext}:

$$\langle E_N(t)|\mathbf{J}(x, t)|E_N(t)\rangle = \langle E_N|\mathbf{j}(x)|E_N\rangle + i\int_0^t dt'\langle E_N|[H_{\text{ext}}(t'), \mathbf{j}(x, t)]|E_N\rangle - ne\mathbf{A}(x, t), \tag{3.24}$$

where $j(x, t) = e^{iH_0t}j(x)e^{-iH_0t}$ is the time-evolved current operator. Using $H_{\text{ext}} = -e\int d^3x\mathbf{A}(x, t)\cdot\mathbf{j}(x)$ and measuring only deviations induced by the applied field from the original ground state of the system (which means we subtract the expectation value in the original ground state from the expectation value in the evolved ground-state), the expectation value of the current becomes

$$\langle E_N(t)|J_i(x, t)|E_N(t)\rangle = \langle E_N|j_i(x)|E_N\rangle + \int_{-\infty}^{\infty} dt' \int d^3x' \sum_j R_{ij}(x - x', t - t')A_j(x', t'), \tag{3.25}$$

where the response function reads

$$R_{ij}(x-x', t-t') = -i\theta(t-t')\langle E_N|[j_i(x,t), j_j(x',t')]|E_N\rangle - ne\delta_{ij}\delta(x-x')\delta(t-t'). \tag{3.26}$$

In the expression of the preceding response function, we have assumed that $A(x, t)$ is proportional to a switch-on term — i.e., $A(x, t) \sim \theta(t)$ — which allows us to place the integral over time from deep past to deep future. We want to calculate the preceding response function in the many-fermion N-body ground state, so $|E_N\rangle = |E_{GS}\rangle$. This zero-temperature result can be extended to nonzero temperatures by averaging over the canonical ensemble of initial states, which we assume occur with a probability $\exp(-\beta E_i)/Z$, where $Z = \sum_i \exp(-\beta E_i)$. The only thing that is modified is the expression for the expectation value of the current commutator:

$$\langle E_N|[j_i(x,t), j_j(x',t')]|E_N\rangle \to \frac{\mathrm{Tr}\{e^{-\beta H_0}[j_i(x,t), j_j(x',t')]\}}{Z}. \tag{3.27}$$

We now want to relate these expressions to the time-ordered response function and then to the finite-temperature Green's functions. This can be done through the fluctuation dissipation theorem, which we explicitly present here.

3.2.1 The Fluctuation Dissipation Theorem

As we saw in the previous section, the linear response to an applied driving force is given by a retarded Green's function of the type: $-i\theta(t)\langle[A(t), B(0)]\rangle$ where the expectation value is taken in the many-body Fermion ground state. We now will work with a canonical ensemble at nonzero temperature T, treating the case $T = 0$ as a limit. Suppose the system has Hamiltonian H that we can diagonalize, obtaining a complete set of eigenstates $|n\rangle$: $H|n\rangle = E_n|n\rangle$. Then,

$$\begin{aligned}\langle A(t)B(0)\rangle &= \frac{1}{Z}\sum_n \langle n|e^{-\beta H}e^{iHt}Ae^{-iHt}B|n\rangle \\ &= \frac{1}{Z}\sum_{n,m}\langle n|Be^{-\beta H}|m\rangle\langle m|e^{iHt}Ae^{-iHt}|n\rangle \\ &= \frac{1}{Z}\sum_{n,m} e^{-\beta E_m + i(E_m-E_n)t}\langle n|B|m\rangle\langle m|A|n\rangle.\end{aligned} \tag{3.28}$$

Fourier-transform the correlation function $J_1(\omega) = \int_{-\infty}^{\infty}\langle A(t)B(0)\rangle e^{i\omega t}dt$ to obtain

$$J_1(\omega) = 2\pi Z^{-1}\sum_{mn} e^{-\beta E_m}\langle n|B|m\rangle\langle m|A|n\rangle\delta(E_m - E_n + \omega) \tag{3.29}$$

(the 2π constant comes from the convention of the definition of the δ function as a Fourrier transform). Similarly, we find for the other time-ordered possibility of the correlation function, $J_2(\omega) = \int_{-\infty}^{\infty}\langle B(0)A(t)\rangle e^{i\omega t}dt$, that

$$J_2(\omega) = 2\pi Z^{-1}\sum_{mn} e^{-\beta E_n}\langle n|B|m\rangle\langle m|A|n\rangle\delta(E_m - E_n + \omega) = e^{-\beta\omega}J_1(\omega). \tag{3.30}$$

We see that J_1 and J_2 are the spectral functions associated with the two correlation functions $\langle A(t)B(0)\rangle$, $\langle B(0)A(t)\rangle$. At zero temperature ($T = 0$), only the terms for which $m = 0$ survive in J_1; for these terms, E_0 is the ground-state energy. Hence, at $T=0$, $J_1(\omega)$ is nonzero only for positive frequencies, and J_2 then turns out to be nonzero for negative frequencies. Our objective for the rest of this section is to relate the time-ordered Green's function to the retarded Green's function. We now consider the retarded Green's function, $G^R(t) = -i\theta(t)\langle[A(t), B(0)]\rangle$, which can be expressed as an inverse Fourier transform of J_1, J_2:

$$G^R(t) = -i\theta(t)\int_{-\infty}^{\infty} J_1(\omega)(1 - e^{-\beta\omega})e^{-i\omega t}\frac{d\omega}{2\pi}. \tag{3.31}$$

We now Fourier-transform the retarded Green's function $G^R(\omega) = \int_{-\infty}^{\infty} G^R(t)e^{i\omega t}\,dt$ to obtain

$$\begin{aligned} G^R(\omega) &= -i\int_{-\infty}^{\infty} J_1(\omega')(1 - e^{-\beta\omega'})\frac{d\omega'}{2\pi}\int_0^{\infty} e^{i(\omega-\omega'+i\eta)t}\,dt \\ &= \int_{-\infty}^{\infty}\frac{J_1(\omega')}{\omega-\omega'+i\eta}(1 - e^{-\beta\omega'})\frac{d\omega'}{2\pi}, \end{aligned} \tag{3.32}$$

where we have added a small damping term, $\eta = 0^+$, to make the integral convergent and well defined. In most of the physical expressions we will be considering, A and B are the same operator. Because of this (and even in the case where A and B would be only Hermitian conjugates of each other and not identical), we have that $J_1(\omega)$ is real. Hence, using a representation of the Dirac delta function, we have

$$\text{Im}(G^R(\omega)) = -\frac{1}{2}(1 - e^{-\beta\omega})J_1(\omega) \tag{3.33}$$

The time-ordered Green's function $G^T(t) = -i\langle T[A(t)B(0)]\rangle$ (T puts the operator $A(t)$ to the left of $B(0)$ if $t > 0$ and to the right if $t < 0$) takes the form

$$G^T(t) = -i\int_{-\infty}^{\infty}(\theta(t)J_1(\omega') + \theta(-t)J_2(\omega'))e^{-i\omega' t}\frac{d\omega'}{2\pi}. \tag{3.34}$$

The Fourier transform of the time-ordered Green's function is

$$G^T(\omega) = \int_{-\infty}^{\infty} J_1(\omega')\left\{\frac{1}{\omega-\omega'+i\eta} - \frac{e^{-\beta\omega'}}{\omega-\omega'-i\eta}\right\}\frac{d\omega'}{2\pi} \tag{3.35}$$

Hence, if $J_1(\omega)$ is real (as is the case if the expectation value is made out of the same operators), we have

$$\text{Re}G^T(\omega) = \mathcal{P}\int_{-\infty}^{\infty}(1 - e^{-\beta\omega'})\frac{J_1(\omega')}{\omega-\omega'}\frac{d\omega'}{2\pi} \tag{3.36}$$

$$\text{Im}G^T(\omega) = -\frac{1}{2}(1 + e^{-\beta\omega})J_1(\omega) \tag{3.37}$$

However, we knew from before that $\text{Im}G^R(\omega) = -\frac{1}{2}(1 - e^{-\beta\omega})J_1(\omega)$, so we have

$$\text{Re}G^R(\omega) = \text{Re}G^T(\omega), \ \text{Im}G^R(\omega) = \tanh\left(\frac{1}{2}\beta\omega\right)\text{Im}G^T(\omega) \tag{3.38}$$

At zero temperature we then have $\text{Im}G^R(\omega) = (\omega)/(|\omega|)\text{Im}G^T(\omega)$. We now make the connection to the electrical conductivity: if $J_1(\omega)$ is the Fourier transform of the current-current correlation function $\langle j(t)j(0)\rangle$, by using $J_\alpha(\omega) = \sigma_{\alpha\beta}(\omega)E_\beta(\omega) = i\omega\sigma(\omega)A_\beta(\omega)$, we have $G^R(\omega) = i\omega\sigma(\omega)$and, hence,

$$J_1(\omega) = -\frac{2}{1 - e^{-\beta\omega}}\omega\sigma(\omega). \tag{3.39}$$

3.2.2 Finite-Temperature Green's Function

We now consider the finite-temperature Green's function. It is important because in this book we primarily compute correlation functions in the formalism of the finite-temperature Green's function. We end up computing quite a few correlation functions, so it is good to set up the formalism on a sound basis. The temperature function

$$G^T(\sigma) = \langle T[A(\sigma)B(0)]\rangle = \theta(\sigma)\langle A(\sigma)B(0)\rangle + \theta(-\sigma)\langle B(0)A(\sigma)\rangle, \tag{3.40}$$

where $A(\sigma) = e^{\sigma H}Ae^{-\sigma H}$, is the imaginary time-evolved operator A. As in the case of the real-time correlation functions, we obtain

$$\langle A(\sigma)B(0)\rangle = Z^{-1}\sum_{mn} e^{-\beta E_m}\langle n|B|m\rangle\langle m|A|n\rangle e^{\sigma(E_m - E_n)},$$

$$\langle B(0)A(\sigma)\rangle = Z^{-1}\sum_{mn} e^{-\beta E_n}\langle n|B|m\rangle\langle m|A|n\rangle e^{\sigma(E_m - E_n)}. \tag{3.41}$$

The finite-temperature Green's functions have a particular periodicity, which we now derive. For bosons (our current operators A and B here have bosonic commutation relations — they are composites of two-fermion operators), we pick $0 < \sigma < \beta$ (or, equivalently, $-\beta < \sigma - \beta < 0$):

$$\begin{aligned} G^T(\sigma - \beta) &= \theta(\beta - \sigma)\langle B(0)A(\sigma - \beta)\rangle \\ &= Z^{-1}\sum_{mn} e^{-\beta E_n - \beta(E_m - E_n)}\langle n|B|m\rangle\langle m|A|n\rangle e^{\sigma(E_m - E_n)} \\ &= Z^{-1}\sum_{mn} e^{-\beta E_m}\langle n|B|m\rangle\langle m|A|n\rangle e^{\sigma(E_m - E_n)} = G^T(\sigma). \end{aligned} \tag{3.42}$$

Due to the periodicity property, we can thus expand $G^T(\sigma)$ in the range $0 \leq \sigma \leq \beta$ as a Fourier series and obtain, for the Fourier coefficients,

$$\begin{aligned} G^T(\omega_m) &= \frac{1}{\beta}\int_0^\beta d\sigma e^{i\omega_m\sigma}G^T(\sigma) \\ &= \frac{1}{\beta Z}\sum_{mn}\langle n|B|m\rangle\langle m|A|n\rangle\frac{e^{-\beta E_n} - e^{-\beta E_m}}{E_m - E_n + i\omega_m}, \end{aligned} \tag{3.43}$$

with $\omega_m = \frac{2\pi m}{\beta}$. The Fourier transform of the finite-temperature Green's function can be immediately expressed in terms of the spectral functions $J_1(\omega)$ and $J_2(\omega)$:

$$G^T(\omega_m) = \frac{-1}{\beta} \int_{-\infty}^{\infty} (1 - e^{-\beta\omega'}) \frac{J_1(\omega')}{i\omega_m - \omega'} \frac{d\omega'}{2\pi}. \tag{3.44}$$

This means $G^T(\omega_m)$ can be obtained directly from $J_1(\omega')$. Because we have earlier proved that

$$G^R(\omega) = \int_{-\infty}^{\infty} (1 - e^{-\beta\omega'}) \frac{J_1(\omega')}{\omega - \omega' + i\eta} \frac{d\omega'}{2\pi}, \tag{3.45}$$

we immediately see that $G^R(\omega)$ can be obtained from $-\beta G^T(i\omega_m)$ by replacing the imaginary variable $i\omega_m$ by $\omega + i\eta$. This *analytic continuation* is rather amazing: $G^T(\omega_m)$ is defined only at a discrete set of points on the imaginary axis, but we can obtain the retarded Green's function from it for *all* the values of ω on the real axis.

3.3 Current-Current Correlation Function and Electrical Conductivity

We now are ready to express the electrical conductivity in first order in linear response as a current-current correlation function through the formalism of finite-temperature Green's functions. The current kernel $R_{ij}(x - x', t - t')$ contains a diamagnetic part, which equals $ne\delta_{ij}\delta(x - x')\delta(t - t')$ and which we neglect. Notice that when we obtained the current operator as the expansion of the tight-binding matrix elements through Peierls substitution, we looked only at the first-order expansion of the $e^{i\int_j^i \mathbf{A}\cdot d\mathbf{l}}$; this also corresponds to neglecting the diamagnetic contribution. The reason for this neglect is not that the diamagnetic contribution is small — it is not — and it is very important in obtaining the electromagnetic response of superconductors as well as the correct response of metals. However, it is diagonal in space indices and, hence, does not contribute to the Hall conductivity or other topological invariants of insulators, which are the focus of our book.

For the Fourier transform of the current, $J_i(k, \omega) = R_{ij}^R(k, \omega)A_j(k, \omega) = i\omega\sigma_{ij}(k, \omega)A_j(k, \omega)$. From our previous exposition of the finite-temperature Green's functions, we can turn the expression of the retarded Green's function into calculating the finite-temperature Green's function. This procedure can actually help with getting the finite-temperature conductivity as well. We can get the retarded Green's function by calculating the finite-temperature Green's function (in Fourier components) and then doing the analytic continuation. We first perform the Fourier transform over the position coordinates; by using the translational invariance of the Hamiltonian and, thus, the Green's function, we obtain

$$R_{ij}^T(q, \sigma - \sigma') = \langle T[j_i(q, \sigma) j_j(-q, \sigma')] \rangle. \tag{3.46}$$

We remember that we want equation (3.46) for $\sigma - \sigma' \geq 0$. For the imaginary time-dependent current operator, $j_i(q, \sigma) = e^{\sigma H} j_i(q) e^{-\sigma H}$, we have

$$j_i(q, \sigma) = \sum_k c_{k+\frac{q}{2},\alpha}^\dagger(\sigma) \frac{\partial h^{\alpha\beta}(k)}{\partial k_i} c_{k-q/2,\beta}(\sigma), \tag{3.47}$$

where $c(\sigma) = e^{\sigma H} c e^{-\sigma H}$. Because the Green's function depends only on $\sigma - \sigma'$, we can shift the origin and put $\sigma' = 0$ to obtain

$$R^T_{ij}(q, \sigma) = \sum_{k,p} \frac{\partial h^{\alpha\beta}(k)}{\partial k_i} \frac{\partial h^{\gamma\theta}(p)}{\partial p_j} \langle T[c^\dagger_{k+q/2,\alpha}(\sigma) c_{k-q/2,\beta}(\sigma) c^\dagger_{p-q/2,\gamma} c_{p+q/2,\theta}] \rangle, \tag{3.48}$$

where $c = c(0)$. By Wick's theorem we have (keeping only the connected part) $\langle T[c^\dagger_{k+q/2,\alpha}(\sigma) c_{k-q/2,\beta}(\sigma) c^\dagger_{p-q/2,\gamma} c_{p+q/2,\theta}] \rangle = \langle c^\dagger_{k+q/2,\alpha}(\sigma) c_{p+q/2,\theta} \rangle \langle c_{k-q/2,\beta}(\sigma) c^\dagger_{p-q/2,\gamma} \rangle$, which is the product of two Green's functions. We now examine the Fourier transform:

$$G^T(p, \sigma) = \sum_n e^{-i\omega_n \sigma} G^T(p, \omega_n), \tag{3.49}$$

where σ is in the interval $(-\beta \leq \sigma \leq \beta)$ and, hence, $\omega_n = \frac{2\pi n}{(2\beta)}$ with $n = 0, \pm 1, \pm 2, \ldots$). We remember that before we had the interval $0 \leq \sigma \leq \beta$. This was because we were dealing with effectively bosonic objects — currents being bilinear in fermionic operators. The fermionic finite-temperature Green's function is antiperiodic with period β (assume $0 \leq \sigma \leq \beta$ and compute using equation (3.42)). The explicit form of the finite-temperature Green's function is

$$G^T(k, \omega_n) = \frac{1}{i\omega_n - h(k)} \tag{3.50}$$

With this, we have the Fourier transform of the $R^T_{ij}(q, \sigma)$:

$$R^T_{ij}(q, \nu_r) = -\frac{1}{\beta^2} \sum_{m,k} \mathrm{Tr} \left[\frac{\partial h(k)}{\partial k_i} G^T(k - \frac{q}{2}, \omega_m) \frac{\partial h(k)}{\partial k_j} G^T(k + \frac{q}{2}, \omega_m - \nu_r) \right]. \tag{3.51}$$

The factor $1/\beta^2$ follows from the Fourier transform. As we said before, we can obtain $R^R_{ij}(k, \omega)$ from $-\beta G^T(\omega_m)$ by the analytic continuation $i\omega_m \to \omega + i\eta$. We are finally capable of writing the conductivity tensor:

$$\sigma_{ij}(q, \omega) = \frac{R^R_{ij}(q, \omega)}{i\omega}, \quad R^R_{ij}(q, \omega) = Q_{ij}(q, i\nu_r \to \omega + i\eta). \tag{3.52}$$

3.4 Computing the Hall Conductance

We are now in a good position to use the machinery already developed to compute the Hall conductance of an insulator. We try to kill two birds with one stone: we try to both compute the Hall conductance and introduce a limit of an insulating Hamiltonian, the flat-band limit. The Hamiltonian in this limit is topologically equivalent to the original Hamiltonian and will be used over and over in the book to perform easy computations of correlation functions and entanglement spectra. For the general case of an arbitrary number of bands, the conductivity formula is very cumbersome: the Green's functions that appear in it contains poles at all the band energies, which, in general, are distinct. Calculations are cumbersome because we must take into account the residues at all the poles. What the flat-band limit accomplishes is to place all the occupied energies of the Hamiltonian at $-E$, whereas all the unoccupied states are placed at energy $+E$ (by convention, E is usually chosen to be 1 in some energy units) while keeping the eigenstates of the system unmodified. Topological properties of the system should not depend on the energies of the occupied bands: the flat-band limit will prove that. What we end up finding is that the Hall conductance does not depend on the energies of the bands — it depends only on the eigenstates. This is the first actual hint of the topological character of the Hall conductance. If, in our adiabatic change of the Hamiltonian, we had found that the Hall conductance depended on the energies of the occupied bands, then it would not have been a topological invariant: by making a small adiabatic transformation (that does not close the gap) of the Hamiltonian, we would obtain different Hall conductance. The fact that we can prove this is not the case is a hint that the Hall conductance is a true topological quantity.

The Hamiltonian $H = \sum_k c_{k\alpha}^\dagger h_k^{\alpha\beta} c_{k\beta}$ is easily solved by diagonalizing the matrix $(h_k)_{\alpha\beta} = h_k^{\alpha\beta}$. This gives us the bands of the Hamiltonian, which disperse. This is a matrix in orbital indices, and we find its eigenvalues:

$$h_k^{\alpha\beta} u_{\beta k}^i = \epsilon_i(k) u_{\alpha k}^i, \tag{3.53}$$

where $u_{\alpha k}^i$ is the α component of the vector u_k^i, which is the ith single-particle eigenstate of energy $\epsilon_i(k)$ of the Hamiltonian at momentum k. We can then diagonalize the matrix h_k:

$$U^\dagger h U = \begin{pmatrix} \epsilon_1(k) & 0 & 0 \\ 0 & \ldots & 0 \\ 0 & 0 & \epsilon_m(k) \end{pmatrix} = \epsilon \tag{3.54}$$

for the m orbitals, or degrees of freedom (spin included), in the system. Here, ϵ is the energy matrix of the system — the matrix with the single-particle energies on the diagonal. The elements of $U_{\alpha i} = u_\alpha^i$, which becomes a unitary matrix because eigenvectors of different energy levels (even if degenerate) are orthogonal, are $U^\dagger U = U_{\alpha i}^* U_{\alpha j} = u_\alpha^{*i} u_\alpha^j = \delta_{ij}$. We then have

$$H = \sum_k \sum_{a=1}^m \epsilon_l(k) \gamma_{k,a}^\dagger \gamma_{k,a}, \qquad \gamma_{k,a} = U_{ai}^\dagger c_{ki}, \qquad c_{ki} = U_{ia}(k) \gamma_{k,a}. \tag{3.55}$$

The unitary transformation U now allows us to perform an adiabatic transformation that helps us compute the Hall conductivity for a generic system.

3.4.1 Diagonalizing the Hamiltonian and the Flat-Band Basis

Because we assumed our system is an insulator, we can put our Fermi level in the insulating gap, or — since we are measuring everything from rescaled Fermi level zero—we can put all the bands separated by a full gap (at all k) from the others below zero energy, so we fill all negative-energy bands. For an insulator with p filled bands (out of m) and with a full gap, we have the following set of inequalities:

$$\epsilon_1(k) \leq \epsilon_2(k) \leq \cdots \leq \epsilon_p(k) < 0 < \epsilon_{p+1}(k) \leq \epsilon_{p+2}(k) \leq \cdots \leq \epsilon_m(k). \tag{3.56}$$

We can now pick two arbitrary energies, ϵ_G and ϵ_E, with the only constraint being $\epsilon_G < 0 < \epsilon_E$ (G and E from occupied ground state and empty). The values ϵ_G and ϵ_E can be constant, say, $\pm 1 eV$. We now *define* the adiabatic interpolation of the energies $\epsilon_1(k)$, ..., $\epsilon_p(k)$ all to ϵ_G and $\epsilon_{p+1}(k)$, ..., $\epsilon_m(k)$ all to ϵ_E:

$$\begin{aligned} E_i(k, t) &= \epsilon_i(k)(1-t) + \epsilon_G t, \quad 1 \leq i \leq p, \\ E_i(k, t) &= \epsilon_i(k)(1-t) + \epsilon_E t, \quad p < i \leq m. \end{aligned} \tag{3.57}$$

The Hamiltonian remains gapped throughout this adiabatic evolution, no bands cross the Fermi level, and, even more, the structure of the band energies (inequalities) remains the same during this adiabatic evolution. This last step is not really important; the only important property is that the Hamiltonian remains gapped for all adiabatic parameter values $t \in [0, 1]$. Because the Fermi level is in the gap and no states cross it during this evolution, we immediately realize that the topological properties (such as Chern numbers) of the interpolated Hamiltonian are identical to the ones of the initial Hamiltonian. This result will actually be substantiated by proving that the Hall conductance does not depend on the energies $E_i(k, t)$ of the interpolation. The Hamiltonian interpolation we consider is

$$h(k, t) = U(k)\mathrm{diag}(E_1(k, t), \ldots, E_p(k, t), E_{p+1}(k, t), \ldots, E_m(k, t))U^{\dagger}(k). \quad (3.58)$$

This Hamiltonian interpolates between the original one and the flat-band system at $t = 1$, — but *with identical eigenstates*. At $t = 1$, we have the nice property that the Hamiltonian is just the sum over the projectors in the occupied states (with coefficient ϵ_G) and the projectors in the empty states (with coefficient ϵ_E):

$$h(k, 1) = \epsilon_G \sum_{a=1}^{p} |a, k\rangle\langle a, k| + \epsilon_G \sum_{\beta=p+1}^{m} |\beta, k\rangle\langle \beta, k|, \quad (3.59)$$

where $|i, k\rangle\langle i, k| = u_n^i u_m^{i*} = P_{nm}^i$ is the matrix having as the n, m element the combination $u_n^i u_m^i$ of the components of the ith energy eigenstate. This matrix obviously projects an eigenstate:

$$P^i u^j = P_{nm}^i u_m^j = u_n^i u_m^{i*} u_m^j = u_n^i \delta_{ij} = u^j \delta_{ij}. \quad (3.60)$$

Because we are dealing with projectors, the following identities are obvious:

$$P_G + P_E = I, \quad P_G P_E = 0 \quad P_G^2 = P_G, \quad P_E^2 = P_E. \quad (3.61)$$

The real advantage of this formalism is that the Green's function of the system assumes a nice form:

$$G(i\omega_m, k) = \frac{P_G(k)}{i\omega_m - \epsilon_G} + \frac{P_E(k)}{i\omega_m - \epsilon_E}, \quad (3.62)$$

as can be proved by direct multiplication plus use of the projector identities:

$$I = (i\omega_m - \epsilon_G P_G - \epsilon_E P_E)\left(\frac{P_G(k)}{i\omega_m - \epsilon_G} + \frac{P_E(k)}{i\omega_m - \epsilon_E}\right). \quad (3.63)$$

We can immediately see that the current of the adiabatic Hamiltonian is

$$\frac{\partial h(k, 1)}{\partial k_i} = \epsilon_G \frac{\partial P_G(k)}{\partial k_i} + \epsilon_E \frac{\partial P_E(k)}{\partial k_i} = (\epsilon_G - \epsilon_E)\frac{\partial P_G(k)}{\partial k_i}. \quad (3.64)$$

The current-current correlator used to compute the conductivity is

$$\begin{aligned} Q_{ij}(i\nu_m) = {} & \frac{1}{V\beta} \sum_{k,n} (\epsilon_G - \epsilon_E)^2 \\ & \times \mathrm{Tr}\left(\frac{\partial P_G(k)}{\partial k_i}\left(\frac{P_G(k)}{i(\omega_n + \nu_m) - \epsilon_G} + \frac{P_E(k)}{i(\omega_m + \nu_m) - \epsilon_E}\right)\frac{\partial P_G(k)}{\partial k_j}\left(\frac{P_G(k)}{i\omega_n - \epsilon_G} + \frac{P_E(k)}{i\omega_n - \epsilon_E}\right)\right), \end{aligned} \quad (3.65)$$

where V is the volume of the system considered.

To perform the projector algebra required for the preceding correlator, we obtain the following set of nice identities involving projectors:

$$P_G(\partial_i P_G) = \partial_i P_G - (\partial_i P_G)P_G, \quad (\partial_i P_E)P_G = -P_E \partial_i P_G, \quad \partial_i P_G = -\partial_i P_E, \tag{3.66}$$

$$(\partial_i P_G)P_G(\partial_j P_G)P_G = (\partial_i P_G)(\partial_j P_G)P_G - (\partial_i P_G)(\partial_j P_G)P_G^2 = 0, \tag{3.67}$$

and an identical expression for $G \to E$. We also have

$$(\partial_i P_G)P_G(\partial_j P_E)P_G = -(\partial_i P_G)(\partial_j P_G)P_E P_G = 0, \tag{3.68}$$

$$(\partial_i P_G)P_E(\partial_j P_G)P_E = -(\partial_i P_G)(\partial_j P_E)P_G P_E = 0, \tag{3.69}$$

$$(\partial_i P_G)P_G(\partial_j P_G)P_E = -(\partial_i P_G)P_G(\partial_j P_E)P_E$$

$$= (\partial_i P_G)(\partial_j P_G)P_E^2 = -(\partial_i P_G)(\partial_j P_E)P_E,$$

$$(\partial_i P_G)P_E(\partial_j P_G)P_G = -(\partial_i P_G)(\partial_j P_E)P_G. \tag{3.70}$$

Of the preceding identities, two are of particular importance, $(\partial_i P_G)P_G(\partial_j P_G)P_G = 0$ and $(\partial_i P_G)P_E(\partial_j P_G)P_E = 0$, because they cancel two out of the four terms of the current-current correlation function. We then need to perform the Matsubara summation. A short crash course on how to do Matsubara summation starts by noticing that the fermionic frequencies $i\omega_m = \frac{(2m+1)\pi}{\beta}$ are the poles of the Fermi distribution function $n_F(z) = \frac{1}{(\exp \beta z)+1}$. We hence have

$$\sum_n f(i\omega_n) = \frac{\beta}{2\pi i}\int_{C_1} dz \frac{1}{e^{\beta z}+1} A(z) = \frac{\beta}{2\pi i}\int_{C_2} dz \frac{1}{e^{\beta z}+1} A(z)$$

$$= \beta \frac{1}{e^{\beta z_i}+1} \text{Res}(A(z, z_i)). \tag{3.71}$$

The contours C_1, C_2 are standard (found in any field theory book) and are chosen as such: C_1 is the counterclockwise contour going vertically to the right and left of the vertical (imaginary) axis, whereas C_2 is the deformed circuit going around the poles of $A(z)$, which are situated at positions z_i. Because we can deform C_1 into C_2, the value of the integral needs to be the same on these two contours.

At zero temperature, the occupation of states $\epsilon_E > 0$ is zero ($n_F(\epsilon_E) = 0$), and the occupation of states $\epsilon_G < 0$ is ($n_F(\epsilon_G) = 1$). We then get

$$Q_{ij}(i\nu_m) = \frac{1}{V}\sum_k -(\epsilon_G - \epsilon_E)^2$$

$$\times \text{Tr}\left(\frac{(\partial_i P_G)(\partial_j P_E)P_E}{\epsilon_G - i\nu_m - \epsilon_E} + \frac{(\partial_i P_G)(\partial_j P_E)P_G}{\epsilon_G + i\nu_m - \epsilon_E}\right). \tag{3.72}$$

We are interested only in the antisymmetric part of the conductivity (the Hall conductivity) and, hence, neglect any symmetric combination in indices i, j to obtain

$$Q_{ij}(i\nu_m) = \frac{1}{V}\sum_k (\epsilon_G - \epsilon_E)^2 Tr\left((\partial_i P_G)(\partial_j P_E)P_E \frac{2i\nu_m}{(i\nu_m)^2 + (\epsilon_G - \epsilon_E)^2}\right). \tag{3.73}$$

Analytic continuation in order to obtain the time-ordered correlator from the temperature-ordered one, $i\nu_m \to \omega + i\delta$, gives, in the DC limit $\omega \to 0$,

$$\begin{aligned} Q_{ij}(\omega \to 0) &= \lim_{\omega\to 0}\frac{1}{V}\sum_k (\epsilon_G - \epsilon_E)^2 \mathrm{Tr}\left((\partial_i P_G)(\partial_j P_E)P_E \frac{2\omega}{(\omega + i\delta)^2 + (\epsilon_G - \epsilon_E)^2}\right) \\ &= \frac{2\omega}{\Omega}\sum_k \mathrm{Tr}[(\partial_i P_G)(\partial_j P_E)P_E] \end{aligned} \tag{3.74}$$

We next massage the trace:

$$\mathrm{Tr}[(\partial_i P_G)(\partial_j P_E)P_E] = \mathrm{Tr}[(\partial_i P_G)(\partial_j P_G)P_G] - \mathrm{Tr}[(\partial_i P_G)(\partial_j P_G)]. \tag{3.75}$$

The last term is again symmetric in i, j, so we neglect it. We have now obtained the correlation function purely in terms of projection operators into the ground-state manifold of occupied bands, $P_G(k)$. Numerically, this is the way we compute the Hall conductance because projectors are *manifestly* gauge invariant, thereby bypassing the need for the gauge smoothing that we referred to in our Berry-phase section. By using the explicit expression of the projectors in terms of Bloch states and after tedious algebra, we find

$$\mathrm{Tr}[(\partial_i P_G)(\partial_j P_G)P_G] = \sum_{\alpha=1}^{m}\langle\partial_i(\alpha, k)|\partial_j|\alpha, k\rangle + \sum_{\alpha,\beta=1}^{m}\langle\alpha, k|\partial_i|\beta, k\rangle\langle\beta, k|\partial_j|\alpha, k\rangle. \tag{3.76}$$

The second term is symmetric in i, j:

$$\sum_{\alpha,\beta=1}^{m}\langle\alpha, k|\partial_i|\beta, k\rangle\langle\beta, k|\partial_j|\alpha, k\rangle = \sum_{\alpha,\beta=1}^{m}\langle\beta, k|\partial_i|\alpha, k\rangle\langle\alpha, k|\partial_j|\beta, k\rangle, \tag{3.77}$$

and we hence neglect it, leading to the antisymmetric part of the $\langle\partial_i(\alpha, k)|\partial_j|\alpha, k\rangle$:

$$\epsilon_{ij}\mathrm{Tr}[(\partial_i P_G)(\partial_j P_G)P_G] = \sum_{\alpha=1}^{m}\left[\langle\partial_i(\alpha, k)|\partial_j|\alpha, k\rangle - \langle\partial_j(\alpha, k)|\partial_i|\alpha, k\rangle\right]. \tag{3.78}$$

The derivatives used here imply gauge smoothing; to be able to differentiate the Bloch wavefunctions, we need to make them smooth over the BZ. This is usually a rather complicated procedure, which is why we use the form of the correlation function in terms of projectors. We hence have, for the Hall conductivity,

$$\sigma_{ij\text{Hall}} = \frac{1}{i\omega}\frac{Q_{ij}(\omega) - i \leftrightarrow j}{2} = \frac{1}{V}\sum_{k}\sum_{a=1}^{m}\langle\partial_i(a,\, k)|\partial_j|a,\, k\rangle - \langle\partial_j(a,\, k)|\partial_i|a,\, k\rangle$$

$$= \frac{1}{\Omega}\sum_{k}\sum_{a=1}^{m} -i(\langle\partial_i(a,\, k)|\partial_j|a,\, k\rangle - \langle\partial_j(a,\, k)|\partial_i|a,\, k\rangle)$$

$$= \int \frac{dk_x dk_y}{(2\pi)^2}\sum_{a=1}^{m} -i(\langle\partial_i(a,\, k)|\partial_j|a,\, k\rangle - \langle\partial_j(a,\, k)|\partial_i|a,\, k\rangle) \tag{3.79}$$

in units of e^2/h. The Hall conductivity is an integral over the filled bands of the Berry curvature. Only the zero-frequency Hall conductivity has topological meaning — nonzero frequency corrections will contain terms related to the energies of the empty and occupied bands.

3.5 Alternative Form of the Hall Response

In the previous several sections, we proved that the Hall conductivity, σ_H, manifests itself in the off-diagonal current response:

$$j_i = \sigma_H \epsilon_{ij} E_j. \tag{3.80}$$

We now obtain a different form of the Hall conductivity that suggests a link between the density of the system and the magnetic field. Through the continuity equation and then through Maxwell's equations, we can relate the Hall conductance to another type of response:

$$\frac{\partial\rho}{\partial t} = -\nabla \cdot j = -\sigma_H\left(\partial_x E_y - \partial_y E_x\right) = \sigma_H \frac{\partial B}{\partial t}, \tag{3.81}$$

and, hence,

$$\sigma_H = \frac{\partial\rho}{\partial B}. \tag{3.82}$$

Equation (3.82) is called the Streda formula. It relates the change in the density of a system as an applied magnetic field is varied. We come back to this form of the Hall conductance in subsequent chapters when we prove the Diophantine equation. Equations (3.80)–(3.82) can be combined together in a covariant way:

$$j^\mu = \sigma_H \epsilon^{\mu\nu\lambda}\partial_\nu A_\lambda. \tag{3.83}$$

The response equations can then obviously be obtained from a coupling $\mathbf{j}\cdot\mathbf{A}$, which, when integrated over space-time, gives the Lagrangian:

$$S_{eff} = \frac{\sigma_H}{2}\int d^2x \int dt A_\mu \epsilon_{\mu\nu\tau}\partial_\nu A_\tau. \tag{3.84}$$

The preceding action is the Chern-Simons field theory of the external field A_μ (different from a fluctuating gauge field because A_μ is a background applied field). Chern-Simons actions will appear extensively later in the book.

3.6 Chern Number as an Obstruction to Stokes' Theorem over the Whole BZ

In the previous sections we have proved that the Hall conductance of a filled band is equal to the integral of the Berry curvature over the BZ. In this section, we show that the Hall conductance of a filled band must be an integer, which is called the Chern number. The Berry curvature is the curl of the Berry gauge field **A**. The BZ is a torus; hence, it has no boundary. An application of Stokes' theorem would then give the Hall conductance as an integral of the Berry gauge field over the boundary of the BZ, but since the latter has no boundary, $\sigma_{xy} = 0$ if $\mathbf{A}(k)$ is well defined in the whole BZ. Nonzero values of the Hall conductance (Chern number) are consequences of the nontrivial structure of the Berry vector potential — in particular, the fact that it has singularities at points in the BZ. A nonzero Chern number (Hall conductance) is a testament to the fact that we cannot choose a global gauge that is continuous and single valued over the entire BZ. Hence, a nonzero Chern number represents an obstruction to the application of Stokes, theorem over the whole BZ. This particular interpretation of the Chern number is useful because it generalizes to the case of the Z_2 invariant of 2-D topological insulators: the Z_2 invariant can be interpreted as an obstruction to Stokes' theorem but only in *half* the BZ.

Under a $U(1)$ gauge transformation, the wavefunction of the ath energy level transforms as

$$|a, k\rangle' = e^{if(k)}|a, k\rangle, \tag{3.85}$$

where we consider $f(k)$ a smooth function over the whole BZ. The corresponding gauge transformation on the Berry potential $A^l(k)$ is

$$\mathbf{A}_a(k)' = \mathbf{A}_a(k) + \nabla f(k). \tag{3.86}$$

Because we performed a gauge transformation, the Hall conductance, which as an observable quantity is gauge invariant, can just as well be calculated in this new gauge. The availability of an arbitrary gauge transformation naively implies that the overall phase of the wavefunction can be chosen arbitrarily by making a suitable gauge transformation. For example, one phase choice is to choose the function $f(k)$ such that it makes the first component of the vector $|a, k\rangle'$ real (for that matter, we could have picked any component, be it first, second, or any other — notice that in an insulator the Bloch state must be at least a two-component vector because an insulator by default must have at least two bands). If the first component of $|a, k\rangle$ is nonzero, we can always pick a phase to gauge-transform and make it real by choosing $e^{if(k)} = ||a, k\rangle_1|/|a, k\rangle_1$. Call this wavefunction ψ_1. Even if we pick it to be real, we still have the ambiguity that we can multiply the wavefunction by ± 1. Still, naively, it seems that we can always find a smooth way of defining the first component — for example, by taking the product of the first component at two very close momenta and requiring that they be positive. The way of defining a single-valued, smooth wavefunction is called finding a smooth gauge.

However, if we could always find a smooth gauge, then the Hall conductance would always vanish, by the preceding argument involving Stokes' theorem. Thus, it hence must be true that there are certain cases in which we cannot pick a smooth gauge for our wavefunction. In our specific gauge-smoothing procedure, it must be that we cannot pick a phase to make the first component real. This happens when the first component of the Bloch part of $|a, k\rangle$ vanishes at some points in the Brillouin zone. We denote the positions in the BZ of the zeros of the first component of the Bloch wavefunction by k_s, with $s = 1, \ldots, N$; around them, we subsequently define small regions (they can be circles or any other shape as long as the regions surround the zeros) by

$$R_s^\epsilon = \{k \in T_{BZ}^2 || k - k_s| < \epsilon, |a, k_s\rangle_1 = 0\}. \tag{3.87}$$

Inside those regions, we cannot pick a smooth wavefunction because our gauge-smoothing procedure fails. Hence, we pick a different phase convention inside these regions: for example, we fix the phase by saying that $|a, k_s\rangle_2$ is real (or pick any other component that does not vanish in that region; we would be extremely unlucky if more than the first component vanishes in that region, but even in that case, we know that a nonvanishing component must exist because the Bloch state cannot be identically zero). Once we pick the phases, the state is completely well defined over the patches that contain the zeros of the first component. Call this wavefunction ψ_2, for which the gauge smoothing just means multiplication by $e^{ig(k)} = ||a, k\rangle_2| / |a, k\rangle_2$. Obviously, ψ_2 is smoothly defined within the small circles (it is not well defined everywhere outside the circles — the second component vanishes outside the circles), whereas ψ_1 is smoothly defined outside the circles and has ambiguities related to the vanishing of its first component in the circles. At the boundary between the two regions, the two wavefunctions are related by a gauge transformation:

$$\psi_2(k) = e^{i(g(k) - f(k))} \psi_1(k) = e^{i\chi(k)} \psi_1(k). \tag{3.88}$$

Hence, the Berry potentials for the two wavefunctions are themselves related by a gauge transformation:

$$\mathbf{A}_2(k) = \psi_2 \partial_{\mathbf{k}} \psi_2 = \psi_1 \partial_{\mathbf{k}} \psi_1 + i \nabla \chi(k) = \mathbf{A}_1(k) + i \nabla \chi(k). \tag{3.89}$$

The Hall conductance is gauge invariant, but if we want to obtain it from integrating the curl of the Berry vector potential over the BZ, we must be able to have smoothly differentiable wavefunctions. We have these by patches, so we write

$$\sigma_{xy} = \frac{e^2}{h} \frac{1}{2\pi i} \left(\int_{T_{BZ}^2 - R_s^\epsilon} \nabla \times \mathbf{A}_1(k) + \int_{R_s^\epsilon} \nabla \times \mathbf{A}_2(k) \right) \tag{3.90}$$

where T_{BZ}^2 is the BZ torus. The Berry vector potentials are now well behaved in each of their respective patches, and we can apply Stokes' theorem to obtain

$$\sigma_{xy} = \frac{e^2}{h} \frac{1}{2\pi i} \left(\int_{\partial(T_{BZ}^2 - R_s^\epsilon)} d\mathbf{k} \cdot \mathbf{A}_1(k) + \int_{\partial(R_s^\epsilon)} d\mathbf{k} \cdot \mathbf{A}_2(k) \right) \tag{3.91}$$

The torus does not have boundary, so we have $\partial(T_{BZ}^2 - R_s^\epsilon) = -\partial(R_s^\epsilon)$ because they have opposite orientation; thus,

$$\begin{aligned} \sigma_{xy} &= \frac{e^2}{h} \frac{1}{2\pi i} \int_{\partial(R_s^\epsilon)} d\mathbf{k} \cdot (\mathbf{A}_2(k) - \mathbf{A}_1(k))) \\ &= \frac{e^2}{h} \frac{1}{2\pi i} \int_{\partial(R_s^\epsilon)} dk \cdot i \nabla \chi(k) = \frac{e^2}{h} n, \end{aligned} \tag{3.92}$$

where

$$n = \frac{1}{2\pi} \oint_{\partial(R_s^\epsilon)} d\mathbf{k} \cdot \nabla \chi(k) \tag{3.93}$$

is the winding number of the gauge transformation on the boundary of the piecewise definition of the wavefunctions. As a simple example, if we make the boundary of the R_s region a perfect circle $\partial(R_s^\epsilon) = k_s + \epsilon e^{i\theta}$, with $\theta \in [0, 2\pi)$, we would have

$$n = \frac{1}{2\pi}\oint_{\partial(R_s^\epsilon)} d(\epsilon e^{i\theta}) \cdot \frac{\partial\chi(k)}{\partial \epsilon e^{i\theta}} = \frac{1}{2\pi}\int_0^{2\pi} d\theta \partial_\theta \chi(k_s + \epsilon e^{i\theta})$$
$$= \frac{1}{2\pi}(\chi(k_s + \epsilon e^{i2\pi - 0^-}) - \chi(k_s + \epsilon)) \tag{3.94}$$

Notice that n has to be an integer because, as we complete a full path around each circle (closed line) R_s, we have the single-valuedness constraint

$$\psi_2(k_s + \epsilon) = e^{i\chi(k_s+\epsilon)}\psi_1(k_s + \epsilon) = \psi_2(k_s + \epsilon e^{i(2\pi - 0^-)})$$
$$= \exp[i\chi(k_s + \epsilon e^{i(2\pi-0^-)})]\psi_1(k_s + \epsilon e^{i(2\pi-0^-)}) = \exp[i\chi(k_s + \epsilon e^{i(2\pi-0^-)})]\psi_1(k_s). \tag{3.95}$$

Hence, upon a full revolution around the point k_s, we necessarily have $\chi(k_s + \epsilon e^{i2\pi-0^-}) - \chi(k_s + \epsilon) = 2m\pi$. Although the phase of the wavefunction is gauge variant, the total vorticity (winding number) is a gauge-invariant quantity. The positions of the vorticities in the BZ can be changed, for example, by picking different components of the Bloch state to gauge smoothen; we can even separate vorticities and create $\pm$ vorticities from vacuum as long as vorticity conservation is maintained — but the sum of all vorticities in the BZ is constant and equal to the Chern number.

This rather abstract discussion of the Chern number as an obstruction to Stokes' theorem in the full BZ as well as the Chern number as a winding of the gauge-gluing functions will be explicitly demonstrated for Dirac fermions at a later stage in the book.

3.7 Problems

1. *Current Operator:* Find the q^2 correction to the current operator $\mathbf{J_q} = \frac{1}{\sqrt{N}}\sum_k c^\dagger_{\mathbf{k}+\mathbf{q}/2} \frac{\partial h_k}{\partial \mathbf{k}} c_{\mathbf{k}-\mathbf{q}/2}$. Is there one?

2. *Hall Conductance and the Jacobian of a Two-Band Model:* Without employing the flattening-band procedure, show that for a two-band Hamiltonian $H = d_i(k)\sigma_i$, the Hall conductance is given by the integral over the BZ of the Jacobian of the map $\hat{d}_i(k) = d_i(k)/d(k)$: $T^2 \to S^2$, where $d(k) = \sqrt{d_i d_i}$. (*Hint*: You need a gapped system to have the target manifold S^2.)

3. *Chern-Simons Equations of Motions:* Obtain the quantum Hall equations of motion by minimizing the Chern-Simons field theory action in equation (3.84).

4

Time-Reversal Symmetry

In the field of topological insulators, we are in the business of unraveling the effect that the presence (or absence) of continuous and discrete symmetries has on the physics of materials. For example, we can define a Chern number as the Hall conductance of a system exhibiting a continuous $U(1)$ symmetry associated with charge conservation.

The most important discrete symmetry for experimentally realizable systems is time-reversal. Time reversal is a transformation that reverses the arrow of time:

$$T : t \to -t. \tag{4.1}$$

Time -reversal (TR) symmetry is a fundamental property: systems behave quite differently depending on whether or not they exhibit time reversal. Until 2005–2006, it was thought that only systems without TR symmetry exhibit interesting physics: if the system is not symmetric upon reversing the motion of the particles, then Hall voltages and other interesting phenomena can occur. In a magnetic field, the particles's trajectory is bent. If we reverse the arrow of time, the particles will not retrace their motion, and the system breaks TR symmetry because of the presence of a magnetic field. Because of the band theory revolution started in 2005 by Kane and Mele, we now understand that analyzing TR-invariant systems can be equally rewarding: even though, as we show next, these systems cannot exhibit Hall effects, they can, nevertheless, exhibit other equally interesting topological phenomena, such as the nontrivial Z_2 topological classification.

In this section, we present a detailed account of TR symmetry for both spinful and spinless particles, of its action on operators and Bloch Hamiltonians, and of its implications on the Berry potential and Berry curvature. We introduce time reversal as a symmetry T of the Hamiltonian H of a system:

$$[H, T] = 0, \tag{4.2}$$

which means that if $|\psi\rangle$ is an eigenstate of H, then so is $T|\psi\rangle$. In many cases $T|\psi\rangle = |\psi\rangle$, but very interesting things happen if the two states are different. This is exactly the difference between integer and half-integer spin particles.

4.1 Time Reversal for Spinless Particles

The TR operator changes only the arrow of time. As such, it leaves the position operator $\hat{x}$ unchanged—in particular, time reversal commutes with any spatial symmetry. However, it flips the sign of the momentum operator $\hat{p}$ because it is proportional to the velocity, a time derivative of a TR-invariant quantity (the position operator):

$$T\hat{x}T^{-1} = x, \qquad T\hat{p}T^{-1} = -\hat{p}. \tag{4.3}$$

We would like to find the representation of the TR operator. By looking at the action of time reversal on the commutator of $\hat{x}$ and $\hat{p}$,

$$T[\hat{x}, \hat{p}]T^{-1} = Ti\hbar T^{-1} = -[\hat{x}, \hat{p}] = -i\hbar, \tag{4.4}$$

we are led to the equation

$$TiT^{-1} = -i. \tag{4.5}$$

The preceding makes it clear that the TR operator must be proportional to the operator of complex conjugation. Such operators are called *anti-unitary* and,unlike operators, do not have eigenvalues. For a particle without spin, the story ends here because the Hilbert space can be made out of scalars; hence,

$$T = K, \tag{4.6}$$

where K is the complex-conjugation operator. In general, however, the TR operator can be represented as

$$T = UK, \tag{4.7}$$

where U is a unitary matrix and K is complex conjugation. We then have

$$T^2 = UKUK = UU^* = U(U^T)^{-1} = \phi. \tag{4.8}$$

Here ϕ has to be a diagonal matrix of phases—the TR operator, applied twice, must get us back to the original state, up to a phase matrix. By using the fact that transpose of a diagonal matrix is the matrix itself, we find

$$U = \phi U^T, \qquad U^T = U\phi, \tag{4.9}$$

and, hence,

$$U = \phi U^T = \phi U \phi. \tag{4.10}$$

This can happen only if $\phi = \pm 1$ and, hence, $T^2 = \pm 1$. This is specific to T being antiunitary — a unitary op can have any phase. If $\phi = -1$, then $U = -U^T$. For spinless particles, $\phi = 1$.

4.1.1 Time Reversal in Crystals for Spinless Particles

We would now like to analyze the consequences that time reversal has for Bloch Hamiltonians. For spinless particles,T leaves the on-site creation operators unchanged, unlike the case for spinful particles. We have

$$Tc_jT^{-1} = c_j, \tag{4.11}$$

where we can add any orbital indices to the creation operators as long as the index is not spin. We then have, after Fourier-transforming,

$$Tc_jT^{-1} = \frac{1}{\sqrt{N}} \sum_k e^{-ikR_j} Tc_k T^{-1} = \frac{1}{\sqrt{N}} \sum_k e^{-ikR_j} c_{-k}. \tag{4.12}$$

In the preceding, the Fourier exponent was complex-conjugated because of the action of the T operator. Hence, the action of the time-reversal operator on the annihilation operator of an electron at momentum k just flips the sign of the momentum: $Tc_kT^{-1} = c_{-k}$. We are now ready to obtain the transformation of a Bloch Hamiltonian under time reversal. For a TR-invariant Hamiltonian, we then have

$$THT^{-1} = \sum_k c_{-k}^\dagger Th(k)T^{-1}c_{-k} = H = \sum_k c_k^\dagger h(k)c_k, \tag{4.13}$$

and we find for the Bloch Hamiltonian $h(k)$,

$$Th(k)T^{-1} = h(-k). \tag{4.14}$$

If $\psi(k)$ is an eigenstate at momentum k, then $T\psi(k)$ is an eigenstate at $-k$. However — and this is crucial — at the special time-reversal invariant momentum points such as $(0, 0, \ldots)$, $(\pi, \pi. \ldots)$ etc. , we *cannot* prove that we have a double degeneracy because for spinless particles we have that $T^2 = 1$ and find (at a generic T invariant momentum $k = 0$)

$$\langle\psi(0)|T\psi(0)\rangle = \sum_m \psi_m(0)^* T\psi_m(0) = \sum_m \psi_m(0)^* \psi_m^*(0) \neq 0, \tag{4.15}$$

so the bands are *not* generically degenerate; in other words, $\psi(k)$ and $T\psi(k)$ can (and usually do) belong to the same band. In other words, a TR-invariant system with spinless particles can very well have just one band. There is no Kramers' theorem, and the states do not need to be doubly degenerate at T-invariant points in the BZ.

4.1.2 Vanishing of Hall Conductance for T- Invariant Spinless Fermions

We show here that spinless systems with T symmetry cannot exhibit nonzero Hall conductance. Although this is obvious from a physical standpoint, we show how to obtain it by brute-force formalism. As an example, we consider a single band $|u(k)\rangle$ below the Fermi level; the generalization to multiple bands is obvious. We have

$$F_{ij}(-k) = -i(\langle\partial_i u(-k)|\partial_j u(-k)\rangle - i \leftrightarrow j). \tag{4.16}$$

But we know that $u(-k) = Tu(k) = u(k)^*$, so

$$\begin{aligned} F_{ij}(-k) &= -i\sum_m (\partial_i u_m(-k)^* \partial_j u_m(-k) - i \leftrightarrow j) \\ &= -i\sum_m (\partial_i u_m(k)\partial_j u_m^*(k) - i \leftrightarrow j) = -F_{ij}(k), \end{aligned} \tag{4.17}$$

which then gives zero upon integration over the BZ.

4.2 Time Reversal for Spinful Particles

We now look at particles with internal angular momentum, or spin, **S**. This requires an extra action of the time-reversal operator; because angular momentum is itself a momentum, it is odd under time reversal

$$T\mathbf{S}T^{-1} = -\mathbf{S}. \tag{4.18}$$

This implies that the spin flips its direction under time reversal. We can represent this action by a rotation by π around some arbitrary axis. Since ancient times, the convention has been to rotate the spin around the y-axis by π. The TR operator must implement this rotation; at the same time, it must be proportional to the complex conjugation operator because its action on the position-momentum commutator remains unchanged regardless of whether the particle has spin or not. With the choice of the rotation axis as y, the form of the TR operator is fixed:

$$T = e^{-i\pi S_y} K. \tag{4.19}$$

We now want to find its square. We assume a standard spin representation in which S_y is purely imaginary. We have

$$\begin{aligned} T \cdot T &= \exp(-i\pi S_y)(K \exp(-i\pi S_y)K) \\ &= \exp(-i\pi S_y)(\exp(i\pi S_y^*)) = \exp(-i2\pi S_y) \end{aligned} \tag{4.20}$$

This result is of fundamental importance: acting with time reversal twice hence rotates the spin by 2π, which for particles with integer spin is equivalent to the identity operator, whereas for particles with half-integer spin, it gives a factor of -1. We remember this is identical to what was obtained in the Berry phase calculation. For spin-$\frac{1}{2}$ particles, the matrix exponential can be easily performed. We pick as spin-$\frac{1}{2}$ the usual $\mathbf{S} = (\hbar/2)(\sigma^x, \sigma^y, \sigma^z)$ and obtain

$$\begin{aligned} e^{-}i\pi\sigma_y/2 &= \sum_{k=0}^{\infty} \frac{1}{k!} \left(\frac{-i\pi\sigma_y}{2}\right)^k \\ &= \cos\left(\frac{\pi}{2}\right) \begin{pmatrix} 1 & 0 \\ 0 & 1 \end{pmatrix} + \sin\left(\frac{\pi}{2}\right) \begin{pmatrix} 0 & -1 \\ 1 & 0 \end{pmatrix} = -i\sigma_y \end{aligned} \tag{4.21}$$

Here we can check explicitly that $T^2 = -1$:

$$T^2 = -i\sigma_y i\sigma_y^* K K = -\sigma_y \sigma_y = -1. \tag{4.22}$$

From $T^2 = -1$, we have that, for half-integer spin particles, $T^{-1} = -T$; hence, for spin-$\frac{1}{2}$,

$$T\mathbf{S}T^{-1} = -i\sigma_y \mathbf{S}^* K i\sigma_y K = -i\sigma_y \mathbf{S}^*(-i)\sigma_y^* K K = \sigma_y \mathbf{S}^* \sigma_y = -\mathbf{S}. \tag{4.23}$$

For the last equality, if the components of $\mathbf{S}$ are x, z, the complex conjugation does nothing, but the matrix σ_y anticommutes with $\sigma_{x,z}$ (Pauli matrices form both a Clifford $\{\sigma_a, \sigma_b\} = 2\delta_{ab}$ and an $SU(2)$ $[\sigma_a, \sigma_b] = i\epsilon_{abc}\sigma_c$ algebra). If the component of $\mathbf{S}$ is σ_y, then taking the complex conjugate changes its sign.

4.3 Kramers' Theorem

We are now in position to prove the most important theorem for a TR-invariant system. For half-integer spin, the $T^2 = -1$ property gives an important theorem called Kramers' theorem: for each energy in a system with an odd number of particles with half-integer spin, there are at least two degenerate states. We will then apply this theorem to systems with an added translational symmetry (Bloch Hamiltonians).

Let $|\psi\rangle$ be a single particle eigenstate of energy E. If we have a TR-symmetric Hamiltonian, then $[T, H] = 0$, and the state $T|\psi\rangle$ is also an eigenstate with energy E. If we can prove that they are actually orthogonal, then the spectrum is doubly degenerate. To show this, we remember that our operator U in $T = UK$ is a unitary operator that satisfies $U = -U^T$ (because $T^2 = -1$); i.e., U is antisymmetric and unitary. Hence,

$$\begin{aligned}\langle\psi|T\psi\rangle &= \sum_{m,n} \psi_m^* U_{mn} K \psi_n \sum_{m,n} \psi_m^* U_{mn} \psi_n^* \\ &= \sum_{m,n} \psi_n^* (-) U_{nm} K \psi_m = -\langle\psi|T\psi\rangle = 0.\end{aligned} \tag{4.24}$$

Note that the $T^2 = -1$ was crucial to proving this theorem. Moreover, we can compute the scattering probability of a state $|\psi\rangle$ into its TR partner $T|\psi\rangle$ for a TR- invariant Hamiltonian:

$$\begin{aligned}\langle T\psi|H|\psi\rangle &= \sum_{m,n,p} (U_{mp} K \psi_p)^* H_{mn} \psi_n = \sum_{m,n,p} (U^\dagger)_{pm} \psi_p H_{mn} \psi_n \\ &= \sum_{m,n,p} (U^\dagger)_{pm} \psi_p (THT^{-1})_{mn} \psi_n = -\sum_{m,n,p,q,r} \psi_p (U^\dagger)_{pm} U_{mr} H_{rq}^* U_{qn}^* \psi_n \\ &= -\sum_{n,p,q} \psi_p H_{pq}^* U_{qn}^* \psi_n = -\sum_{m,n,p} U_{mp}^* \psi_p H_{mn} \psi_n.\end{aligned} \tag{4.25}$$

So, we have proved that $\langle T\psi|H|\psi\rangle = -\langle T\psi|H|\psi\rangle = 0$. If we had a set of n excitations and we were to calculate the probability that n left-movers are scattered back to n right-movers, we would get a factor of $(-1)^n$. We can immediately show that for an even number of excitations, the scattering process between the Kramers' pairs does not vanish, but for an odd number of excitations it does.

All these results could have much more easily been obtained by noting the following property, which holds for any two eigenstates $|\psi\rangle$ and $|\phi\rangle$:

$$\begin{aligned}\langle T\phi|T\psi\rangle &= \sum_{m,p,r} (U_{mp} K \phi_p)^* U_{mr} K \psi_r \\ &= \sum_{m,p,r} U_{mp}^* \phi_p U_{mr} \psi_r^* = \sum_{p,r} \phi_p \delta_{pr} \psi_r^* = \langle\psi|\phi\rangle.\end{aligned} \tag{4.26}$$

This is independent of whether $T^2 = 1$ or -1 and holds as long as T is antiunitary. By choosing $|\psi\rangle = T\,|\phi\rangle$, we obtain the preceding properties.

4.4 Time-Reversal Symmetry in Crystals for Half-Integer Spin Particles

We now want to obtain the action of the T operator on Bloch Hamiltonians of spin -$\frac{1}{2}$ particles. As before, by Fourier transform, the Hamiltonian has a spectrum labeled as $\mathbf{k}$:

$$H = \sum_{\mathbf{k}} c_{\mathbf{k}\alpha\sigma}^\dagger h_{\alpha,\beta}^{\sigma,\sigma'}(\mathbf{k}) c_{\mathbf{k}\beta\sigma'}, \tag{4.27}$$

where α, β are the orbital indices and σ, σ' are the spin indices. We first have to find the action of the TR operator on the Hilbert space. The Hilbert space is a tensor product of states created by the on-site operators $c_{ja\sigma}$, $c^\dagger_{j,a\sigma}$. We want to prove

$$Tc_{ja\uparrow}T^{-1} = c_{ja\downarrow}, \quad Tc_{ja\downarrow}T^{-1} = -c_{ja\uparrow}. \tag{4.28}$$

Notice the relative minus sign present in the the second equation. This is a rather subtle equation: if we naively take $c_{ja\downarrow}$ and plug it into the second equation, we get $T^2c_{ja\uparrow}T^{-2} = -c_{ja\uparrow}$. We might be inclined to think that because $T^2 = T^{-2} = -1$, we get a contradiction. Of course, this is not true because we have to look at the operator equation as acting on states of the defined particle number. When we do that, we immediately recognize that T^{-2} and T^2 necessarily act on states of different fermion number (i.e., one even, the other odd, or viceversa), and hence only one of them is -1, the other being 1 (because even fermion number states necessarily have integer spin whereas odd fermion number states necessarily have half-integer spin). For example, assume that — because we know that the TR operator flips spin— the following two relations are true: $Tc_\uparrow T^{-1} = Ac_\downarrow$ and $Tc_\downarrow T^{-1} = Bc_\uparrow$; also; assume A, B are ± 1. (It is trivial to prove the generalized condition that A, B are phases. A, B can at most be phases because we want to maintain canonical commutation relations.) We now want to look at the action of the operator $Tc_\uparrow$ on the singly occupied state $c^\dagger_\uparrow|0\rangle$:

$$\begin{aligned} Tc_\uparrow c^\dagger_\uparrow|0\rangle &= T|0\rangle = |0\rangle^* = Ac_\downarrow Tc^\dagger_\uparrow|0\rangle = AT^{-1}Tc_\downarrow Tc^\dagger_\uparrow|0\rangle \\ &= AT^{-1}c_\uparrow T^2 c^\dagger_\uparrow|0\rangle = -ABT^{-1}|0\rangle = -AB|0\rangle^*. \end{aligned} \tag{4.29}$$

Hence, we see that $AB = -1$. The rest is gauge convention, but the product of AB must be negative unity. Thus, we have

$$Tc_{ja\sigma}T^{-1} = i\sigma_{\sigma\sigma'}c_{ja\sigma'}. \tag{4.30}$$

We also have $Tc^\dagger_{ja\uparrow}T^{-1} = c^\dagger_{ja\downarrow}$, $Tc^\dagger_{ja\downarrow}T^{-1} = -c^\dagger_{ja\uparrow}$, which gives

$$Tc^\dagger_{ja\sigma}T^{-1} = c^\dagger_{ja\sigma'}i(\sigma^y)^T_{\sigma'\sigma}. \tag{4.31}$$

We are interested in the transformation of the "Bloch" Hamiltonian, so we Fourier-transform the coefficients $c_{ja\sigma} = \frac{1}{\sqrt{N}}\sum_{\mathbf{k}} e^{i\mathbf{k}\cdot\mathbf{R_j}}c_{\mathbf{k}a\sigma}$ Applying the T operator gives

$$\begin{aligned} Tc^\dagger_{ja\sigma}T^{-1} &= \tfrac{1}{\sqrt{N}}\textstyle\sum_{\mathbf{k}} e^{i\mathbf{k}\cdot\mathbf{R_j}}Tc^\dagger_{\mathbf{k}a\sigma}T^{-1} = \tfrac{1}{\sqrt{N}}\textstyle\sum_{\mathbf{k}} e^{-i\mathbf{k}\cdot\mathbf{R_j}}Tc^\dagger_{-\mathbf{k}a\sigma}T^{-1} \\ &= c^\dagger_{ja\sigma'}i(\sigma^y)^T_{\sigma'\sigma} = \tfrac{1}{\sqrt{N}}\textstyle\sum_{\mathbf{k}} e^{-i\mathbf{k}\cdot\mathbf{R_j}}c^\dagger_{\mathbf{k}a\sigma'}i(\sigma^y)^T_{\sigma'\sigma}, \end{aligned} \tag{4.32}$$

and hence $Tc^\dagger_{\mathbf{k}a\sigma}T^{-1} = c^\dagger_{-\mathbf{k}a\sigma'}i(\sigma^y)^T_{\sigma'\sigma}$. An identical logic takes us to $Tc_{\mathbf{k}a\sigma}T^{-1} = i\sigma^y_{\sigma\sigma'}c_{-\mathbf{k}a\sigma'}$. We now want to obtain the transformation of the Bloch Hamiltonian:

$$\begin{aligned} H = THT^{-1} &= \sum_{\mathbf{k}} Tc^\dagger_{\mathbf{k}a\sigma}h^{\sigma,\sigma'}_{a,\beta}(\mathbf{k})c_{\mathbf{k}\beta\sigma'}T^{-1} \\ &= \sum_{\mathbf{k}} c^\dagger_{-\mathbf{k}a\sigma''}i(\sigma^y)^T_{\sigma''\sigma}Th^{\sigma,\sigma'}_{a,\beta}(\mathbf{k})T^{-1}i\sigma^y_{\sigma'\sigma'''}c_{-\mathbf{k}\beta\sigma'''} \\ &= \sum_{\mathbf{k}} c^\dagger_{\mathbf{k}a\sigma''}i(\sigma^y)^T_{\sigma''\sigma}Th^{\sigma,\sigma'}_{a,\beta}(-\mathbf{k})T^{-1}i\sigma^y_{\sigma'\sigma'''}c_{\mathbf{k}\beta\sigma'''}. \end{aligned} \tag{4.33}$$

We hence obtain the transformation of the Bloch Hamiltonian, $h^{\sigma'',\sigma'''}_{a,\beta}(\mathbf{k}) = i(\sigma^y)^T_{\sigma''\sigma}Th^{\sigma,\sigma'}_{a,\beta}(-\mathbf{k})T^{-1}i\sigma^y_{\sigma'\sigma'''}$. Because $h^{\sigma,\sigma'}_{a\beta}(-\mathbf{k})$ is a number, time reversal acts on it as complex conjugation: $Th^{\sigma,\sigma'}_{\alpha,\beta}(-\mathbf{k})T^{-1} = (h^{\sigma,\sigma'}_{\alpha,\beta}(-\mathbf{k}))^*$, and we obtain the final form of the TR action on the Hamiltonian:

$$h^{\sigma'',\sigma'''}_{\alpha,\beta}(\mathbf{k}) = i(\sigma^y)^T_{\sigma''\sigma}(h^{\sigma,\sigma'}_{\alpha,\beta}(-\mathbf{k}))^* i\sigma^y_{\sigma'\sigma'''}. \tag{4.34}$$

Without matrix indices, it is written in the compact form

$$Th_{\alpha,\beta}(\mathbf{k})T^{-1} = h_{\alpha,\beta}(-\mathbf{k}), \tag{4.35}$$

where the TR operator acts only on the spin indices of the Hamiltonian (i.e., T is diagonal in orbital indices — which is the reason we have applied time reversal on each orbital index of the Hamiltonian). We hence see that the time reversal takes the Bloch Hamiltonian at $\mathbf{k}$ and transforms it into the Bloch Hamiltonian at $-\mathbf{k}$. For a Bloch wavefunction of our Hamiltonian at momentum $\mathbf{k}$, $|u_I(\mathbf{k})\rangle$ with energy $E_{\mathbf{k}}$,

$$h(\mathbf{k})|u_I(\mathbf{k})\rangle = E_{\mathbf{k}}|u_I(\mathbf{k})\rangle, \tag{4.36}$$

then the wavefunction $T|u_I(\mathbf{k})\rangle$ is an eigenstate of the same Hamiltonian at momentum $-\mathbf{k}$, with energy $E(-\mathbf{k}) = E(\mathbf{k})$:

$$h(-\mathbf{k})T|u_I(\mathbf{k})\rangle = Th(\mathbf{k})T^{-1}T|u_I(\mathbf{k})\rangle = TE(\mathbf{k})|u_I(\mathbf{k})\rangle = E(\mathbf{k})T|u_I(\mathbf{k})\rangle, \tag{4.37}$$

where we have used the fact that the energy is a real number.

At special points in the BZ, which are invariant (mod a reciprocal lattice vector) under time reversal (such as $\mathbf{k}_{\text{special}} \in \{(0, 0), (\pi, 0), (0, \pi), (\pi, \pi)\}$ in two dimensions and similar ones in three dimensions — we denote these special points as $\mathbf{G}/2$ in any dimension), the Bloch Hamiltonian is invariant under the TR transformation: $Th_{\alpha\beta}(\mathbf{G}/2)T^{-1} = h_{\alpha\beta}(\mathbf{G}/2)$; hence, $\psi(\mathbf{G}/2)$ and $T\psi(\mathbf{G}/2)$ have the same energy, resulting in a double degeneracy at these special points in the Brillouin zone. The double degeneracy is guaranteed by the fact that we know that the two states are orthogonal due to Kramers' theorem.

Let us review the bidding up to this point: for spin-$\frac{1}{2}$ (or any half-integer spin) TR-invariant Hamiltonians, we have shown that states come in Kramers' degenerate pairs. When we introduce translational symmetry in the problem, the quantum states are indexed by the momentum quantum number. As a result, the Kramer's degeneracy generically is split into states at momentum $\mathbf{k}$ and $-\mathbf{k}$. Only states at special momenta ($\mathbf{k} \equiv -\mathbf{k}$) maintain a double degeneracy at the same $\mathbf{k}$ when translational symmetry is present. As a corollary, we now see that the minimal model for a time-reversal invariant topological insulator of half-integer spin electrons has to have at least four bands: a two-band model would always be a metal because a double degeneracy must exist between the bands at the TR- invariant momenta.

4.5 Vanishing of Hall Conductance for T-Invariant Half-Integer Spin Particles

In section 4.1.2, we showed that the Hall conductance is zero for spinless (or integer spin) fermions with TR invariance. Although technically more cumbersome, an identical statement can be proved for half-integer fermions with time reversal. We now show that the Berry curvature for points k and $-k$ in the BZ are the opposites of each other. For a TR-invariant system, because there are at least two occupied bands, I and II, we have

$$F(k_x, k_y) = (-i\langle\partial_{k_x}u_I(k)|\partial_{k_y}u_I(k)\rangle - x \leftrightarrow y) + \text{I} \leftrightarrow II. \tag{4.38}$$

A system with more than two occupied bands can always be reduced to sets of two occupied bands. This is the same as saying that the manifold of occupied bands is $U(2)^{N_{\text{occupied}}/2}$ — i.e., any other degeneracies besides the ones imposed by TR symmetry can be lifted; TR symmetry requires double degeneracies at TR-invariant momenta. Hence, the manifold for two occupied bands is $U(2)$. After some rather tedious algebra, we finally obtain the following relations between the bands:

$$\langle \partial_{-k_x} u_I(-k) | \partial_{-k_y} u_I(-k) \rangle - x \leftrightarrow y = -\langle \partial_{k_x} u_{\text{II}}(k) | \partial_{k_y} u_{II}(k) \rangle - x \leftrightarrow y. \tag{4.39}$$

A similar calculation yields

$$\langle \partial_{-k_x} u_{II}(-k) | \partial_{-k_y} u_{II}(-k) \rangle - x \leftrightarrow y = -\langle \partial_{k_x} u_I(k) | \partial_{k_y} u_I(k) \rangle - x \leftrightarrow y, \tag{4.40}$$

which immediately shows that the added curvature for these two bands satisfies

$$F(k_x, k_y) = -F(-k_x, -k_y). \tag{4.41}$$

This, in turn, forces the Chern number to vanish when integrated over the full BZ. In a future section, we will analyze the transformation of the field strength through the more "modern" language of the sewing matrix — which will allow us to generalize this to other point-group symmetries.

4.6 Problems

1. *Time Reversal for Spin-$\frac{3}{2}$*: Find a representation for the TR operator for spin-$\frac{3}{2}$ particles.

2. *Properties Under Time Reversal of Magnetic Fields and Spin-Orbit Coupling:* Show that a spin in a magnetic field is odd under time reversal, whereas spin-orbit coupling is even under time reversal.

3. *TR-Breaking Lattice Hamiltonian*: Find the TR properties of the Hamiltonian; $h(\mathbf{k}) = \sin(k_x)\sigma_x + \sin(k_y)\sigma_y + M\sigma z$ in the case where σ are (a) orbital (isospin) matrices and (b) spin matrices. Then generalize to a Hamiltonian $h(\mathbf{k}) = d_i(\mathbf{k})\sigma_i$.

4. *Vanishing Chern Number of a TR-Invariant Hamiltonian*: From the properties under time reversal of the two-band Hamiltonian $h(\mathbf{k}) = d_i(\mathbf{k})\sigma_i$ found in the previous problem, deduce directly that the Chern number vanishes if time reversal is respected, irrespective of whether the σ_i are orbital (isospin) or spin matrices.

5. *Eigenvalues of Spatial Symmetries in a TR-Invariant Hamiltonian*: Show that time reversal commutes with spatial symmetries, such as inversion. Then show that in spinful system (spin-$\frac{1}{2}$), the Kramers doublets have the same eigenvalues under spatial symmetries at high-symmetry points in the BZ.

5

Magnetic Field on the Square Lattice

Historically, the first system exhibiting topological behavior was the quantum Hall effect. When placed in a magnetic field large enough that Landau-level quantization becomes important, electrons exhibit a quantized Hall effect, in which the Hall conductance is an integer in units of the quantum of conductance, e^2/h. This effect can be analyzed through different means; in the continuum limit we obtain Landau levels, with their massive degeneracy of states—and the Hall conductance measures the number of Landau levels that are occupied. However, a better understanding of the problem is obtained by understanding the problem from a lattice perspective, which is the subject of the current chapter. We present a detailed exposition of the problem of a 2-D lattice pierced by a uniform magnetic field and learn about magnetic translation generators, the magnetic translation group, the Hofstadter problem, the Diophantine equation, explicit gauge fixing, and Hall conductance on the square lattice.

5.1 Hamiltonian and Lattice Translations

We first look at spinless electrons and want to analyze the simplest nontrivial, tight-binding Hamiltonian:

$$H = T_x + T_y + \text{h.c.}, \tag{5.1}$$

where T_x, T_y are the (covariant) translation operators by one lattice constant in the x- and y-directions, which, when a magnetic field is applied, take the covariant form

$$T_x = \sum_{m,n} c^\dagger_{m+1,n} c_{m,n} e^{i\theta^x_{m,n}}, \quad T_y = \sum_{m,n} c^\dagger_{m,n+1} c_{m,n} e^{i\theta^y_{m,n}}. \tag{5.2}$$

Since the Hamiltonian contains only nearest-neighbor terms, the phase factors can be consistently chosen as the integral of the external vector potential over the bond linking the nearest neighbors (as we learned in the Peierel's substitution of equation (3.9a):

$$\theta^x_{mn} = \frac{e}{\hbar}\int_m^{m+1} \mathbf{A}\cdot d\mathbf{x}, \quad \theta^y_{mn} = \frac{e}{\hbar}\int_n^{n+1} \mathbf{A}\cdot d\mathbf{y}. \tag{5.3}$$

We introduce the lattice derivatives:

$$\Delta_x f_{mn} = f_{m+1,n} - f_{m,n}, \quad \Delta_y f_{mn} = f_{m,n+1} - f_{m,n}. \tag{5.4}$$

The lattice curl of the phase factors is related to the the flux per plaquette ϕ_{mn}:

$$\begin{aligned} \text{rot}_{mn}\theta &= \Delta_x \theta^y_{mn} - \Delta_y \theta^x_{mn} = \theta^y_{m+1,n} - \theta^y_{m,n} - \theta^x_{m,n+1} + \theta^x_{m,n} \\ &= \frac{e}{\hbar}\int_{\text{unit cell}} \mathbf{A}\cdot d\mathbf{l} = 2\pi \frac{e}{h}\int B\, dS = 2\pi\phi_{mn}, \end{aligned} \tag{5.5}$$

where ϕ_{mn} is the number of flux quanta in units of h/e. The lattice Hamiltonian has a $U(1)$ gauge symmetry:

$$c_i \to U_i c_i, \quad e^{i\theta_{ij}} \to U_i e^{i\theta_{ij}} U_j^{-1}, \quad |U_i| = 1 \quad \forall j \in (m, n). \tag{5.6}$$

Under the preceding transformation, it is clear that the Hamiltonian remains invariant. The transformation on the hopping phases is just a gauge transformation on A. The spectrum of the current problem is the famous Hofstadter butterfly. In the weak field limit $\phi \to 0$, it gives the Landau-level structure, which we will soon see. Before that, we want to analyze its symmetries.

First, we look at the symmetries of this Hamiltonian. We immediately see that the covariant translation operators do not commute; for example, when we act on a single-particle state at site (m, n), $|\psi_{mn}\rangle = c^\dagger_{m,n}|0\rangle$, we find

$$T_x T_y |\psi_{ij}\rangle = T_x c^\dagger_{i,j+1} e^{i\theta^y_{ij}} |0\rangle = e^{i\theta^x_{i,j+1} + i\theta^y_{ij}} c^\dagger_{i+1,j+1}|0\rangle, \tag{5.7}$$

$$T_y T_x |\psi_{ij}\rangle = e^{i\theta^x_{ij} + i\theta^y_{i+1,j}} c^\dagger_{i+1,j+1}|0\rangle, \tag{5.8}$$

$$T_y T_x |\psi_{mn}\rangle = e^{i2\pi\phi_{mn}} T_x T_y |\psi_{mn}\rangle. \tag{5.9}$$

Even in an external, constant magnetic field, we see that the Hamiltonian is not translationally invariant with the original translation operators because the two do not commute with each other (and hence do not commute with the Hamiltonian, which is a sum of the two). Even though the magnetic field is translationally invariant, the gauge potential is not. A gauge transformation is required to make the Hamiltonian translationally invariant, but we will see that we cannot maintain the translational symmetry of the original lattice. We can find operators that commute with the Hamiltonian, $\hat{T}_x, \hat{T}_y$—they are called the magnetic translation operators and are defined by

$$\hat{T}_x = \sum_{m,n} c^\dagger_{m+1,n} c_{m,n} e^{i\chi^x_{m,n}}, \quad \hat{T}_y = \sum_{m,n} c^\dagger_{m,n+1} c_{mn} e^{i\chi^y_{m,n}}. \tag{5.10}$$

We could have guessed the form of these operators on physical grounds: they have to be one-body operators because we are solving a one-body Hamiltonian. We also knew that they have to have phases that are different from the original ones, because those do not commute. We find the phases χ by requiring that the operators commute with the Hamiltonian, which is a sum of the translation operators in the x- and y - directions. For $\hat{T}_x$,

$$[T_x, \hat{T}_x] = \sum_{mn} c^\dagger_{m+2,n} c_{m,n} e^{i(\chi^x_{mn} + \theta^x_{m+1,n})} [e^{i(\chi^x_{m+1,n} + \theta^x_{m,n} - \chi^x_{mn} - \theta^x_{m+1,n})} - 1]. \tag{5.11}$$

To obtain zero, we hence require

$$\Delta_x \chi^x_{mn} = \Delta_x \theta^x_{mn}. \tag{5.12}$$

We also want another commutator to vanish:

$$[\hat{T}_x, T_y] = 0, \tag{5.13}$$

which, once worked out, imposes the constraint

$$\Delta_y \chi^x_{mn} = \Delta_x \theta^y_{mn} = \Delta_y \theta^x_{mn} + 2\pi\phi_{mn}. \tag{5.14}$$

We similarly have

$$[\hat{T}_y, T_x] = 0, \quad [\hat{T}_y, T_y] = 0, \tag{5.15}$$

which gives the mirror constraints to the preceding constraints:

$$\Delta_x \chi^y_{m,n} = \Delta\theta^y_{mn} \quad (= \Delta_x \theta^y_{m,n} - 2\pi\phi_{mn}), \tag{5.16}$$

$$\Delta_y \chi^y_{m,n} = \Delta_y \theta^y_{m,n}. \tag{5.17}$$

The constraints can be solved, with the solutions

$$\chi^x_{mn} = \theta^x_{mn} + 2\pi n\phi_{m,n}, \quad \chi^y_{mn} = \theta^y_{mn} - 2\pi m\phi_{mn}. \tag{5.18}$$

We have now found operators that commute with H but do not commute between themselves:

$$[\hat{T}_x, \hat{T}_y] \neq 0. \tag{5.19}$$

To obtain the commutators of the two preceding operators,we use a useful trick: we take the action of the operators on the single particle state at site (i, j).

$$\begin{aligned} \hat{T}_x\hat{T}_y c^\dagger_{i,j} |0\rangle &= \sum_{mn} e^{i(\chi^x_{mn}+\chi^y_{ij})} c^\dagger_{m+1,n} c_{m,n} c^\dagger_{i,j+1} |0\rangle = e^{i(\chi^x_{i,j+1}+\chi^y_{ij})} c^\dagger_{i+1,j+1} |0\rangle, \\ \hat{T}_y\hat{T}_x c^\dagger_{i,j} |0\rangle &= \sum_{mn} e^{i(\chi^y_{mn}+\chi^x_{ij})} c^\dagger_{m,n+1} c_{m,n} c^\dagger_{i+1,j} |0\rangle = e^{i(\chi^y_{i+1,j}+\chi^x_{ij})} c^\dagger_{i+1,j+1} |0\rangle. \end{aligned} \tag{5.20}$$

They do not commute. In general, we can find a combination of the translation operators that does commute. This combination depends on the gauge that we pick. We select the Landau gauge $A_y = Bx = 2\pi\phi m$, where ϕ is the uniform flux per plaquette. Hence, $\theta^y_{mn} = 2\pi\phi m(n+1-n) = 2\pi\phi m$, and we have

$$\hat{T}_x\hat{T}_y = e^{i2\pi\phi}\hat{T}_y\hat{T}_x. \tag{5.21}$$

By successive application of $\hat{T}_x$ to the preceding commutation, we find that $\hat{T}^q_x\hat{T}_y = e^{i2q\pi\phi}\hat{T}_y\hat{T}^q_x$, and, in general, no power of the magnetic translation operators commutes with the other translation operator. However, for the special case of rational flux per plaquette, $\phi = p/q$, where p, q are relatively prime, we then find that

$$\hat{T}^q_x\hat{T}_y = \hat{T}_y\hat{T}^q_x. \tag{5.22}$$

We hence have two operators, $\hat{T}^q_x, \hat{T}_y$, which commute between themselves and commute with the Hamiltonian; hence, they define a new set of good quantum numbers. In the gauge chosen here, $\hat{T}_y = \sum_{m,n} c^\dagger_{m,n+1} c_{m,n}$ is the noncovariantized translation operator, whereas $\hat{T}_x = \sum_{m,n} c^\dagger_{m+1,n} c_{m,n} e^{i2\pi n\phi}$. The action of these two operators on single-particle states is that of translation by one lattice constant in the y-direction and by q lattice constants in the x-direction:

$$\hat{T}_y c_{i,j} |0\rangle = c^\dagger_{i,j+1} |0\rangle , \quad \hat{T}^n_x c^\dagger_{i,j} |0\rangle = e^{i2\pi j\phi} c^\dagger_{i+n,j} |0\rangle,$$

$$\hat{T}^q_x c^\dagger_{i,j} |0\rangle = e^{i2\pi q\phi} c^\dagger_{i+q,j} |0\rangle = e^{i2\pi p} c^\dagger_{i+q,j} |0\rangle = c^\dagger_{i+q,j} |0\rangle. \tag{5.23}$$

The new translational unit cell in the x-direction is called the magnetic unit cell, and it is q times larger than the usual unit cell. Because $\hat{T}^q_x$ and $\hat{T}_y$ commute with the Hamiltonian, it means that the eigenstates have quantum numbers under these operators. Since these operators perform translations, it is then clear that we can index the state by momentum quantum numbers. The Bloch conditions then become

$$H \left|\mathbf{k}\right\rangle = E(\mathbf{k}) \left|\mathbf{k}\right\rangle \quad \hat{T}^q_x \left|\mathbf{k}\right\rangle = e^{ik_x qa} \left|\mathbf{k}\right\rangle \quad \hat{T}_y \left|\mathbf{k}\right\rangle = e^{ik_y} \left|\mathbf{k}\right\rangle, \tag{5.24}$$

where, because the unit cell is q times larger in the x-direction, the magnetic BZ is q times smaller: $0 \le k_x \le 2\pi/q, \quad 0 \le k_y \le 2\pi$. From here we can prove that the Landau-level problem on a lattice has a q-fold degeneracy at different wavevectors, i.e., the bulk bands have the same energy at q different wavevectors on the y-axis. Assume $\left|k_x, k_y\right\rangle$ is an eigenstate of the Hamiltonian. Because the Hamiltonian commutes with $\hat{T}_x$, $\hat{T}_y$, we have that $\hat{T}_x \left|k_x, k_y\right\rangle$ is also an eigenstate of the Hamiltonian. However, because $\hat{T}_{x,y}$ do not commute between themselves, $\hat{T}_x \left|k_x, k_y\right\rangle$ cannot be an eigenstate of the Hamiltonian at the same wavevector. Instead,

$$\hat{T}_y \hat{T}_x \left|k_x, k_y\right\rangle = e^{-i2\pi\phi} \hat{T}_x \hat{T}_y \left|k_x, k_y\right\rangle = e^{i(k_y - 2\pi\phi)} \hat{T}_x \left|k_x, k_y\right\rangle. \tag{5.25}$$

The eigenvalue under $\hat{T}_y$ of $\hat{T}_x \left|k_x, k_y\right\rangle$ is $k_y - 2\pi\phi$, which leads us to conclude that $\hat{T}_x \left|k_x, k_y\right\rangle \sim \left|k_x, k_y - 2\pi\phi\right\rangle$. The eigenstates $\left|k_x, k_y - 2\pi\phi\right\rangle$ and $\left|k_x, k_y\right\rangle$ have identical energy. Because $\phi = p/q$ are rational and p, q are relatively prime, we have that the spectrum is q-fold degenerate, corresponding to the application of $\hat{T}_x$ q times.

5.2 Diagonalization of the Hamiltonian of a 2-D Lattice in a Magnetic Field

We would like to make use of the newly learned fact that the true unit cell is q times larger than the initial lattice unit cell to diagonalize the Hamiltonian. Let us choose the Landau gauge from before, $A_y = 2\pi\Phi m$. The expanded Hamiltonian then is

$$H = \sum_{m,n} -t_a c^\dagger_{m+1,n} c_{m,n} - t_b c^\dagger_{m,n+1} c_{m,n} e^{i2\pi\Phi m} + \text{h.c.} \tag{5.26}$$

Fourier-transform $c_{m,n} = \frac{1}{(2\pi)^2} \int_{-\pi}^{\pi} dk_x \int_{-\pi}^{\pi} dk_y e^{ik_x m + ik_y n} c_{k_x,k_y}$, where for now $-\pi \le k_x, k_y \le \pi$, and require $c_{k_x+2\pi j, k_y+2\pi l} = c_{k_x,k_y}$. In a sign that we have not used the full translational symmetry of the model, the Fourier-transformed Hamiltonian has—due to the magnetic field—coupling between different k-sectors and is *not* diagonal in k:

$$H = -\int_{-\pi}^{\pi} \frac{dk_x}{2\pi} \int_{-\pi}^{\pi} \frac{dk_y}{2\pi} \left[t_a \cos(k_x) c^\dagger_{k_x,k_y} c_{k_x,k_y} + t_b e^{-ik_y} c^\dagger_{k_x+2\pi\Phi,k_y} c_{k_x,k_y} + \text{h.c.} \right]. \tag{5.27}$$

The Fourier transform of the Hamiltonian mixes $(k_x, k_y) \rightarrow (k_x \pm 2\pi\Phi, k_y)$. To diagonalize our Hamiltonian, we need to find a momentum space where there is no mixing. If $\Phi = p/q$, where p and q are relatively prime, we see that the Hamiltonian can be broken in several sectors by making the Brillouin zone q times smaller than the initial one in the x-direction (corresponding to an enlarged magnetic unit cell made out of q initial unit cells):

$$H = \frac{1}{(2\pi)^2} \int_{-\pi/q}^{\pi/q} dk_x^0 \int_{-\pi}^{\pi} dk_y \hat{H}_{k_x^0,k_y}, \tag{5.28}$$

where

$$\begin{aligned} \hat{H}_{k_x^0,k_y} = & \sum_{n=0}^{q-1} \{-2t_a \cos(k_x + 2\pi\Phi n) c^\dagger_{k_x+2\pi\Phi n,k_y} c_{k_x+2\pi\Phi n,k_y} \\ & - t_b(e^{-ik_y} c^\dagger_{k_x+2\pi\Phi(n+1),k_y} c_{k_x+2\pi\Phi n,k_y} + e^{ik_y} c^\dagger_{k_x+2\pi\Phi(n-1),k_y} c_{k_x+2\pi\Phi n,k_y}). \end{aligned} \tag{5.29}$$

To get from equation (5.27) to equation (5.29), we made $k_x = k_x^0 + 2\pi\Phi n$. Note that this partition of the k-space can happen only if p and q are relatively prime. Only in this case is the covering $k_x^0 + 2\pi\Phi n$ able to reproduce the whole initial $-\pi \leq k_x \leq \pi$. For example, for the case $p = 2, q = 3$, we find $k_x^0 \in [-\pi/3, \pi/3]$, $k_x^0 + 2\pi\Phi \in [\pi, 5\pi/3] = [-\pi, -\pi/3]$, and $k_x^0 + 4\pi\Phi \in [7\pi/3, 9\pi/3] = [\pi/3, \pi]$—thereby showing that $k_x^0 + 2\pi\Phi n$ covers the entire range $-\pi$ to π. However, if p, q are not relatively prime, this is not possible: $p = 2, q = 2$, $k_x^0 \in [-\pi/2, \pi/2]$, $k_x^0 + 2\pi = [-\pi/2, \pi/2]$, so you can never sample the sectors $[-\pi, \pi/2]$, $[\pi/2, \pi]$

The Hamiltonian in equation (5.27) had states mixing (k_x, k_y) with $(k_x \pm 2\pi\Phi, k_y)$, which was in the same BZ. In the Hamiltonian in equation (5.29) no k_{x1}^0 mix with another k_{x2}^0 when both are in the $[-\pi/q, \pi/q]$ reduced BZ. If they did mix, it would mean that $k_{x1}^0 + 2\pi\Phi n_1 = k_{x2}^0 + 2\pi\Phi n_2$ and so $k_{x1}^0 - k_{x2}^0 \geq 2\pi\Phi$, which is impossible because, at most, $k_{x1}^0 - k_{x2}^0 = 2\pi/q$. The price paid, as we know from the general arguments of the magnetic translation group, is that the magnetic unit cell is made up of q plaquettes in the x-direction and the magnetic BZ is q times smaller than the non magnetic one.

The Hamiltonian $\hat{H}_{k_x^0,k_y}$ in equation (5.29) is, by itself, a q-site Hamiltonian. The Schrodinger equation

$$\hat{H}(k_x^0, k_y) |\psi\rangle = E_{k_x^0,k_y} |\psi\rangle \tag{5.30}$$

hence reduces to solving a 1-D tight-binding model on a 1-D lattice chain (in momentum space $k_x^0 + 2\pi\Phi n$, $n = 0, 1, 2, \ldots, q - 1$). The single-particle energies are obtained by expanding into single-particle states at each lattice point m,

$$|\psi\rangle = \sum_{m=0}^{q-1} a_m c^\dagger_{k_x^0+2\pi\Phi m,k_y} |0\rangle, \tag{5.31}$$

where $|0\rangle$ is the vacuum, and diagonalizing. The eigenvalue and eigenstate equation is

$$-2t_a \cos(k_x^0 + 2\pi\Phi n) a_n - t_b(e^{-ik_y} a_{n-1} + e^{ik_y} a_{n+1}) = E_{k_x^0,k_y} a_n, \tag{5.32}$$

which is to be solved subject to the boundary conditions $\psi_{j+q} = \psi_j$. This is called the Harper equation. Its solutions give the famous Hofstadter butterfly. This equation has a duality that we will use in the future. Let us make the transformation

$$a_j = \sum_{l=1}^{q} e^{i2\pi\Phi jl} x_l, \tag{5.33}$$

with $j = 1, \ldots, q$, and obtain an equation for the x_l:

$$-2t_b \cos(k_y + 2\pi\Phi l)x_l - t_a(e^{-ik_x^0} x_{l-1} + e^{ik_x^0} x_{l+1}) = E_{k_x^0,k_y} x_l. \tag{5.34}$$

This is the same equation as equation (5.32) but with the transformation $t_a \leftrightarrow t_b$, $k_x^0 \leftrightarrow k_y$. This transformation is a gauge choice, which amounts to choosing $\theta^x_{m,n} = 2\pi\Phi n$.

To diagonalize equation (5.32), it is convenient to make the transformation $a_j = b_j e^{-ik_y j}$, which renders the equation in a simple form:

$$-t_b b_{j-1} - t_b b_{j+1} - 2t_a \cos(k_x^0 + 2\pi\Phi j)b_j = E_{k_x^0,k_y} b_j, \tag{5.35}$$

in which the k_y dependence has disappeared. However, it could not have disappeared forever—indeed, it is still present in the boundary condition, which now becomes

$$b_{j+q} = e^{ik_y q} b_j. \tag{5.36}$$

Because the boundary condition is the only place where the dependence on k_y appears and because it depends only on $k_y \cdot q$, we recover the q-fold periodicity of the spectrum that we had obtained on general grounds in the discussion on the magnetic translation group:

$$E(k_x^0, k_y) = E\left(k_x^0, k_y + \frac{2\pi n}{q}\right), \quad n = 1, \ldots, q-1. \tag{5.37}$$

The energies are the solution of the vanishing of the determinant:

$$\text{Det}(M(E, k_x^0, k_y)) = \text{Det} \begin{vmatrix} v_1 - E & -t_b & 0 & 0\ldots & -t_b e^{-iqk_y} \\ -t_b & v_2 - E & -t_b & 0\ldots & 0 \\ \ldots & \ldots & \ldots & \ldots & \ldots \\ 0 & \ldots & -t_b & v_{q-1} - E & -t_b \\ -t_b e^{iqk_y} & 0\ldots & 0 & -t_b & v_q - E \end{vmatrix} = 0, \tag{5.38}$$

where $v_j = -2t_a \cos(k_x^0 + 2\pi\Phi j)$. Because it is a $q \times q$ matrix, the Hamiltonian will have q bands.

5.2.1 Dependence on k_y

We continue our attempts to analytically find more information about the magnetic field problem on the square lattice by first analyzing the dependence of the characteristic polynomial on the k_y momentum. It is easy to prove that the dependence on k_y in the determinant of equation (5.38) comes only through the term $2t_b^q \cos(qk_y)$: we start by expanding the preceding determinant above in pivots in the first row and focus only on the terms in the expansion that contain k_y. In the first-row expansion, only the pivots 12 and $1q$ multiply determinants of $q-1 \times q-1$ matrices that depend on k_y. The $1q$ pivot is e^{-iqk_y} and multiplies the determinant:

$$\text{Det}\begin{vmatrix} -t_b & v_2-E & -t_b\ 0\ldots & 0 \\ \ldots & \ldots & \ldots\ \ldots & \ldots \\ 0 & \ldots & -t_b\ \ldots & v_{q-1}-E \\ -t_b e^{iqk_y} & 0 & \ldots\ 0 & -t_b \end{vmatrix} = 0. \tag{5.39}$$

This determinant can be expanded through pivots in its first column, which has only two non zero elements. We need to pick only the element 11 of the matrix in equation (5.39), which was the element $21 = -t_b$ of the original matrix (eq. (5.38)). The second pivot, $-t_b e^{iqk_y}$ (the $q1$ element of the original matrix (eq. (5.38)). does not count for the purpose of determining the k_y dependence of the characteristic polynomial in equation (5.38) because it is the Hermitian conjugate of e^{iqk_y}, and so the result is independent of k_y. After these two expansions, under the pivots $1q * 21 = -t_b e^{-ik_y q}(-t_b)$ the matrix has become upper diagonal, so the determinant is the product of the diagonal elements of the remaining matrix, all of which are $-t_b$. We thus get a contribution $(-1)^{1+q}(-1)(-1)^{q-2} t_b^q e^{-ik_y q} = t_b^q e^{-ik_y q}$, where between the minus signs, $(-1)^{q-1}$ comes from the expansion in the $1q$ pivot and -1 comes from the 21 pivot of the original matrix, whereas $(-1)^{q-2}$ comes from the diagonals with $-t_b$. Because we obviously must get something unitary, the real dependence on k_y is obtained by adding the conjugate, and we get $2t_b^q \cos(k_y q)$. However, the duality just obtained immediately tells us that we must have a similar form for $k_y \to k_x^0$ and $t_b^q \to t_a^q$. Hence, the characteristic polynomial must take the form

$$F(E) - 2t_b^q \cos(qk_y) + 2t_a^q \cos(qk_x^0) = 0, \tag{5.40}$$

where $F(E)$ is a function of the energy *only* and *not* of k_x^0, k_y. It is also very easy to show that, because $\cos(k_x^0 + 2\pi\Phi j)$ is uniformly distributed on the periodic system of sites $k_x^0 + 2\pi\Phi j$, it turns out that $F(E)$ is an even function of the energy E if q is even and is an odd function if q is odd. Hence, for even q, the spectrum is symmetric with respect to $E = 0$, whereas for odd q, the $E = 0$ level is at the center of the middle band. The characteristic polynomial is of order E^q, and it contains only terms of powers E^q, E^{q-2}, E^{q-4}. This is easy to see because the terms E^{q-1}, E^{q-3}, ... necessarily contain odd functions of v_j, which are annihilated when summed over the whole circle. There are $q-1$ gaps in the system, excluding the band bottom and band top.

5.2.2 Dirac Fermions in the Magnetic Field on the Lattice

We now obtain one of the interesting results of the late 1980s. Xiao-Gang Wen first proved that the solution of the magnetic field on a lattice problem contains q Dirac fermions at $E = 0$ for q even. For $t_a = 0$, by doing tedious pivot expansion, we obtain

$$
\begin{aligned}
\mathrm{Det}(M(E, k_x^0, k_y))|_{E=0,t_a=0} &= F(0) - 2t_b^q \cos(qk_y) \\
&= t_b \begin{vmatrix} -t_b & -t_b & 0 & 0 & \dots & 0 & 0 \\ 0 & 0 & -t_b & 0 & 0 & \dots & 0 \\ 0 & -t_b & 0 & -t_b & 0 & 0 & \\ 0 & 0 & -t_b & 0 & -t_b & 0 & 0 \\ \dots & \dots & \dots & \dots & \dots & \dots & \dots \\ 0 & 0 & 0 & 0 & -t_b & 0 & -t_b \\ -t_b e^{iqk_y} & 0 & \dots & 0 & 0 & -t_b & 0 \end{vmatrix} \\
&\quad +(-1)^q t_b e^{-iqk_y} \begin{vmatrix} -t_b & 0 & -t_b & 0 \dots & 0 & 0 \\ 0 & -t_b & 0 & -t_b & 0 \dots & 0 \\ \dots & \dots & \dots & \dots & \dots & \dots \\ 0 & 0 & \dots & -t_b & 0 & -t_b \\ 0 & 0 & \dots & 0 & -t_b & 0 \\ -t_b e^{iqk_y} & 0 & \dots & 0 & 0 & -t_b \end{vmatrix} \\
&= -t_b^2 \begin{vmatrix} 0 & -t_b & 0 & 0 & \dots & 0 \\ -t_b & 0 & -t_b & 0 & 0 & \\ 0 & -t_b & 0 & -t_b & 0 & 0 \\ \dots & \dots & \dots & \dots & \dots & \dots \\ 0 & 0 & 0 & -t_b & 0 & -t_b \\ 0 & \dots & 0 & 0 & -t_b & 0 \end{vmatrix} + (-1)^{q+1} t_b^2 2\cos(qk_x) t_b^{q-2} (-1)^{q-2},
\end{aligned}
\tag{5.41}
$$

and we find

$$
F(0) = -2t_b^2 \begin{vmatrix} 0 & -t_b & 0 & 0 & \dots & 0 \\ -t_b & 0 & -t_b & 0 & 0 & \\ 0 & -t_b & 0 & -t_b & 0 & 0 \\ \dots & \dots & \dots & \dots & \dots & \dots \\ 0 & 0 & 0 & -t_b & 0 & -t_b \\ 0 & \dots & 0 & 0 & -t_b & 0 \end{vmatrix} . \tag{5.42}
$$

After some more algebraic manipulation, we can compute the determinant and find

$$F(0) = 2t_b^q(-1)^{q/2} \quad \text{if} \quad t_a = 0. \tag{5.43}$$

By duality, we then have

$$F(0) = (-1)^{q/2}(t_a^q + t_b^q). \tag{5.44}$$

This equation shows us that we have zero-energy solutions (with q even) for momenta that satisfy the equation

$$(-1)^{q/2}(t_a^q + t_b^q) = 2t_a^q \cos(qk_x^0) + 2t_b^q \cos(qk_y). \tag{5.45}$$

This gives the momenta at which the Dirac nodes occur:

$$q = 4n \to k_y = \frac{2\pi m}{q}, \quad k_x^0 = 0,$$

$$q = 4n + 2 \to k_y = \frac{(2m+1)\pi}{q}, \quad k_x^0 = \frac{\pi}{q} = -\frac{\pi}{q}. \tag{5.46}$$

In either cases, there are exactly q zero modes in the system, which we call Dirac nodes [36]. They are formed when the symmetric bands around $E = 0$ touch. The dispersion around these nodes is of Dirac form:

$$E = \pm\gamma q\sqrt{t_a^q (k_x^0)^2 + t_b^q k_y^2}, \tag{5.47}$$

with γ a constant number.

5.3 Hall Conductance

The time has come to compute the Hall conductance of the square-lattice problem in a magnetic field. This can be computed in two ways, each of which gives a unique physical insight into the problem. The Diophantine equation approach can be used to compute the Hall conductance through high-level, generic arguments without a tight-binding model and gives a general argument as to why the Hall conductance must be quantized in a gap. The Harper equation, on the other hand, uses the specifics of the square-lattice nearest-neighbor-hopping problem to compute the Hall conductance in an anisotropic limit (the full calculation in the isotropic limit is too hard to attempt). This second approach makes explicit issues such as gauge fixing over the full BZ, which were mentioned earlier in the book.

5.3.1 Diophantine Equation and Streda Formula Method

Quantized Hall conductance happens only in bulk insulating systems. Consider a generic insulator and suppose that we place the Fermi level in the rth gap; that is, there are r bands below the Fermi level. In a magnetic field, we have three positive integers $r, p, q > 0$. A mathematical fact, the Darboux theorem, tells us that three positive integers always satisfy

$$r = qs_r + pt_r, \tag{5.48}$$

where s_r, t_r are integers and $|t| \leq q/2, 0 \leq r \leq q$. The subscript r indicates that integers s_r and t_r depend on the value of r. The latter condition is physically justified by the fact that in the magnetic lattice problem just studied, we have $q-1$ gaps, which—along with the band top and bottom—give the aforementioned limits on r. Equation (5.48) is called the *Diophantine equation*. To obtain the Hall conductance, we can now use the earlier-derived Streda formula:

$$\sigma_{xy} = \frac{e^2}{\hbar}\frac{1}{V}\frac{\partial \rho}{\partial B}, \tag{5.49}$$

where ρ is the total density of states (not density) below the gap, V is the unit-cell volume (area), and BV is the flux per unit cell (the magnetic field multiplied by the unit-cell area). The density of states of the occupied bands is equal to the number of bands occupied divided by the total number of bands, which we know to be q. Hence, we divide the Diophantine equation by q:

$$\rho = \frac{r}{q} = s_r + \frac{p}{q} t_r = s_r + \frac{BV}{2\pi} t_r, \tag{5.50}$$

where we have used the fact that the flux per unit cell is $BV = p/q$. We thus find, for the Hall conductance,

$$\sigma_{xy} = \frac{e^2}{h} t_r, \tag{5.51}$$

which is a unique number that comes out of the Diophantine equation. The preceding derivation cannot help but leave the reader feeling that we have obtained a great deal of information (the Hall conductance) without putting in a lot (for example, we have not specified a model, any conditions, etc). Of course, there are several caveats. First, we must assume that the density of states is a differentiable function of the magnetic field B. Second, we must show that s_r and t_r are independent of B to take the derivative. This can actually be made plausible because they are integers, and integers do not change dramatically upon infinitesimal changes of parameters. The only way they can change is by the system becoming gapless and bands intersecting.

As we will also see by direct calculation, t_r is the Hall conductance. Then the rth band, which is between the $(r-1)$st and rth gap, carries integral Hall conductance I_r:

$$1 = q(s_r - s_{r-1}) + p(t_r - t_{r-1}) = qJ_r + pI_r. \tag{5.52}$$

In the general case, the values of t_r vary wildly as we move the Fermi level r, for generic values of p and q, at p/q relatively larger. The case we will primarily consider is going to be $p = 1$. For this case, we have $0 \leq r = qs_r + t_r \leq q$. We have two cases.

1. q even: For $r < q/2$ (which means all the bands with negative energy), we see if we take $|s_r| > 0$, we have that, because $|t_r| \leq q/2$, there is no way to satisfy the identity, because $|r - qs_r| > q/2$ if $s_r > 0$; so, for $r < q/2$, $s_r = 0$ and $t_r = r$. This means that the Hall conductance carried by each band is $I_r = t_r - t_{r-1} = 1$. For $r > q/2$, we have that $|r - qs_r| > q/2$ if $s_r = 0, s_r \leq 0$, or $s_r \geq 2$, so the only possibility is $s_r = 1$, which gives $t_r = r - q$ and, again, Hall conductance $I_r = 1$. The gap $r = q/2$ separates the bands $q/2 - 1$ from the band $q/2 + 1$, but as we now know, for the case q even, this gap actually closes at q points in the BZ (the Dirac fermions), and we cannot talk about the Hall conductance of the bands $q/2 \pm 1$ separately, but we have to talk about them together. Hence the two bands, symmetric at the middle of the energy dispersion, carry a Hall conductance

$$\sigma_{xy} = t_{q/2+1} - t_{q/2-1} = q/2 + 1 - q - (q/2 - 1) = -(q-2). \tag{5.53}$$

2. q odd: In this case, we have an identical characterization of the bands, but with different limits on the r. For $r \leq \frac{q-1}{2}$, we have $s_r = 0$, $t_r = r$, whereas for $r > \frac{q-1}{2}$, we have $s_r = q$, $t_r = r - q$. It is easy to see that the $r = \frac{q-1}{2}$ gap corresponds to $s_r = 0$ because for $s_r = 1$, we would have $t_r = r - q = -\frac{q+1}{2}$ incompatible with $|t_r| \geq q/2$. We thus obtain that for the band between the $\frac{q-1}{2}$ and the $\frac{q+1}{2}$ gaps—i.e., for the center band, we have: $\sigma_{xy} = \frac{q+1}{2} - q - \frac{q-1}{2} = -(q-1)$.

So far we have showed that the Diophantine equation allows for the determination of the Hall conductance. In fact, we can adopt the inverse point of view and show that the existence of a quantized Hall conductance implies the existence of a Diophantine equation, which then determines the value of the Hall conductance. This seems almost miraculous. We write the

Hall conductance in the form $\sigma_{xy} = -\frac{e^2}{h} t_r$ when the Fermi level is in the gap with t_r, an unknown integer guaranteed to be invariant from topological arguments. Because energy gaps are stable under small perturbations and persist under small variations of the magnetic field, we can assume that t_r does not change under a magnetic field change (as long as it is small) and integrate the Streda formula to obtain

$$\rho = \text{constant} + \frac{e}{h} V B t_r, \tag{5.54}$$

where $V = |\mathbf{a} \times \mathbf{b}|$ is the area of a unit cell. When the flux per unit cell is p/q, the area of the magnetic BZ is $\frac{(2\pi)^2}{Vq}$, q times smaller than that of the original BZ in the absence of the magnetic field. The density of electrons in a single band is $\frac{1}{Vq}$. When there are r bands below the Fermi energy (i.e., the Fermi level is in the rth gap), we have $\rho = r/q$, and hence

$$r = \text{constant} \cdot q + p t_r. \tag{5.55}$$

Because r, p, and q are integers, const $\cdot\, q$ is also an integer; if const $= m/n$ then we have that $q = kn$, where k is integer. However, we can imagine changing the magnetic field and, hence, changing q. We can change the magnetic field by keeping p constant, and we can change B by very little (for example, by changing q by 1 when q is very large) so that we do not close any gaps. But if q changes, then it cannot remain a multiple of n; hence, $n = 1$ and const is an integer (which we call s_r), and we have rediscovered the Diophantine equation.

5.4 Explicit Calculation of the Hall Conductance

We will now try to compute the Hall conductance of the bands in the square-lattice model in a magnetic field through direct methods. This computation was sketched first in the original Thouless, Kohmoto, Nightingale, and den Nijs (TKNN) paper [37]; we now present an expanded and pedagogical description of it. In order to explicitly compute the Hall conductance, we look at the anisotropic limit—the full isotropic limit is inaccessible. In the fully anisotropic limit, we start from a series of 1-D uncoupled wires in the x-direction; this is the $t_b = 0$ limit of the Harper equation:

$$-2t_a \cos(k_x^0 + 2\pi\phi j)\psi_j = E(k_x^0, k_y)\psi_j, \tag{5.56}$$

which has the solutions

$$E_m(k_x^0, k_y) = -2t_a \cos(k_x^0 + 2\pi\phi m), \quad \psi_j = \delta_{j,m}. \tag{5.57}$$

For the eigenstate of energy m, we know that the wavefunction is exactly 1 on the mth site of the q-site lattice and zero otherwise. In this limit, the dispersion is flat with respect to the k_y momentum, and there are many level crossings on the k_x^0-axis. This is easy to understand: as $t_b = 0$, the Hamiltonian does not feel the presence of the magnetic field. In the gauge we chose, $A_y = Bx$, the magnetic field entered through Peierls substitution in the hopping in the y-direction, which we now have chosen to be zero. As such, our Hamiltonian has translational invariance in the x-direction identical to one unit cell: we do not need to enlarge our unit cell to q times the original one because the magnetic field is zero. The dispersion in the full BZ is that of a single band, $-2t_a \cos(k_x)$. However, k_x^0 is in the magnetic BZ—the dispersion here is a folding of the single-band dispersion in the full BZ. To consolidate our understanding, let us,

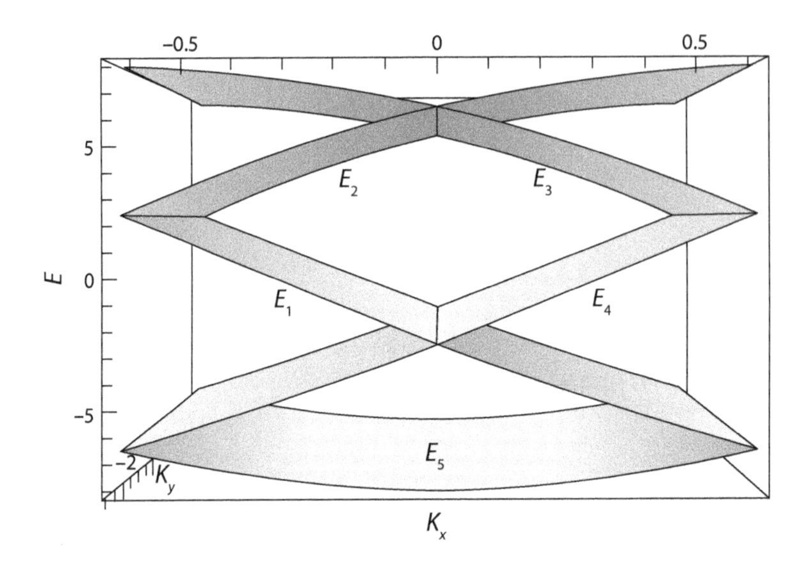

Figure 5.1. Band dispersion in the weak coupling limit ($t_b = 0$) of the Harper equation for $p = 1$ and $q = 5$. The k_y-direction is degenerate.

for example, choose the case $p = 1$ and $q = 5$. The five bands have the following degeneracies: $E_5(-\pi/5) = E_1(-\pi/5)$, $E_5(\pi/5) = E_4(\pi/5)$, $E_4(-\pi/5) = E_2(-\pi/5)$, $E_2(0) = E_3(0)$. These dispersions are shown in figure 5.1. Notice that the dispersion is just the folded dispersion of a single band described by the $-2t_a \cos(k_x)$ hopping term (which is the only one present in the case $t_b = 0$) in the full BZ $[-\pi, \pi]$ folded into the magnetic BZ $[-\pi/q, \pi/q]$.

As we understand the behavior of the system in the fully anisotropic limit, we now would like to understand what happens as we turn on the t_b hopping. If t_b is small, we can analyze the system through perturbation theory. In perturbation theory, it is pretty clear what will happen—gaps will open at the degeneracy points in the magnetic BZ (remember that the fully isotropic limit has q bands with gaps between them). We need to understand the order in perturbation theory in which gaps will open. As such, we need to understand which bands actually cross in the magnetic BZ.

Because there do not exist couplings between k_x^0 in the same BZ, the condition for degeneracy between bands m_1 and m_2 becomes

$$k_x^0 + \frac{2\pi p}{q} m_1 = -(k_x^0 + \frac{2\pi p}{q} m_2) \mod 2\pi \tag{5.58}$$

with $m_{1,2}$ integers. This condition with a $+$ is just periodicity of the cosine function and does not make sense. Let us show this: assume $k_x^0 + \frac{2\pi p}{q} m_1 = k_x^0 + \frac{2\pi p}{q} m_2 \mod (2\pi)$, which means $s = \frac{p(m_1 - m_2)}{q} \in \mathcal{Z}$. Since $1 \leq m_1 \leq m_2 \leq q$, we have $|m_1 - m_2| < q$; but this is impossible, because p and q are relatively prime. Hence, to get an integer s, it must be that $m_1 - m_2 = lq$ with $l \in \mathcal{Z}$. The only solution to that equation is then $m_1 = m_2$, which is not a condition for a degeneracy. From the preceding, it also obvious that degeneracies can happen only at $k_x^0 = -\pi/q, 0, \pi/q$ in the magnetic BZ and at wavevectors

$$k_x^0 = \frac{\pi}{q}(-p(m_1 + m_2) + qs) \in \left[-\frac{\pi}{q}, \frac{\pi}{q}\right]. \tag{5.59}$$

The range of the magnetic BZ uniquely determines the integers m_1, m_2 for which there are band crossings. For our example—$p = 1$, $q = 5$—we have

$$m_1 = 1, \quad m_2 = 2 \to k_x^0 = -3 + 5s \in [-1, 1] \to \text{Impossible}, \tag{5.60}$$

$$m_1 = 1, \quad m_2 = 3 \to k_x^0 = -4 + 5s \in [-1, 1] \to s = 1, \ k_x^0 = \frac{\pi}{q}, \tag{5.61}$$

$$m_1 = 1, \quad m_2 = 4 \to k_x^0 = -5 + 5s \in [-1, 1] \to s = 1, \ k_x^0 = 0, \tag{5.62}$$

$$m_1 = 1, \quad m_2 = 5 \to k_x^0 = -6 + 5s \in [-1, 1] \to s = 1, \ k_x^0 = -\frac{\pi}{q}. \tag{5.63}$$

As we said, crossings can happen only at $0, \pm\pi/q$; let us look first at the crossing at $k_x^0 = 0$. They involve bands m_1, m_2, which satisfy

$$(m_1 + m_2)p = qs \to s = xp, \quad x \in \mathcal{Z},$$
$$xq = m_1 + m_2 \in [3, 2q - 1] \to x = 1 \to m_1 + m_2 = q, \tag{5.64}$$

where we used the fact that $1 \leq m_1, m_2 \leq q$ and $m_1 \neq m_2$ (for degenerate distinct bands) to show that $m_1 + m_2 \in [3, 2q - 1]$. Hence, for any p we find that the bands m_1, m_2 that cross at $k_x^0 = 0$ must satisfy $m_1 + m_2 = q$, which can also be seen from figure 5.1 in our example situation.

For $t_a > 0$, the minimum energy band is the $m = q \equiv 0$ band of energy $E_q(k_x^0) = -2t_a \cos(k_x^0)$; this is because $k_x^0 \in [-\pi/q, \pi/q]$, centered around 0, has the maximum value of the cosine. All the other ms then fill in (for p, q relatively prime) the full $[-\pi, \pi]$ interval in the unfolded BZ and, hence have lower values of the cosine, i.e., higher energies. In general, we can easily find the solutions for which $-(m_1 + m_2)p + qs = \pm 1$ to find the intersections at $\pi/q, -\pi/q$

Adding a weak hopping perturbation t_b will now open gaps at the degeneracy points of the bands. First, we want to find a relation between the number of the gap counting from the band bottom up and the number m_1, m_2 of the bands that used to intersect at the point where the gap is opened. In other words, we would like to know which bands hybridize to give rise to the gap denoted by the integer r. In the $t_b = 0$ case, we refer to the rth intersection between the bands to be the point where the rth gap would open upon the inclusion of weak t_b. For any p, q, the lowest band is always $m = q \equiv 0$ of $E_0(k_x^0) = -2t_a \cos(k_x^0)$, which is a convex function of k_x^0 in the interval $[-\pi/q, \pi/q]$. We then have that the rth intersection is at $k_x^0 = \pm\pi/q$ for odd r and $k_x^0 = 0$ for even r (see fig 5.1 for an example). We pick the convention that for the fully empty system, the Fermi level is in the zeroth gap. For $k_x^0 = \pi/q$, we have that $m_1 + m_2 = (q-1) \bmod q$, whereas for $k_x^0 = -\pi/q$, we have $m_1 + m_2 = (q + 1) \bmod q$. Together with the knowledge that the lowest-energy band is $m = q \equiv 0$, and with the condition $1 \leq m_1 \neq m_2 \leq q$, this equation fully determines the succession of gaps.

Let us be clearer with an example: at $k_x^0 = \pi/q$, the $r = 1$ intersection is made by the bands q and $q - 1$, whereas at $k_x^0 = -\pi/q$, the $r = 1$ intersection is made by the bands q and 1. Notice that bands $q - 1$ and 1 go on to intersect at $k_x^0 = 0$ ($r = 2$ intersection) as they should because $m_1 + m_2 = q$ intersect at $k_x^0 = 0$. The $q - 1$ band continues to $k_x^0 = -\pi/q$, where it intersects the second band $q - 1 + 2 = q + 1$, whereas the first band goes on to $k_x^0 = \pi/q$, where it intersects the $(q - 2)$nd band at the $r = 3$ intersection. The second and $(q - 2)$nd band zigzag and intersect

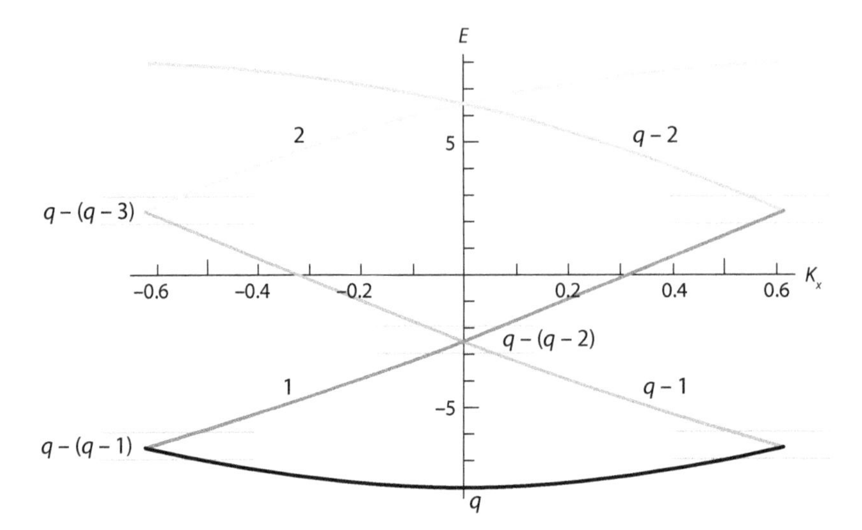

Figure 5.2. Relation between the gap number (intersection number) and the bands that intersect in the case $p = 1$.

again at k_x^0 ($r = 4$ intersection) to meet up with the $(q - 3)$rd and third band, respectively, to form the $r = 5$ intersection, and so on, as clearly shown in figure 5.2.

We thus see there is a simple relation between the gap number r and the bands m_1, m_2 that intersect to then split and form the gap

$$r = q - |m_1 - m_2|. \tag{5.65}$$

Obviously, for $k_x^0 = \pm\pi/q$, this formula gives the same number of the gap as can be seen, for example, in equation (5.64). Also, for the first gap at π/q, we have to take $m_1 = q \equiv 0$ for the formula to work, but this is, of course, obvious because we can number the bands from 1 to q or from 0 to $q - 1$; in the latter case, the qth band is exactly equal the zeroth band.

We would now like to make the connection with the Hall conductance. We will show that there is an integer in the problem that has the same values as the Hall conductance when the Fermi level stands in the rth gap. This integer equals the order in perturbation theory at which the rth gap opens. We then prove that this integer is, indeed, the Hall conductance by direct evaluation of the Chern number. Denote

$$k_x = k_x^0 + 2\pi\frac{p}{q}m_1, \quad k_x' = k_x^0 + 2\pi\frac{p}{q}m_2, \tag{5.66}$$

with $m_1 \neq m_2$. The degeneracy condition for two bands is $k_x = -k_x'$. We have $k_x - k_x' = \frac{2\pi p}{q}(m_1 - m_2)(\text{mod } 2\pi) = \frac{2\pi p}{q}t_r$ with $t_r \leq q/2$ an integer (which, in anticipation, we have denoted by the same symbol as in the Diophantine equation—it will turn out to be the same number). To show we can always make $|t_r| \leq q/2$ we remark that if $t_r = q/2 + x$, $x > 0$, we have

$$k_x - k_x' = \frac{2\pi p}{q}\left(\frac{q}{2} + x\right) = \pi p + 2\pi\frac{p}{q}x \ (\text{mod} 2\pi) = 2\pi\frac{p}{q}\left(-\left(\frac{q}{2} - x\right)\right), \tag{5.67}$$

so $t_r' = -(q/2 - x)$ within the wanted bounds. We apply this to our spectrum: if $|m_1 - m_2| \leq q/2$ (which denotes the upper part of the spectrum, i.e., intersections (future gaps) $r \geq q/2$), we have $t_r = -|m_1 - m_2|$ and, hence, $r = q + t_r$. For $|m_1 - m_2| \geq q/2$ (which denotes the upper part

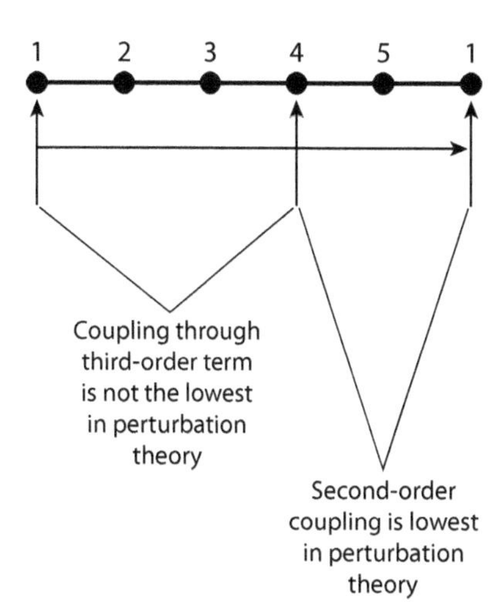

Figure 5.3. Example of perturbation theory coupling in the case $p = 1$. Sites 1 and 4 are coupled to second order in perturbation theory.

of the spectrum,—i.e., degeneracies, where gaps open—$r \leq q/2$), we have $t_r = q - |m_1 - m_2|$, and, hence, $r = q - |m_1 - m_2| = t_r$. This is exactly the solution to the Diophantine equation in the previous section. Note that t_r is the smallest of either $-|m_1 - m_2|$ or $q - |m_1 - m_2|$. Thus, we have obtained the solution of the Diophantine equation. All that remains is to relate the integer t_r to the Hall conductance of the system.

We probe even further into the gap opening in perturbation theory. Because $\psi_j^m = \delta_{jm}$ (the amplitude of the band of energy E_m is fully localized on the site $j = m$), turning on the term $-t_b\left(e^{-ik_y}\psi_{j-1} + e^{ik_y}\psi_{j+1}\right)$, which couples only adjacent sites, means that if bands m_1 and m_2 intersect, then we need to apply perturbation theory a number $\min(|m_1 - m_2|, q - |m_1 - m_2|)$ times to be able to hybridize the bands and open a gap. In other words, due to the perturbation being the nearest neighbor in sites and due to the fact that the bands that intersect have their wavefunctions localized on the sites equal to the band number, we have to apply perturbation theory several times to couple far-away sites with a nearest-neighbor coupling, even though the two bands might be degenerate. For intersecting bands m_1, m_2, a gap will open only in the t_rth order in perturbation theory. Of course, $t_r = \min(|m_1 - m_2|, q - |m_1 - m_2|)$ as we showed before; a physical way of understanding this is to notice that because of the periodic boundary conditions, the maximum distance between two sites can be of order $q/2$. A coupling of order greater than $q/2$ can be made less than $q/2$ by coupling through the periodic boundary condition, as in figure 5.3. Note that this is important because it means that the bands at π/q and $-\pi/q$, which are actually different in number (i.e., the bands that become degenerate at $-\pi/q$ are different bands from the ones that become degenerate at $-\pi/q$), will have identical Hamiltonians because they have the same $|m_1 - m_2|$ up to a shift by q lattice sites.

Let us recap what we have showed. We have analyzed the problem in the anisotropic limit, whereupon the introduction of a small y-hopping opens gaps in the spectrum, and have found a formula relating the gap number with the quantum numbers of the bands that hybridize as well as with the order in perturbation theory in which the bands hybridize. We now show how to compute the Hall conductance (Chern number) when the Fermi level is in one of the gaps. We also show that the Hall conductance equals the order t_r in perturbation theory at which the bands hybridize to open a gap.

The perturbation theory will couple (hybridize) bands m_1, m_2 close to the degeneracy points and open gaps, as in figure 5.4. Away from the degeneracy points, because the hybridization varies as $\left(\frac{t_b}{\Delta E}\right)^{t_r}$, the bands will still retain their character—i.e., all components of the bands except $j = m_{1,2}$ will be negligible. Let $\psi_{m_1} = a$, $\psi_{m_2} = b$. The coupling between these bands is $\Delta e^{-ik_y t_r}$, where $\Delta \sim t_b^{t_r}$. We expand around the unperturbed energies of the two bands at the

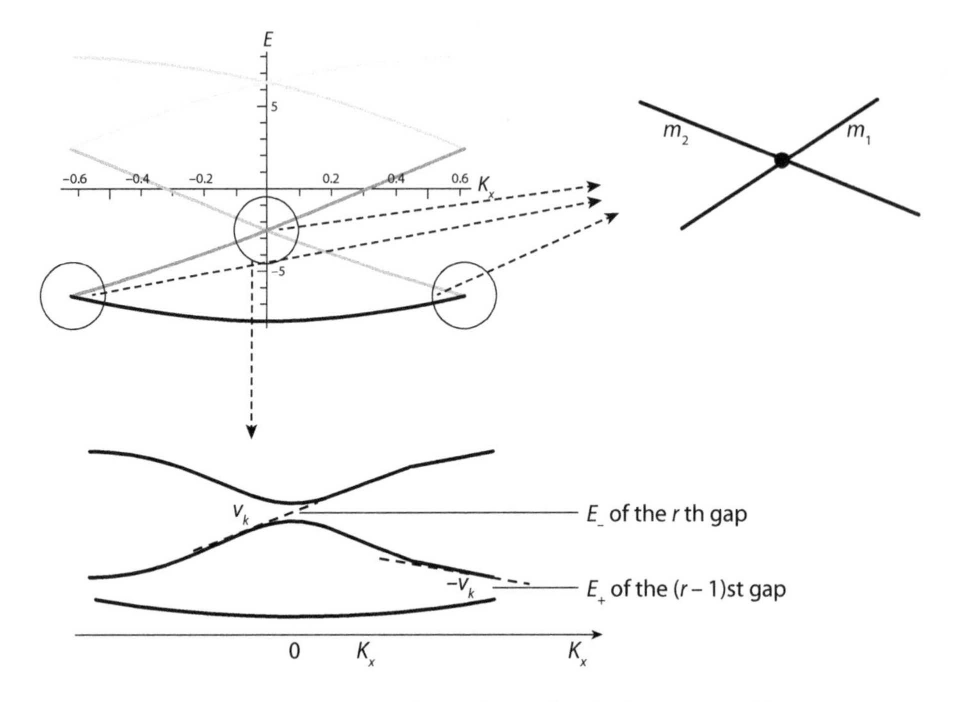

Figure 5.4. Gap opening in the weak-coupling-limit Harper problem.

degeneracy point:

$$\epsilon_{m_1} = vk_x, \quad \epsilon_{m_2} = -vk_x, \tag{5.68}$$

where the two velocities are actually equal due to an inversion symmetry present in the folded-zone model and $k_x = k_x^0 - K_x^0$, where K_x^0 is the degeneracy point $(= \pm\pi/q,\, 0)$. The Hamiltonian, shifted by the energy of the degeneracy point, becomes

$$\begin{pmatrix} \epsilon & \Delta e^{-ik_y t_r} \\ \Delta e^{ik_y t_r} & -\epsilon \end{pmatrix} \begin{pmatrix} a \\ b \end{pmatrix} = E \begin{pmatrix} a \\ b \end{pmatrix}, \tag{5.69}$$

where $\epsilon = vk_x$. We can make a gauge transformation $b = b' e^{ik_y t}$, in which case we have the equation

$$\begin{pmatrix} \epsilon & \Delta \\ \Delta & -\epsilon \end{pmatrix} \begin{pmatrix} a \\ b' \end{pmatrix} = E \begin{pmatrix} a \\ b' \end{pmatrix}, \tag{5.70}$$

with gap 2Δ and energies and eigenvalues parametrized by an angle θ:

$$E_+ = \sqrt{\epsilon^2 + \Delta^2}; \quad (a, b') = (\cos\theta, \sin\theta), \quad \sin 2\theta = \frac{\Delta}{E_+}, \tag{5.71}$$

$$E_- = -\sqrt{\epsilon^2 + \Delta^2}; \quad (a, b') = (-\sin\theta, \cos\theta), \quad \sin 2\theta = \frac{\Delta}{|E_-|}. \tag{5.72}$$

This parametrization guarantees orthogonality of the two wavefunctions. Away from the gap, for $\epsilon >> \Delta$ (in the middle of the band, we do not want to go far away to hit the other degeneracy point), we have, for the lower band $E_- \approx -|\epsilon|$,

$$\begin{pmatrix} \epsilon & 0 \\ 0 & -\epsilon \end{pmatrix} \begin{pmatrix} a \\ b' \end{pmatrix} = -|\epsilon| \begin{pmatrix} a \\ b' \end{pmatrix}. \tag{5.73}$$

Without loss of generality, let the rth gap be at $K_x^0 = 0$; then, as k_x^0 passes 0, ϵ goes from less than 0 to greater than 0, and the wavefunction goes from (for $\epsilon >> \Delta$)

$$\epsilon < 0 \to (a, b') = (a, 0); \quad \epsilon > 0 \to (a, b') = (0, b'), \tag{5.74}$$

with $|a| = |b'| = 1$ for proper normalization; in our θ parametrization,

$$\begin{aligned} &\epsilon < 0 \to \theta = \frac{\pi}{2} \to (a, b') = (-1, 0); \\ &\epsilon > 0 \to \theta = 0 \to (a, b') = (0, 1) \to (a, b) = (0, e^{ik_y t_r}), \end{aligned} \tag{5.75}$$

We have assumed that the rth gap opens at the origin $K_x^0 = 0$, so ϵ $(= vk_x^0) <> 0$ corresponds to $k_x^0 <> 0$. For this gap, the band *below* it is the band of energy E_-. However, the band travels to the point $-\pi/q$, at which it becomes the band of energy E_+. If we just blindly put $\epsilon = -v(k - \pi/q)$, then we would get the wrong answer. This is because at π/q, the band b has energy ϵ before the gap opening. So with the original basis, the Hamiltonian at π/q is

$$\begin{pmatrix} -\epsilon & \Delta \\ \Delta & \epsilon \end{pmatrix} \begin{pmatrix} a \\ b' \end{pmatrix} = E \begin{pmatrix} a \\ b' \end{pmatrix}, \tag{5.76}$$

with $\epsilon = -v(k - \pi/q)$. Because we are now looking at the band above the $(r-1)$st gap, we have energy E_+ and the form of the eigenstates:

$$\begin{aligned} &\epsilon < 0 \to \theta = 0 \to (a, b') = (1, 0), \\ &\epsilon > 0 \to \theta = \frac{\pi}{2} \to (a, b') = (0, 1) \to (a, b) = (0, e^{ik_y t_{r-1}}). \end{aligned} \tag{5.77}$$

Where we have used the form of the eigenstates described here in terms of θ, notice that $(a, b') = (1, 0)$, not $(a, b') = (-1, 0)$, as we had in the previous case. This is due to the fact that the energy band has changed from E_- to E_+. We have, of course, neglected the constant term in the Hamiltonian (relative to the previous gap) because it does not do anything to the eigenstates. For $k < \pi/q$, $\epsilon > 0$; as k_x^0 passes through π/q, the wavefunction changes from $\epsilon = -(k_x^0 - \pi/q) > 0$ to $\epsilon < 0$.

To be more explicit, the following sequence of events happens as k_x moves through the magnetic BZ starting, for example, from the middle of a band at, say, $k_x^0 = -\frac{\pi}{2q}$ and moving toward increasing k_x^0 following the band: we first encounter the $k_x^0 = 0$ gap, where the band we are interested in has the energy E_- (of the Hamiltonian around the $k_x^0 = 0$ degeneracy point), where the energy is $\epsilon = v_F k$, and we go from $\epsilon < 0$ to $\epsilon > 0$—or, in wavefunction space, from $(-1, 0)$ to $(0, e^{ik_y t_r})$. We then travel more and reach the π/q point, where $\epsilon = (k_x^0 - \pi/q)v_F$, but we now have to look at the E_+ branch of the energy (of the Hamiltonian around the $k_x^0 = \pi/q$ degeneracy point), and in the eigenstates we move from $(0, e^{ik_y t_{r-1}})$ to $(1, 0)$ when crossing $k_x^0 = \pi/q$ (the bands at these points are gapped). These wavefunctions are well defined over the whole BZ, but we notice for that to occur, we have some k_x-dependent phases that make

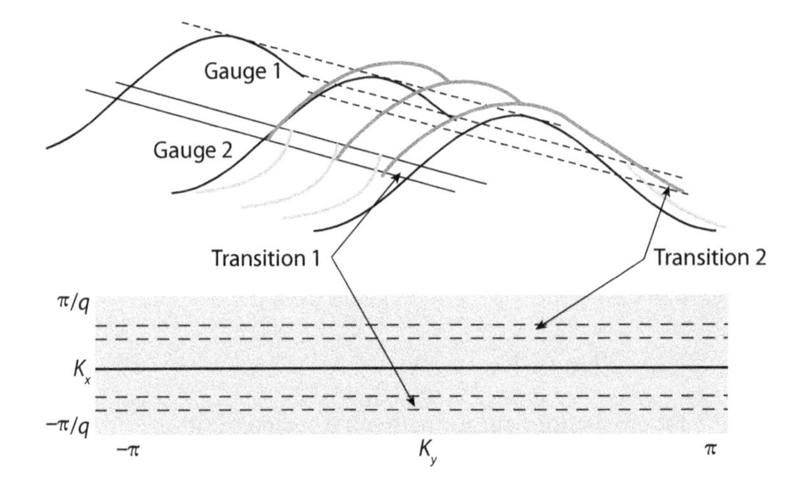

Figure 5.5. Transition regions on every band in the Harper problem.

the wavefunction change. This is an indication that a total phase is not possible, because the wavefunction changes character from having the second component equal to zero to having the first component equal to zero.

In the middle of the band, the two wavefunctions meet, and we have a transition function between them (see fig. 5.5). Going from $-\pi/q$ to 0, we encounter a transition between wavevectors (1, 0) to (−1, 0), for which the transition function (at transition 1 in fig. 5.5) is simply $U_1 = -1$. However, once we move past $k_x^0 = 0$, our wavefunction changes to $(0, e^{ik_y t_r})$; then, close to the point π/q, it changes again to $(0, e^{ik_y t_{r-1}})$. Hence, at transition 2, the transition function is $(0, e^{ik_y(t_r - t_{r-1})})$. These changes in the character of the bands are easy to understand—the eigenstates must change from one to another as we go through the two degeneracy points, but the change picks up phases due to the Dirac equation. Because the wavefunction is regular across each region, we can apply Stokes' theorem for the Hall conductance of *each magnetic band*:

$$\begin{aligned}\sigma_{xy} &= \frac{1}{2\pi i}\int_{-\pi/q}^{\pi/q} dk_x \int_0^{2\pi} dk_y \left(\partial_{k_x}\psi^*\partial_{k_y}\psi - \partial_{k_y}\psi^*\partial_{k_x}\psi\right) \\ &= \frac{1}{2\pi i}\int_{-\pi/q}^{\pi/q} dk_x \int_0^{2\pi} dk_y (\nabla_k \times (\psi^* \nabla_k \psi)) \cdot \hat{z} \\ &= \frac{1}{2\pi i}\left(\int_{\text{region 1}} d\mathbf{k}\cdot[\psi^*\nabla\psi] + \int_{\text{region 2}} d\mathbf{k}\cdot[\psi^*\nabla\psi]\right), \end{aligned} \tag{5.78}$$

where ψ is the wavefunction of the state we are considering. The boundaries of integrations 1 and 2 are exemplified in figure 5.6. The only contribution comes from the part of the domain with the k-dependent transition function (even if the function were k-dependent in the other regions, we would still get the same thing because different contributions would cancel out; for example, the parts of the domain at $k_y = \pm\pi$ and any k_x cancel within the same domain, whereas the parts where the transition function is −1, which are in the lower-half BZ, also cancel because they are not momentum dependent):

$$\int_{\text{region 1}} d\mathbf{k}\cdot[\psi^*\nabla\psi] = \int_{-\pi}^{\pi} e^{-ik_y t_r}\nabla_y e^{ik_y t_r} = 2\pi i t_r, \tag{5.79}$$

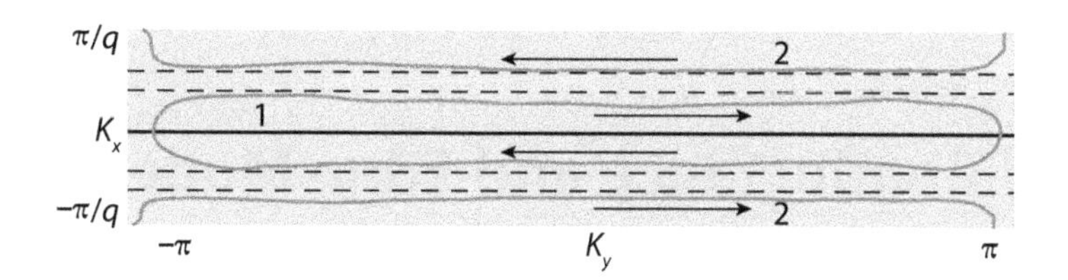

Figure 5.6. Integration domains 1 and 2 used in the text.

$$\int_{\text{region 2}} d\mathbf{k} \cdot [\psi^* \nabla \psi] = \int_{\pi}^{-\pi} e^{-ik_y t_{r-1}} \nabla_y e^{ik_y t_{r-1}} = -2\pi i t_{r-1}. \tag{5.80}$$

Hence the Hall conductance from one band is, as advertised,

$$\sigma_{xy} = t_r - t_{r-1}. \tag{5.81}$$

This completes our proof. The Hall conductance of all bands below the rth gap is $\sigma_{xy} = t_r$ (in units of e^2/h), thereby reaching the same conclusion as in the Diophantine equation.

5.5 Problems

1. *Magnetic Translation Operators in the Symmetric Gauge:* Find the commutation relation for the magnetic translation operators in the symmetric gauge (the one given in the text was explicitly constructed in the Landau gauge). Then find their action on the single-particle states, and show that the Landau-level problem has a q-fold degeneracy, just like in the Landau gauge.

2. *Numerical Implementation of the Hofstadter Problem:* Numerically implement the Hofstadter Hamiltonian on a finite-size lattice. Compute the eigenstates with periodic boundary conditions; then, form the projector into each band. Finally, use the Chern number formula in terms of the projector into the band to obtain the Chern number for each band, and verify that it gives the same result as the Diophantine equation.

3. *Continuum Limit of the Hofstadter Problem:* Show that in the low field limit, when the flux per plaquette is p/q, $q \to \infty$, the Landau-level eigenstates in the continuum on the torus (θ functions—please look them up if you are unfamiliar with their expressions) are the solutions to the Hofstadter problem. The continuum lowest-Landau-level eigenstates are the solutions of the lowest band in the Hofstadter model—the bans whose Chern number equals unity.

4. *Winding Number of the Position Operator:* On the lattice, the position operator has to be periodic with the periodicity of the lattice. Such an operator is $X = \sum_{j=1}^{N} e^{i\Delta_x j} |j\rangle \langle j|$, where $\Delta_x = \frac{2\pi}{N_x}$. When in a two-dimensional system, we can imagine the system as made out of coupled one-dimensional systems (wires) in the x-direction. Then, the X operator will be y-dependent, $X(k_y) = \sum_{j=1}^{N} e^{i\Delta_x j} |j, k_y\rangle \langle j, k_y|$. We now project the $X(k_y)$ operator into a specific band, using the projector P. Compute the eigenvalues of $PX(k_y)P$ as a function of k_y for a band in the Hofstadter model and for a trivial insulator $H = \sigma_z$, and show that the phase of the eigenstates winds from 0 to 2π as k_y goes from 0 to 2π in the Hofstadter model but not in the trivial insulator. The winding number is identical to the Chern number of the band. Show this numerically and then analytically.

6

Hall Conductance and Edge Modes: The Bulk-Edge Correspondence

We have so far computed the Hall conductance of lattice electrons in a magnetic field. The system was an insulator (we were always able to place the Fermi level in the gap), and we used periodic boundary conditions in real space. We obtained a nonzero Hall conductance, indicating transport, even though the system has a gap. How can this be possible? The answer lies in the presence of edge states. Even though the system is a bulk insulator, when edges are made in the system (such as by the application of contacts, which are needed in order to measure anything), there will be gapless edge modes traversing the bulk gap: the system is a bulk insulator but an edge Fermi liquid. In this chapter, we show that the presence of edges is an unescapable conclusion of multiple factors: one of them, as described later, is that the system transports electrons from one edge to another during a flux insertion—and there must be a band of states on each edge ready to receive/donate such electrons. We also present a direct calculation of the spectrum with open boundary conditions (which means we have edges) based on a transfer matrix formalism due to Hatsugai [38]. Gapless edge states, as it turns out, are a fundamental aspect of many topological insulators; indeed, the triviality or nontriviality of such an insulator usually depends on the presence or absence of edge states.

However, the conclusion that gapless edge states must exist can be reached by the following simple argument: put two insulators, each with different values of the Hall conductance, close to each other, so that they share a boundary (which in D dimensions is a $(D-1)$-D space). Away from the boundary, the two insulators extend to $\pm\infty$, respectively. The boundary between insulators preserves $U(1)$ charge symmetry, which is the symmetry necessary to define a Hall conductance (all our Hamiltonians so far contained only $c^\dagger c$ terms; no $c^\dagger c^\dagger$ superconducting terms were present). Because we know that the Hall conductance is an integer that characterizes the phase of the system, and because we know that the Hall conductance *cannot* be changed unless the bulk gap collapses and reopens again, the conclusion is that the boundary region that connects two insulators with different Hall conductances must have a gap-closing-and-reopening point somewhere on it—i.e., it must have an edge mode crossing the Fermi level. Otherwise, the whole space would be gapped, which by default would mean that the Hall conductance in the whole space would be the same. This type of an argument applies to a boundary between any two topologically different insulators, as long as the boundary respects the symmetry that protects the bulk-insulating state.

6.1 Laughlin's Gauge Argument

The original argument for the presence of edge states in the quantum Hall effect is due to Robert Laughlin in the early 1980s. In a seminal paper, Laughlin argued that the presence of edge modes is an inescapable consequence of transverse quantized transport in an insulator. This is even more remarkable because at that time it was widely accepted that delocalized states cannot exist in two dimensions, whereas Laughlin's argument says that delocalized states must exist. Of course, in retrospect, we now know that systems with nonzero Hall conductance have delocalized bulk states that carry the Chern number. These bulk states, which can be deep in the valence band, rise up in energy close to an edge due to the edge-confining potential and become the gapless edge modes. In the following, we present a modern-day version of the Laughlin gauge argument.

We consider a two-dimensional material (quantum well) with a magnetic field perpendicular to it. We also consider periodic boundary conditions in the y-direction but place edges on the sample in the x-direction. Because the boundary conditions are periodic in the y-direction, we can devise a thought experiment in which we "wrap" the sample in the y-direction into a cylinder, with x being parallel to the axis of the cylinder. Along the cylinder, k_y is a good quantum number. Through the cylinder, parallel to the x-axis, insert a flux, Φ. This flux is different from the flux generated by the magnetic field, which is "radial" to the cylinder. The flux, Φ, is used to "control and change" the k_y momentum: the flux enters the Hamiltonian as $k_y \rightarrow k_y + 2\pi\frac{\Phi}{L_y}$, where L_y is the circumference of the cylinder. At the two edges of our cylindrical sample, we attach leads and put a voltage V_x perpendicular to k_y between the leads: this, of course, is equivalent to creating a potential difference between the two edges. We wish to relate the current carried around the cylinder, j_y, to the potential drop from one edge to the other. This is the Hall effect. The current operator can now be related to the change of the Hamiltonian as a function of the flux thread through the cylinder:

$$j_y = \frac{\partial H}{\partial k_y} = \frac{L_y}{2\pi}\frac{\partial H}{\partial \Phi}. \tag{6.1}$$

We take the expectation value of this current in the instantaneous ground state of the system:

$$\langle\psi| \, j_y \, |\psi\rangle = \frac{\partial \langle\psi| \, H \, |\psi\rangle}{\partial \Phi} - \langle\psi| \, H\partial_\Phi \, |\psi\rangle - (\partial_\Phi \, \langle\psi|)H \, |\psi\rangle = \frac{\partial E}{\partial \Phi}, \tag{6.2}$$

where we have used the fact that $|\psi\rangle$ is normalized and, hence, $\partial\langle\psi|\psi\rangle/\partial\Phi = 0$. This is the Byers-Yang formula. We discretize this formula by assuming a linear flux threading (we thread the flux Φ very slowly):

$$I_y = \frac{\Delta E}{\Delta \Phi}. \tag{6.3}$$

Here, ΔE is the energy change in the process of flux insertion. We will see that this change has to do with the potential across the Laughlin cylinder. This is the current *along* the cylinder, i.e., azimuthal current along the translationally invariant direction. If the system is a bulk insulator, which is what we focus on in this book, not much will happen except at the edges of the system. Even in this case, nothing will happen unless the system is topologically nontrivial and has a Hall conductance. The presence of edges is not just a gimmick: in the absence of edges, there is no way to measure voltage—any time we attach leads to a system, we make an edge. If we put edges on the system, then the system is still an insulator deep in the bulk, but close to the edge, the confining electric field perpendicular to the two edges $E_x(L/2)$, $E_x(-L/2)$ (the field coming from the boundary condition) forces the bands to band upwards and cross the Fermi level. Assume there are n bands crossing the Fermi level (where n could even be zero).

We now want to take the flux threaded through the Laughlin cylinder to be exactly a flux quanta. The reason why we want exactly one flux quanta is because, in this case, we know exactly what happens to the system: in an adiabatic insertion, the momentum of *all* the occupied states changes by $2\pi/L_y$. This is exactly the momentum level spacing in the y-direction. Very close to the edge, after an adiabatic flux insertion, every single band that crosses the Fermi level now has one occupied momenta *above* the Fermi level on (say) the right edge and one unoccupied momenta *below* the Fermi level on the left edge of the sample. But the two edges had a potential difference of V_x between them, which means that the energy we are required to provide during the flux insertion (the change in energy ΔE) is equal to the energy required to move n particles through the potential difference:

$$\Delta E = neV_x, \tag{6.4}$$

where n is the number of bands crossing the Fermi level in a continuous system. Because $\Delta\Phi = 1$ (in units of the fundamental flux), we have

$$\sigma_{xy} = I_y/V_x = n \tag{6.5}$$

(in units of e^2/h). This equation, based only on gauge invariance, is true even in a disordered system, as long as we realize that a flux insertion of a unit flux will create edge particle-hole excitations of the system.

6.2 The Transfer Matrix Method

From the Laughlin gauge argument, we have seen that the existence of Hall conductance is intimately linked to the existence of edge modes: the system is a bulk insulator but an edge metal. In the quantum Hall effect, the edge modes have to be chiral (with the chirality given by the magnetic field), which means there has to be a preferred direction on an edge. The number of chiral edge modes on one edge (and antichiral on the opposite edge) has to be equal to the Hall conductance. To gain insight into the existence of edge modes, we now diagonalize the Hamiltonian in a Laughlin geometry – periodic in one direction and open in the other direction See Figure 6.1. We also thread flux, Φ, through the cylinder, which adds to the momentum, k_y. A magnetic field is placed on the lattice, which gives flux $\phi = p/q$ through each plaquette. In the Landau gauge, we can choose the vector potential in the y-direction to depend only on x so we can make the physical situation compatible with periodic boundary conditions in the y-direction.

The flux, Φ, threaded through the cylinder corresponds to each y-bond having an extra phase $2\pi\Phi/L_y$. We again choose the Landau gauge in which $A_y = 2\pi\phi m$, in which case the phases $\theta_{m+1,n;m,n} = 0$ and $\theta_{m,n+1;m,n} = 2\pi\phi m$. The tight-binding Hamiltonian is, as before,

$$H = -t_x \sum_{m,n} c^\dagger_{m+1,n} c_{m,n} - t_y e^{i2\pi(\Phi/L_y} \sum_{m,n} c^\dagger_{m,n+1} e^{i2\pi\phi m} c_{m,n} + \text{h.c.} \tag{6.6}$$

We now consider two edges on the system, one at $m = L_x$ (right after the $L_x - 1$ site) and one at $m = 0$ (right before the first site). As such, the problem has $L_x - 1$ sites in the x-direction. Because we broke translational invariance in the x-direction by the edges, we can Fourier-transform only in the y-direction:

$$c_{m,n} = \frac{1}{L_y} \sum_{k_y} e^{ik_y n} c_m(k_y). \tag{6.7}$$

The y-momentum takes the discrete values $k_y = 2\pi n_y/L_y$, with $n_y = 0, 1, ..., L_y - 1$. The Hamiltonian becomes

$$H = \sum_{m,k_y} \left[-t_x c^\dagger_{m+1,k_y} c_{m,k_y} - t_y e^{i2\pi(\Phi/L_y)} c^\dagger_{m,k_y} e^{-ik_y + i2\pi\phi m} c_{m,k_y} + \text{h.c.} \right]. \tag{6.8}$$

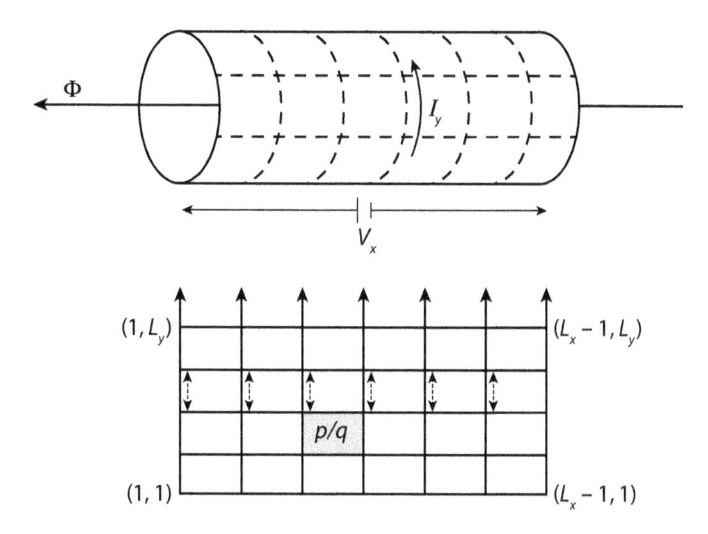

Figure 6.1. Lattice problem with open-boundary conditions (edge states) and flux threading. The arrow indicates the magnetic phases—they are solely in the y-direction but depend on the x-position, per the Landau gauge.

Again, considering a one-particle state, $|\psi(k_y, \Phi)\rangle = \sum_m \psi_m(k_y, \Phi)c^{\dagger}_{m,k_y}|0\rangle$, we can solve the Schrodinger equation to obtain the *difference* equation:

$$-t_x(\psi_{m+1}(k_y, \Phi) + \psi_{m-1}(k_y, \Phi)) - 2t_y \cos\left(k_y - 2\pi\frac{\Phi}{L_y} - 2\pi\phi m\right)\psi_m(k_y, \Phi)$$
$$= E\psi_m(k_y, \Phi) \tag{6.9}$$

This is Harper's equation *in real space.* It is the Fourier transform with respect to the x-momentum k_x of Harper's equation (5.32), which we previously solved to get the periodic conditions problem in the previous chapter. The preceding equation is no longer translationally invariant by a unit cell in the x-direction due to the existence of $2\pi\phi m$ in the cosine. However, for $\phi = p/q$, it is quasi-periodic with a longer period of q lattice sites in the x-direction. This is, of course, the magnetic unit cell we have already discussed. We divide the preceding by t_x and represent the equation in a matrix form:

$$\begin{pmatrix} \psi_{m+1}(\epsilon, k_y, \Phi) \\ \psi_m(\epsilon, k_y, \Phi) \end{pmatrix} = \tilde{M}_m(\epsilon, k_y, \Phi)\begin{pmatrix} \psi_m(\epsilon, k_y, \Phi) \\ \psi_{m-1}(\epsilon, k_y, \Phi) \end{pmatrix}, \tag{6.10}$$

where $\epsilon = E/t_x$. This equation relates the wavefunction at advanced sites, $m+1$, m to the wavefunction at the previous sites, m, $m-1$. This is called the transfer matrix formalism, and it was applied to the magnetic field lattice problem by Hatsugai [38] . The site-dependent matrix $\tilde{M}_m(\epsilon, k_y, \Phi)$ is

$$\tilde{M}_m(\epsilon, k_y, \Phi) = \begin{pmatrix} -\epsilon - 2\frac{t_y}{t_x}\cos(k_y - 2\pi\frac{\Phi}{L_y} - 2\pi\phi m) & -1 \\ 1 & 0 \end{pmatrix}. \tag{6.11}$$

We drop the explicit functional dependence on k_y, Φ—it is trivial to restore it when needed. Because the system has a natural periodicity in the x-direction via translations by a magnetic unit cell, we take the full length of the system, L_x, to be an integer multiple of (commensurate

Figure 6.2. Lattice in the Hatsugai problem. The wavefunction amplitude dies on sites 0 and L_x.

with) q; $L_x = ql$, with $l \in \mathcal{Z}$. In this case, several things can be obtained analytically, whereas in the incommensurate case only numerical solutions are available. In this case, we can define a q-site transfer matrix, which will be our building block for the edge-state theory:

$$M(\epsilon) = \tilde{M}_q(\epsilon)\tilde{M}_{q-1}(\epsilon) \cdots \tilde{M}_1(\epsilon) \equiv \begin{pmatrix} M_{11}(\epsilon) & M_{12}(\epsilon) \\ M_{21}(\epsilon) & M_{22}(\epsilon) \end{pmatrix}, \tag{6.12}$$

where the last equality defines the components of the 2×2 matrix $M(\epsilon)$. This matrix has not been found analytically and explicitly for any values of p, q but it would be useful for academic purposes if a closed-form solution for at least some of the p, q were found. Obviously, $M_{11}(\epsilon)$, $M_{12}(\epsilon)$, $M_{21}(\epsilon)$ and $M_{22}(\epsilon)$ are polynomials in the energy ϵ of degree q, $q-1$, $q-1$ and $q-2$, respectively. In terms of this matrix, we have the transfer equation

$$\begin{pmatrix} \psi_{q+1}(\epsilon, k_y, \Phi) \\ \psi_q(\epsilon, k_y, \Phi) \end{pmatrix} = M(\epsilon, k_y, \Phi) \begin{pmatrix} \psi_1(\epsilon, k_y, \Phi) \\ \psi_0(\epsilon, k_y, \Phi) \end{pmatrix}, \tag{6.13}$$

which transports the wavefunction all through the magnetic unit cell. Because the system is periodic with q, we have

$$\tilde{M}_q(\epsilon)\tilde{M}_{q-1}(\epsilon) \cdots \tilde{M}_1(\epsilon) = \tilde{M}_{2q}(\epsilon)\tilde{M}_{2q-1}(\epsilon) \cdots \tilde{M}_{q+1}(\epsilon) = M(\epsilon), \tag{6.14}$$

and so on, which means that for the total length of our commensurate chain, we have the following transfer matrix formula:

$$\begin{pmatrix} \psi_{L_x+1}(\epsilon, k_y, \Phi) \\ \psi_{L_x}(\epsilon, k_y, \Phi) \end{pmatrix} = (M(\epsilon, k_y, \Phi))^l \begin{pmatrix} \psi_1(\epsilon, k_y, \Phi) \\ \psi_0(\epsilon, k_y, \Phi) \end{pmatrix}. \tag{6.15}$$

So far we have done everything except specify the boundary conditions. We now impose open boundary conditions, which, according to figure 6.2, means that at sites 0, L_x there is vanishing wavefunction amplitude, and after that the lattice repeats itself. Basically, the boundary conditions are identical to taking a periodic chain from 0 to $2L_x \equiv 0$ and then slowly turning off the hopping between sites $L_x - 1$ and L_x and from sites 0 to 1 to make two $L_x - 1$ chains, each with the boundary conditions

$$\psi(L_x) = \psi(0) = 0. \tag{6.16}$$

The boundary condition imposed on the preceding transfer matrix equation states that

$$\begin{pmatrix} \psi_{L_x+1}(\epsilon, k_y, \Phi) \\ 0 \end{pmatrix} = (M(\epsilon, k_y, \Phi))^l \begin{pmatrix} \psi_1(\epsilon, k_y, \Phi) \\ 0 \end{pmatrix}. \tag{6.17}$$

This condition requires that

$$[M^l(\epsilon)]_{21} = 0. \tag{6.18}$$

$[M^l(\epsilon)]_{21}$ is a polynomial in ϵ of degree $L_x - 1$; hence, it has $L_x - 1$ real roots. These roots are the eigenvalues of the 1-D lattice Hamiltonian and determine the full energy spectrum of the problem. Some of these eigenvalues belong to eigenstates that are in the bulk and some of them belong to eigenstates localized on the edge of the sample (edge states).

6.3 Edge Modes

The eigenvalue in equation (6.18) can be satisfied if the q-site matrix M has the property

$$M_{21}(\epsilon) = 0. \tag{6.19}$$

This is so because the product of upper-diagonal matrices is also upper diagonal. We write the $q - 1$ solutions of equation (6.19) as μ_j, $j = 1, \ldots, q - 1$ and order them in ascending order $\mu_i < \mu_j, i < j$. We can prove that this matrix does not have degeneracies. The transfer matrix equation gives, for a translation of qk sites,

$$\begin{pmatrix} \psi_{qk+1}(\mu_j, k_y, \Phi) \\ 0 \end{pmatrix} = \begin{pmatrix} [M_{11}(\mu_j)]^k & \# \\ 0 & \# \end{pmatrix} \begin{pmatrix} \psi_1(\mu_j, k_y, \Phi) \\ 0 \end{pmatrix}, \tag{6.20}$$

where each # is a quantity that cannot be obtained analytically. Notice the presence of M_{11}^k in the first spot of this matrix, which would not have happened if $M_{21} \neq 0$. Hence,

$$\frac{\psi_{qk+1}}{\psi_1} = (M_{11}(\mu_j))^k. \tag{6.21}$$

Because we want a normalized wavefunction $\sum_{m=1}^{L_x-1} |\psi_m|^2 = 1$, we see that the preceding equation implies that the eigenfunctions are localized on the two different edges or in the bulk, depending on the value of $M_{11}(\mu_j)$.

$$\begin{aligned} &|M_{11}(\mu_j)| < 1 \rightarrow \frac{\psi_{qk+1}}{\psi_1} \rightarrow 0 \rightarrow \text{localized at site 1—left edge.} \\ &|M_{11}(\mu_j)| > 1 \longrightarrow \frac{\psi_{qk+1}}{\psi_1} \rightarrow \infty \rightarrow \text{localized at site } L_x - 1\text{—right edge.} \\ &M_{11}(\mu_j)| = 1 \rightarrow \left|\frac{\psi_{qk+1}}{\psi_1}\right| = 1 \rightarrow \text{degenerate with, merges into bulk.} \end{aligned} \tag{6.22}$$

We have then found how to obtain the edge modes. The rest of the eigenvalues belong to eigenstates that give the bulk modes. Hence, the roots of $[M^l]_{21}(\epsilon) = 0$—but not those of $M_{21}(\epsilon) = 0$—belong to the bulk states.

6.4 Bulk Bands

If we put periodic boundary conditions in both x- and y-directions, the bands disperse in the magnetic BZ with respect to both k_x^0 and k_y. Because the length of the magnetic unit cell is equal to q, we have to take L_x to be a multiple of q in the periodic boundary condition case, which is also one of the reasons why we chose it to be a multiple of q even in the open-boundary-condition case. Then, if instead of doing a Fourier transform and plotting the bands versus both k_x^0, k_y, we only Fourier-transform y to k_y, we can simply diagonalize the quasi-1-D lattice

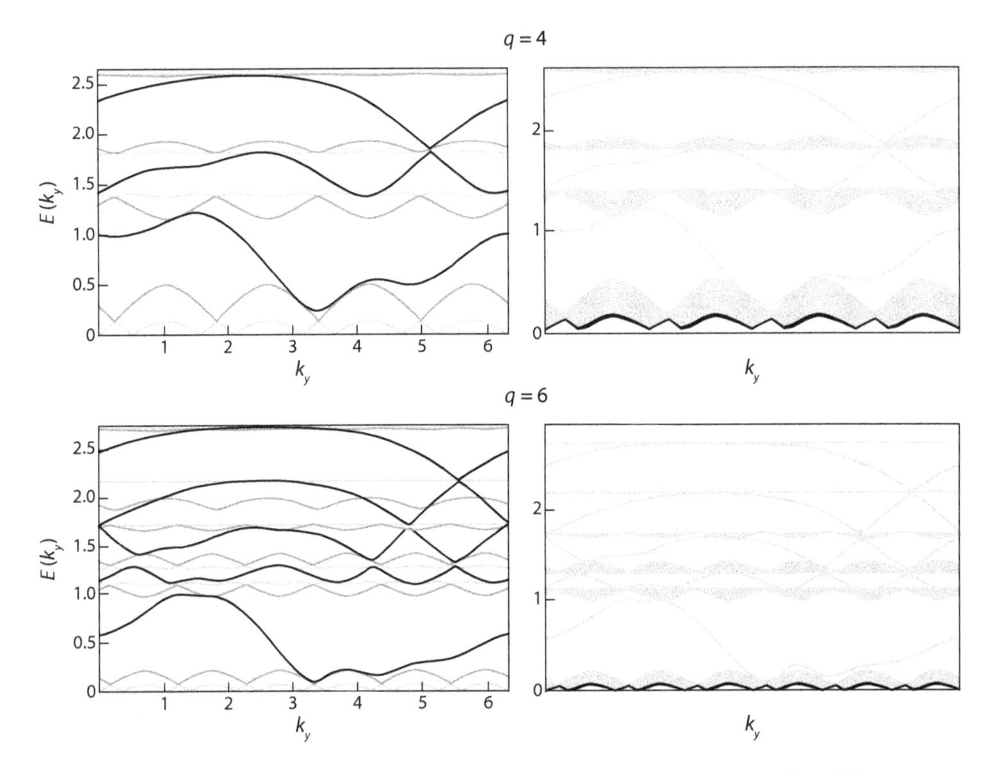

Figure 6.3. Example of bulk band edges for graphene $q = 4, 5$ (graphene has different band counting, which we will show in the next chapter, but the band filling is the same). Left: bands are obtained from the band-edge condition $\Delta = \pm 2$ (at each k_y). Right: bands are obtained by diagonalization of the problem of $L_x = 500$ sites. The edge states are seen to connect bulk bands in the band gap.

Hamiltonian at each k_y to obtain L_x/q energies. This is the same number of energies that we would obtain if we Fourier-transformed on k_x^0 because the spacing between two k_x^0 momenta is $2\pi/L_x$. Equivalently, we could have Fourier-transformed in x and y, calculated the energy spectrum for all allowed values of k_x^0 and k_y in the first BZ, and then—for every value of k_y—collapsed *all* the k_x^0 values onto the same axis, i.e., the constant k_y line. This will make the bands (at each k_y) "fat" because, for a single k_y, there is sizable k_x^0 dispersion, at least in the strong field limit. This collapse of all the k_x^0 for each k_y will have (for each band in the k_x^0, k_y space) upper and lower energy bounds, which will disperse with k_y. An example of this is presented in figure 6.3.

The important point of the preceding paragraph is that we can obtain the bulk bands by looking at the system with *periodic* boundary conditions. In particular, we can obtain the bulk bands from the particular Bloch (periodicity) condition they satisfy, which, in this case—because the unit cell has length q—becomes

$$\psi_{m+q}(\epsilon) = \rho(\epsilon)\psi_m(\epsilon) \tag{6.23}$$

for m any lattice site. The Bloch condition says that the proportionality $\rho(\epsilon)$ constant must be a phase $|\rho(\epsilon)| = 1$. If it had not been a phase, the wavefunction would grow or shrink exponentially as we make the system very large, and because for periodic boundary conditions, the system does not have edges, the wavefunction would become unnormalizable, i.e., infinite or vanishing, respectively. Applying the preceding condition to $m = 1$, we find

$$\begin{pmatrix} \psi_{q+1}(\epsilon, k_y, \Phi) \\ \psi_q(\epsilon, k_y, \Phi) \end{pmatrix} = M \begin{pmatrix} \psi_1(\epsilon, k_y, \Phi) \\ \psi_0(\epsilon, k_y, \Phi) \end{pmatrix} = \rho(\epsilon) \begin{pmatrix} \psi_1(\epsilon, k_y, \Phi) \\ \psi_0(\epsilon, k_y, \Phi) \end{pmatrix}. \tag{6.24}$$

Notice ρ is an eigenvalue of M; as such, it satisfies the equation

$$\rho^2 - \mathrm{Tr}[M]\rho + \mathrm{Det}[M] = 0. \tag{6.25}$$

It is trivial to prove, by direct calculation, that $\mathrm{Det}[M] = \prod_{i=1}^{q} \mathrm{Det}[\tilde{M}_i] = \prod_{i=1}^{q} 1 = 1$. Following Hatsugai, we call $\mathrm{Tr}[M(\epsilon)] = \Delta(\epsilon)$. The solution for ρ is easy to find and takes completely different forms if $\Delta^2 <, > 4$:

$$\Delta^2 \leq 4 \rightarrow \rho = \frac{\Delta \pm i\sqrt{4 - (\Delta)^2}}{2}, \tag{6.26}$$

which is of Bloch form. It is easy to show that, in this case, $|\rho| = 1$. These are the bulk bands to which we return in a moment.

When $\Delta^2 > 4$, the states are the edge modes. This can be seen through direct calculation of powers of ρ, which is not very inspiring. Another way of seeing that $\Delta^2 > 4$ implies edge modes is to note that the edge-mode condition $M_{21}(\epsilon) = 0$ and the identity $\mathrm{Det}(M) = 1$ (valid for both edge and bulk states) show that at energies of the edge states, we have $M_{11}M_{22} = 1$, which, in turn, implies the following:

$$\text{Edge modes: } (M_{21} = 0;\ M_{11}M_{22} = 1) \rightarrow \Delta = M_{11} + \frac{1}{M_{11}} \rightarrow \Delta^2 \geq 4. \tag{6.27}$$

We now turn to the bulk bands. Their energies all satisfy $\Delta^2 \leq 4$ (for each k_y). Because Δ is a polynomial of degree q in the energy ϵ, the equation $\Delta^2 = 4$ has $2q$ solutions λ_i, which can be placed as the lower and upper edges of the q "fat" bands that we previously mentioned:

$$\epsilon \in [\lambda_1, \lambda_2], \ldots, [\lambda_{2j-1}\lambda_{2j}], \ldots, [\lambda_{2q-1}, \lambda_{2q}], \quad \lambda_i \leq \lambda_j, i < j. \tag{6.28}$$

The lower and upper boundaries of the bulk-band regions are λ_{2j-1} and λ_{2j}. Everything in between them gets filled when we diagonalize the problem on a large L_x chain. The band edges, λ_i, come from $\Delta = \pm 2$. This is rather clear because for $\Delta^2 > 4$, we have gaps or, at most, edge modes, whereas for $\Delta \leq 4$, we have bands. So, the edges of the bands must come from $|\Delta|^2 = 4$ if the function $\Delta(\epsilon)$ is to be smooth, which it is because it is a polynomial in ϵ. Solutions for which both $M_{21} = 0$ have the extra property that they give the points at which the edge modes touch the bulk bands. These, of course, have $M_{11} = \pm 1$.

The Laughlin gauge argument has taught us that the Hall conductance equals the number of chiral states localized on one edge. However, from the exact diagonalization results presented in figure 6.4 we can see that the edge states on a single edge can be both chiral (have positive slope at some k_y) and antichiral (have negative slope at some other k_y) while belonging to the same edge. This happens only in the high-field limit, in which the bulk bands have strong dispersion; in the weak-field limit, the bulk bands become very thin and effectively nondispersive (in the limit of the weak field, they are Landau levels), and the edge modes on one given (left or right) edge are either all chiral or all antichiral. When the edge mode contains both chiral and antichiral components on one edge, the Hall conductance is given by the linking number of the edge with the constant Fermi level in the gap. An example of how

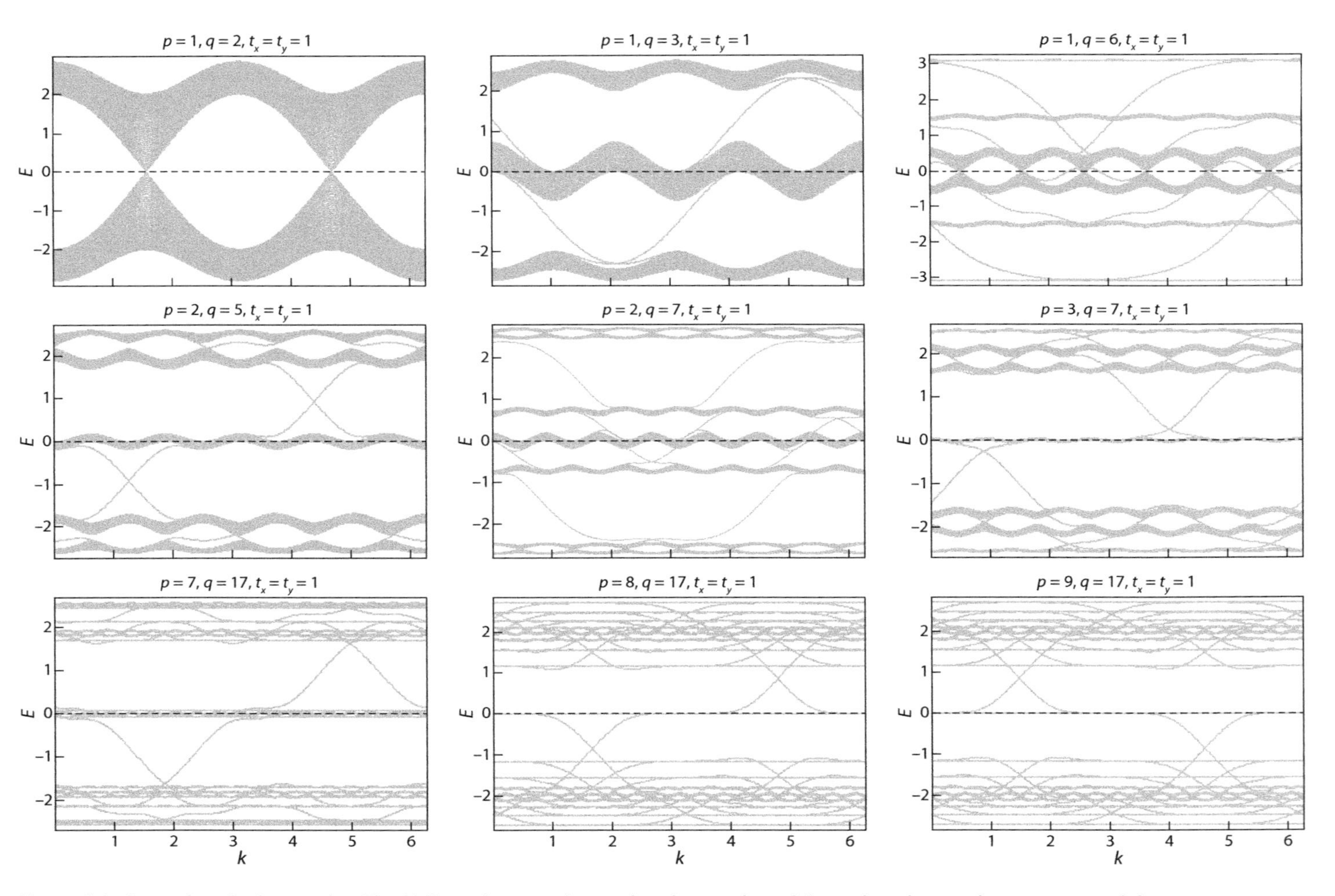

Figure 6.4. Examples of edge modes. The Hall conductance is equal to the number of times the edge mode wraps around the gap.

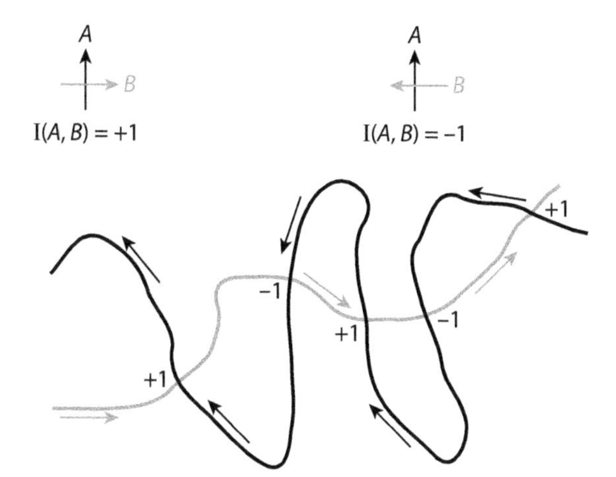

Figure 6.5. Linking number example.

to compute linking numbers is given in figure 6.5. The linking number has an orientation, as seen in figure 6.5. The Hall conductance is identical to the linking number by virtue of Laughlin's argument.

6.5 Problems

1. *Chern Number and Hall Conductance from Bulk and Edge:* Through a numerical diagonalization procedure, obtain the first eigenstates of the Harper problem of a magnetic field with flux $\frac{2\pi}{5}$ on the square lattice and compute the Chern number of each of the bands. Then, through the transfer matrix method, obtain the open-boundary-condition spectrum and count the number of edge modes in each of the gaps in the spectrum. Show that the sum of the Chern numbers of the bands below a certain gap equal the number of edge modes in the gap.

2. *Nonmonotonicity of Hall Conductance in the Hofstadter Problem:* Test the validity of the Diophantine equation giving the Hall conductance versus the number of edge modes obtained in numerical diagonalization of the transfer matrix method with open-boundary conditions for different fluxes that are of the form $\frac{2\pi p}{q}$, with $p \neq 1$. Show that in this case, the sequence of Hall conductances when the Fermi level is placed in the first, second, third, etc., gap can be nonmonotonic (when $p = 1$, remember the Hall conductance was monotonically growing until half-filling and then monotonically decreasing until all the bands were filled).

3. *Magnetic Field of Irrational Flux per Plaquette:* Through the exact diagonalization of large systems (which can be done in the noninteracting single-particle case), solve the open-boundary condition of a system with flux $\frac{2\pi}{\sqrt{2}}$ per unit cell for a large number of sites (in this case, the problem cannot be solved analytically as in the Harper equation because the flux per unit cell is not a rational number; hence, enlarging the unit cell does not help make the flux an integer). Plot the energies, and look at the eigenstates: Are there gaps in the system? Are there eigenstates localized at the edge of the sample? What is the Hall conductance in these (if any) gaps?

7

Graphene

In the previous chapters, we have primarily investigated phenomena on the square lattice. In this chapter, we switch gears and look at another two-dimensional lattice, the hexagonal, or honeycomb, lattice. This is not just a simple variation of our work so far. The hexagonal lattice is important for of both experimental and theoretical reasons. Experimentally, the hexagonal lattice is realized in graphene, an amazing material of fundamental importance and interest. The synthesis of graphene sheets via the "scotch tape" method was awarded with the Nobel Prize in Physics for 2010. Graphene is currently a popular subject of research and is likely to remain so for many years. From a theoretical point of view, graphene is also interesting: it has two sites per unit cell (A-sites and B-sites), so the minimal model for graphene has to be a two-band model. It turns out that the hexagonal lattice with nearest-neighbor hopping is a semimetal with gapless Dirac fermions at the Fermi level. Gaps for these Dirac fermions can be opened in different ways, and the insulator obtained can exhibit fascinating properties, such as a zero-field quantized Hall effect (i.e., the Chern insulator or quantum anomalous Hall-effect state).

We focus on graphene in this long chapter, We show that the nearest-neighbor-hopping model of graphene has Dirac nodes and then ask what symmetries protect the nodes from opening a gap. We learn that in two spatial dimensions, both inversion and TR symmetry are needed to keep Dirac fermions gapless. These two conditions endow Dirac fermions with a "reality" constraint, and we relate their gaplessness to the Wigner–vonNeumann classification. We then keep inversion and time reversal intact but break the C_3 symmetry of the graphene nearest-neighbor model and show that, although Dirac nodes are *locally* stable, they are not globally stable: two Dirac nodes can annihilate and open a gap. This can happen, however,only if the two nodes have opposite "vorticity," which is the same effect the Berry phase accumulated around a constant-energy contour encircling the Dirac point. We hence learn that the Berry phase is a vorticity that keeps Dirac nodes locally stable. We then move on to show that gapless graphene has edge modes that link the two Dirac nodes. We obtain analytic solutions for these modes and then argue that the opening of different types of gaps at the Dirac nodes gives rise to different types of insulators. This provides a simple and heuristic way of understanding the Chern insulator presented in a later chapter of this book.

7.1 Hexagonal Lattices

The graphene lattice, shown in figure 7.1, has the following translation vectors:

$$\mathbf{a}_1 = \frac{a}{2}(3, \sqrt{3}), \quad \mathbf{a}_2 = \frac{a}{2}(3, -\sqrt{3}), \tag{7.1}$$

where a is the bond length. The vectors give rise to the reciprocal lattice vectors

$$\mathbf{b}_1 = \frac{2\pi}{3a}(1, \sqrt{3}), \quad \mathbf{b}_2 = \frac{2\pi}{3a}(1, -\sqrt{3}), \tag{7.2}$$

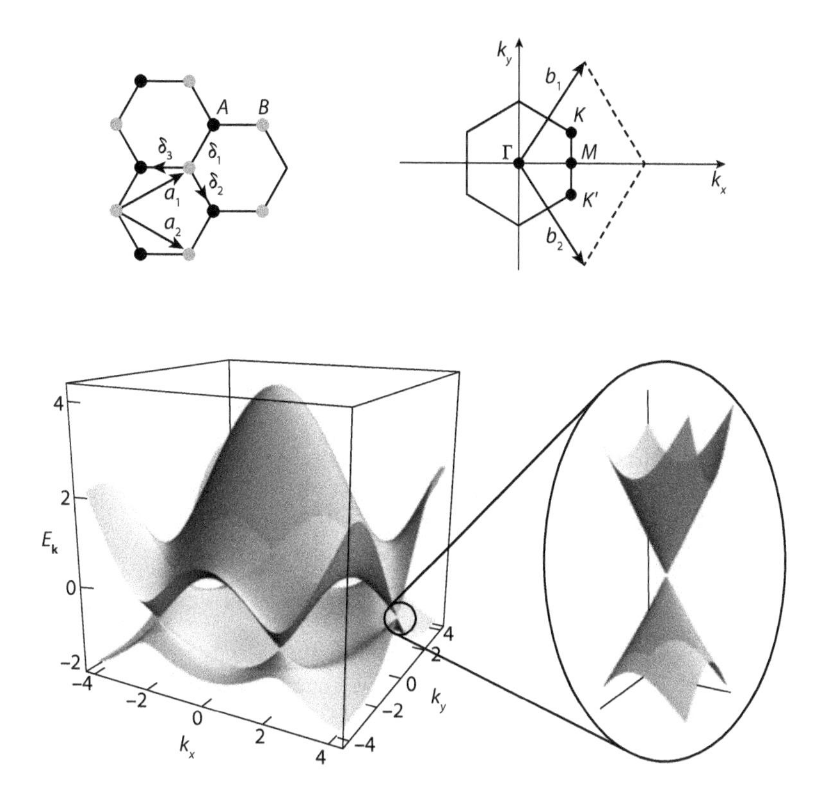

Figure 7.1. The graphene lattice, primitive vectors, BZ and dispersion.

which satisfy $\mathbf{b}_i \cdot \mathbf{a}_j = 2\pi\delta_{ij}$. Notice that the lattice has a two-site unit cell because A-sites (see fig. 7.1) can be moved into A-sites only by the primitive wavevectors. In practice, the BZ can be taken to be the parallelogram formed by the wavevectors $\mathbf{b}_1, \mathbf{b}_2$, although in most work on the subject, the BZ is taken to be the hexagon with the two vertices at points (K, K'), where

$$\mathbf{K} = \frac{2\pi}{3a}\left(1, \frac{1}{\sqrt{3}}\right), \quad \mathbf{K}' = \frac{2\pi}{3a}\left(1, -\frac{1}{\sqrt{3}}\right). \tag{7.3}$$

These two points are TR partners of each other, which is clear in the parallelogram BZ but which can easily be shown to be the case in the hexagonal BZ as well because $-\mathbf{K} = \mathbf{K}' - \mathbf{b}_1 - \mathbf{b}_2$.

The simplest tight-binding Hamiltonian for graphene contains hoppings to nearest- neighbor sites. It is a 2×2 Hamiltonian, which is easy to diagonalize in the (c_A, c_B) basis of second quantized operators for sites A, B. For hopping t_1, t_2, t_3 along the $\boldsymbol{\delta}_1, \boldsymbol{\delta}_2, \boldsymbol{\delta}_3$ bonds (see fig. 7.1) the Hamiltonian Fourier- transformed in momentum space is

$$H = \sum_k \begin{pmatrix} c_{Ak}^\dagger & c_{Bk}^\dagger \end{pmatrix} \begin{pmatrix} 0 & \sum_{i=1}^3 t_i e^{i\mathbf{k}\cdot\boldsymbol{\delta}_i} \\ \sum_{i=1}^3 t_i e^{-i\mathbf{k}\cdot\boldsymbol{\delta}_i} & 0 \end{pmatrix} \begin{pmatrix} c_{Ak} \\ c_{Bk} \end{pmatrix}. \tag{7.4}$$

The distances between the three nearest neighbors of one of the sites is

$$\boldsymbol{\delta}_1 = \frac{a}{2}(1, \sqrt{3}), \quad \boldsymbol{\delta}_2 = \frac{a}{2}(1, -\sqrt{3}), \quad \boldsymbol{\delta}_3 = a(-1, 0). \tag{7.5}$$

Unfortunately, in this basis, the Hamiltonian is not in Bloch form, i.e., $h(\mathbf{k}+\mathbf{G}) \neq h(\mathbf{k})$, where $G = b_1, b_2$, because $\mathbf{b}_i \cdot \boldsymbol{\delta}_j \neq 2\pi\delta_{ij}$. We could work in this basis and forego Bloch form, but this would not be consistent with results in the past chapters and would also preempt the automatic identification of time reversal invariant points $\mathbf{G}/2$ and $-\mathbf{G}/2$, up to reciprocal lattice wavevectors. We decide to restore Bloch form by making a gauge transformation of the preceding Hamiltonian on B-sites,

$$c_{Bk} \to c_{Bk} e^{i\mathbf{k}\cdot\delta_3}, \tag{7.6}$$

to get the Bloch Hamiltonian $H = \sum_{\mathbf{k}} c_{\mathbf{k}}^{\dagger} h(k) c_{\mathbf{k}}$, where

$$h(k) = \begin{pmatrix} 0 & -t_a e^{i\mathbf{k}\cdot\mathbf{a}_1} - t_b e^{i\mathbf{k}\cdot\mathbf{a}_2} - t_c \\ -t_a e^{-i\mathbf{k}\cdot\mathbf{a}_1} - t_b e^{-i\mathbf{k}\cdot\mathbf{a}_2} - t_c & 0 \end{pmatrix}. \tag{7.7}$$

The gauge transformation that makes the Hamiltonian of Bloch form corresponds to a different gauge choice of orbitals in the graphene lattice.

7.2 Dirac Fermions

The simple tight-binding graphene Hamiltonian displays the interesting physics of massless Dirac fermions. For the case of isotropic-hopping matrix elements $t_a = t_b = t_c = 1$, if we expanded around the points $\mathbf{K}$, $\mathbf{K}'$, at $\mathbf{k} = \mathbf{K} + \boldsymbol{\kappa}$, with $\kappa \ll K$, we find that the Hamiltonian has the Dirac form

$$h(\mathbf{K} + \boldsymbol{\kappa}) = \kappa_x \sigma_x + \kappa_y \sigma_y. \tag{7.8}$$

Expanding around $\mathbf{K}'$ to first order, we obtain another Dirac Hamiltonian:

$$h(\mathbf{K}') = H(-\mathbf{K} + \boldsymbol{\kappa}) = -\kappa_x \sigma_x + \kappa_y \sigma_y. \tag{7.9}$$

There are no other independent Dirac nodes in the problem. The presence of these nodes renders graphene to be a semimetal, with fundamentally different properties from an insulator, because low-energy excitations are always present in such a system.

The question one should now ask is whether these Dirac points are stable to perturbations. Our expansion of the Hamiltonian has showned us the existence of two Dirac fermions. However, the Hamiltonian we used was by no means generic. For example, it contained only nearest-neighbor coupling with C_3 symmetry, and it did not allow for different on-site energies of the A and B-sites in the unit cell, etc. Would adding small perturbations to the graphene lattice result in the gapping of these Dirac fermions? What are the perturbations we are allowed to add while keeping the system a semimetal? What kind of perturbations open a gap? For all these questions, we need to look at the symmetries of graphene.

7.3 Symmetries of a Graphene Sheet

Two main symmetries characterize the hexagonal lattice with identical atoms on A and B-sites. The first symmetry is time reversal, which is present regardless of the spatial distribution of hopping matrix elements, as long as they are real. The second symmetry, inversion, is present if hoppings in the lattice are symmetric upon inversion with the inversion center either in the middle of the unit cell bond or in the middle of the hexagonal lattice: it is clearly present

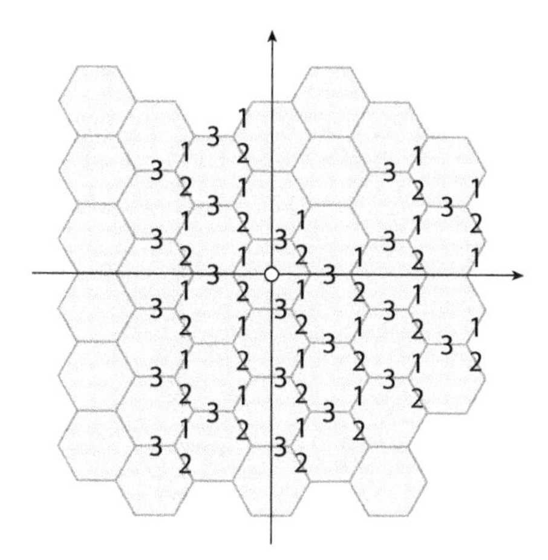

Figure 7.2. t_1, t_2, t_3-hopping model of graphene and inversion center (open dot in the center).

in the isotropic hopping scenario. We analyze these symmetries and their consequences on the physics of Dirac fermions. The effects of the C_3 symmetry on the Hamiltonian are then analyzed.

7.3.1 Time Reversal

As previously discussed, time reversal for spinless fermion Bloch Hamiltonians takes the form

$$Th(\mathbf{k})T^{-1} = h(-\mathbf{k}) \longrightarrow h(\mathbf{k})^* = h(-\mathbf{k}). \tag{7.10}$$

If we impose TR invariance on the system, this symmetry fixes the form of the Hamiltonian at K' once the Hamiltonian at K is known

$$h(\mathbf{K}+\boldsymbol{\kappa})^* = h(-\mathbf{K}-\boldsymbol{\kappa}) = (\kappa_x\sigma_x + \kappa_y\sigma_y)^* = \kappa_x\sigma_x - \kappa_y\sigma_y$$

$$\longrightarrow h(-\mathbf{K}+\boldsymbol{\kappa}) = -\kappa_x\sigma_x + \kappa_y\sigma_y. \tag{7.11}$$

Notice that T by itself does not protect the Dirac fermions from opening a gap as long as the gap at points K, K' has special properties: if at K we have a Bloch Hamiltonian $h(\mathbf{K}+\boldsymbol{\kappa}) = \kappa_x\sigma_x + \kappa_y\sigma_y + m\sigma_z$, whereas at K' we have the Bloch Hamiltonian $h(-\mathbf{K}-\boldsymbol{\kappa}) = h(\mathbf{K}+\boldsymbol{\kappa})^* = \kappa_x\sigma_x - \kappa_y\sigma_y + m\sigma_z$, the system preserves time reversal. This type of mass, which preserves T but opens a gap in the Dirac spectrum, physically represents a different on-site energy for the atoms on the A-sites and B-sites. It corresponds to adding a momentum-independent term $m\sigma_z$ in the full Bloch Hamiltonian. For example, the Hamiltonian of boron nitride, (BN) a system that also has a hexagonal lattice structure but, as the name suggests, as different atoms on the A- and B-sites, has this form and is a gapped insulator.

7.3.2 Inversion Symmetry

The graphene lattice has another symmetry which is spatial inversion (see fig. 7.2 for the t_1, t_2, t_3-hopping model and inversion center). If the hopping and on-site matrix elements do not change upon the action of the inversion operator

$$I : (x, y) \rightarrow (-x, -y), \tag{7.12}$$

we have that the Hamiltonian also has inversion symmetry. Inversion is a unitary operator. It does not contain the complex conjugation operator, and it leaves the $[x, p]$ commutator unchanged, due to the fact that both x, p change under inversion, unlike in the T case, in which only p transformed under inversion. Under inversion with respect to the middle of the unit cell, the second quantized operators transform as

$$Ic_{i,A}I^{-1} = c_{-i,B}, \qquad Ic_{i,B}I^{-1} = c_{-i,A}, \tag{7.13}$$

or, in compact form,

$$Ic_{i,a}I^{-1} = \sigma^x_{a,a'}c_{-i,a'}, \tag{7.14}$$

where $a = A, B$. The Fourier component of the second quantized operators transforms as

$$Ic_{\mathbf{k},a}I^{-1} = \sigma^x_{a,a'}\frac{1}{\sqrt{N}}\sum_j e^{-i\mathbf{k}r_j}c_{ja'} = \sigma^x_{aa'}c_{-\mathbf{k},a'}. \tag{7.15}$$

For the inversion-symmetric second quantized graphene Hamiltonian with $[H, I] = 0$, we find the transformation on the Bloch Hamiltonian:

$$\begin{aligned} IHI^{-1} &= \sum_{\mathbf{k}} \sigma^x_{\theta a}c^\dagger_{-\mathbf{k}\theta}Ih_{\alpha\beta}(\mathbf{k})I^{-1}\sigma^x_{\beta,\delta}c_{-\mathbf{k}\delta} = \sum_{\mathbf{k}} \sigma^x_{\theta a}c^\dagger_{-\mathbf{k}\theta}h_{\alpha\beta}(\mathbf{k})\sigma^x_{\beta,\delta}c_{-\mathbf{k}\delta} \\ &= \sum_{\mathbf{k}} \sigma^x_{\theta a}c^\dagger_{\mathbf{k}\theta}h_{\alpha\beta}(-\mathbf{k})\sigma^x_{\beta,\delta}c_{\mathbf{k}\delta}. \end{aligned} \tag{7.16}$$

If inversion is a good symmetry, then $IHI^{-1} = H$ and, hence,

$$h(\mathbf{k}) = \sigma_x h(-\mathbf{k})\sigma_x. \tag{7.17}$$

In this equation, the matrix representation of the inversion operator is σ_x in the graphene lattice. We will later see other representations, such as σ_z.

By itself, inversion also does not guarantee the stability of the Dirac points. If at K we open a gap m, $h(\mathbf{K} + \boldsymbol{\kappa}) = \kappa_x\sigma_x + \kappa_y\sigma_y + m\sigma_z$, then we can have a perfectly inversion-symmetric Hamiltonian if the Hamiltonian at the K' point is $h(-\mathbf{K} - \boldsymbol{\kappa}) = \sigma_x(\kappa_x\sigma_x + \kappa_y\sigma_y + m\sigma_z)\sigma_x = \kappa_x\sigma_x - \kappa_y\sigma_y - m\sigma_z$.This type of mass, however, cannot come from a momentum-independent term (in the full Bloch Hamiltonian). Indeed, a full lattice realization of such a mass term was first found by Haldane in the first example of a topological insulator, the Chern insulator. We analyze it in chapter 8.

7.3.3 Local Stability of Dirac Points with Inversion and Time Reversal

The reason that inversion and time reversal separately do not protect the Dirac fermions is because they both link a generic $\mathbf{k}$ to $-\mathbf{k}$, thereby not really imposing any constraints on *the Hamiltonian at a generic* $\mathbf{k}$. However, when both these symmetries are present and when they are used together, they relate $\mathbf{k}$ to $\mathbf{k}$ and impose constraints on the form of the Bloch Hamiltonian at each $\mathbf{k}$ separately:

$$h(\mathbf{k}) = \sigma_x h(-\mathbf{k})\sigma_x = h^*(-\mathbf{k}) \longrightarrow h(\mathbf{k}) = \sigma_x h^*(\mathbf{k})\sigma_x. \tag{7.18}$$

The first equality uses I, and the second uses T. The equation (29) is the matrix representation of the operator product of inversion and time reversal, $(TI)h(\mathbf{k})(TI)^{-1} = h(\mathbf{k})$). For a generic, two-level Hamiltonian of Bloch form,

$$H = d_i(\mathbf{k})\sigma_i + \epsilon(\mathbf{k})I_{2\times 2}. \tag{7.19}$$

T and inversion impose the conditions (we drop the explicit $\mathbf{k}$-dependence)

$$d_i\sigma_i + \epsilon = \sigma_x(d_i\sigma_i + \epsilon)^*\sigma_x = d_x\sigma_x + d_y\sigma_y - d_z\sigma_z + \epsilon. \tag{7.20}$$

We see that as a result of the combined inversion and time reversal, we obtain

$$d_z(\mathbf{k}) = -d_z(\mathbf{k}) = 0. \tag{7.21}$$

As such, if both inversion and time reversal are respected, no σ_z term can arise, and the Dirac points are locally stable. What does this mean for a Dirac Hamiltonian? If, around the point $\mathbf{K}$ in the BZ, the Hamiltonian is $k_x\sigma_x + k_y\sigma_y$ (where k is the deviation from $\mathbf{K}$), we are not allowed to add a σ_z-term to the Hamiltonian by any perturbation. Although σ_x-and σ_y-terms are not forbidden by time reversal and inversion, their only result, *if small*, is to shift the Dirac points in momentum space. Explicitly,

$$H' = k_x\sigma_x + k_y\sigma_y + a_1\sigma_x + a_2\sigma_y, \tag{7.22}$$

and as long as a_1 and a_2 are small, H' has nodes at $k_x = -a_1$ and $k_y = -a_2$. Hence, with inversion and time reversal, a single Dirac node is *locally* stable—no small perturbation can open a gap. Large perturbations can and do open a gap, but the gap-opening mechanism is very different.

As long as the Dirac fermions at $\mathbf{K}$ and $-\mathbf{K}$ are forbidden to open a gap by T and inversion, they carry a vortex in the Bloch wavefunction, which is the same as saying that their Berry phase $\int A_i(\kappa)d\kappa_i$ (where $A_i(\kappa)$ is the Berry potential and the integral is performed over any closed contour enclosing the Dirac point) equals $\pm\pi$. Because the matrix σ_z is forbidden (whether multiplied by a constant term or even by an even or odd power of k), we thus have that the Hamiltonian at both $\mathbf{K}$ and $\mathbf{K}'$ can be written as a matrix: $H(k) = \kappa_i A_{ij}\sigma_j$. The Berry phase of such a Dirac fermion is given by π sign(Det(A)), which we be prove later. Note that the lack of σ_z is important because it allows us to write the Hamiltonian with both $i, j = 1, 2$ rather than $H(\kappa) = \kappa_i A_{ia}\sigma_a$, with $a = 1, 2, 3$ — in this case A_{ia} would not be a square matrix and defining a determinant (vorticity) would be impossible. The fact that vorticities can be defined when the dimension of the BZ (in this case, 2) is equal to the codimension of the crossing (in this case, also 2 because we have two Pauli matrices) is no accident. The Wigner–vonNeumann classification says that a generic crossing has codimension 3—we need to tune three parameters to obtain a degeneracy because there exist three anticommuting Pauli matrices whose coefficients need to be tuned to zero. However, the BZ provides for two parameters (k_x, k_y), which are tuned automatically, leaving us with only one tunable parameter needed to obtain a degeneracy. This would be the Dirac fermion mass, which would need to be fine-tuned to vanish. But, if time reversal and inversion are present, then the matrix that would couple to this remaining tunable parameter cannot exist, and degeneracies can happen without tuning—i.e., they are locally stable. The presence of inversion and time reversal is said to impose a reality condition on the Hamiltonian (the Hamiltonian can be made real by a gauge transformation).

For the Dirac fermion at $\mathbf{K}$ we have $A_{11} = A_{22} = 1$, whereas for the one at $\mathbf{K}'$, we have $A_{11} = -A_{22} = -k$. Hence, the Berry phase, which is πsign(Det(A)), is opposite for the two fermions. In graphene, the Dirac fermion at $\mathbf{K}$ carries Berry phase π, whereas the Dirac fermion at $\mathbf{K}' = -\mathbf{K}$ carries Berry phase $-\pi$. Because the wavefunctions have vorticity, the nodes cannot be removed by themselves. For example, for the Dirac fermion at **K**, the wavefunction has a vortex (which gives it a Berry phase π) and the inclusion of extra terms that break the C_3 symmetry (for example,the simplest way is to make $t_1 \neq t_2 \neq t_3$) but respect T and I can only move the Dirac point away from **K** but cannot open a gap.

7.4 Global Stability of Dirac Points

The stability of the Dirac nodes proved in the previous section is valid for *any* perturbation that respects T and I, as long as it is small. For example, we can break C_3 and make the hopping on the bonds different, or we can add second and third nearest-neighbor hoppings. The second nearest-neighbor hopping does nothing because it is diagonal in the sublattice space (couples identical sublattices).What happens if the perturbation is large? What happens if we add large, arbitrary-range hopping terms in our Bloch Hamiltonian? It is clear that the Hamiltonian can be gapped. For example, pick $t_a = t_b = 0$ in our graphene model, i.e., make a model of very anisotropic graphene with nonzero hopping in only one direction. The energy levels would then be $\pm t_c$, thereby giving a fully gapped Hamiltonian. How did this happen? It turns out that Dirac modes can and will open a gap by coming together and annihilating at a TR-symmetric point; at this point, the dispersion has to be quadratic in one direction. This section analyzes this aspect of the problem. We show that if the Hamiltonian has C_3 symmetry (on top of I and T), then the Dirac nodes are *globally* stable, and their position is fixed at the **K** and **K**$'$ points in the BZ. If, however, C_3 symmetry is broken, then the Dirac nodes can move off the **K**, **K**$'$ points—and, upon large-enough perturbations, the two Dirac nodes can meet up and annihilate at a TR-invariant point in the BZ.

7.4.1 C_3 Symmetry and the Position of the Dirac Nodes

We now prove that, in the case where the Hamiltonian has the three symmetries T, I, and C_3, the position of the Dirac nodes does not change and stays at

$$\mathbf{K} = \frac{2\pi}{3a}\left(1, \frac{1}{\sqrt{3}}\right), \quad \mathbf{K}' = \frac{2\pi}{3a}\left(1, -\frac{1}{\sqrt{3}}\right). \tag{7.23}$$

As stated previously, T and I guarantee the absence of σ_z terms in the Hamiltonian. However, with just T and I, the hopping can be *anisotropic*, the simplest example being a model with different nearest-neighbor hoppings. An extra C_3 symmetry, i.e., rotation by $\frac{2\pi}{3}$ around a hexagon center in the lattice or around the points A or B, adds extra constraints to the system. The C_3 symmetry is manifest by the fact that the hopping matrix elements must be invariant upon the cyclic change of

$$\boldsymbol{\delta}_1 \to \boldsymbol{\delta}_2 \to \boldsymbol{\delta}_3 \to \boldsymbol{\delta}_1, \tag{7.24}$$

where, as previously defined, $\boldsymbol{\delta}_1 = \frac{a}{2}(1, \sqrt{3})$, $\boldsymbol{\delta}_2 = \frac{a}{2}(1, -\sqrt{3})$, $\boldsymbol{\delta}_3 = a(-1, 0)$ are the $A \to B$ bond vectors. This C_3 transformation is obvious if we take the $\frac{2\pi}{3}$ rotation center to be either the center of the hexagon, one of the A-sites, or one of the B-sites. Note that this symmetry implies equal hopping parameters for the nearest-neighbor hopping $A \to B$, but, in general, for nth next nearest-neighbor hopping, it does *not* imply equal hopping parameters. For example, for the hopping matrix element from site A to the third-nearest-neighbor site B (there are two third-nearest neighbor sites B, as given in fig. 7.3), we have perfect C_3 symmetry even though

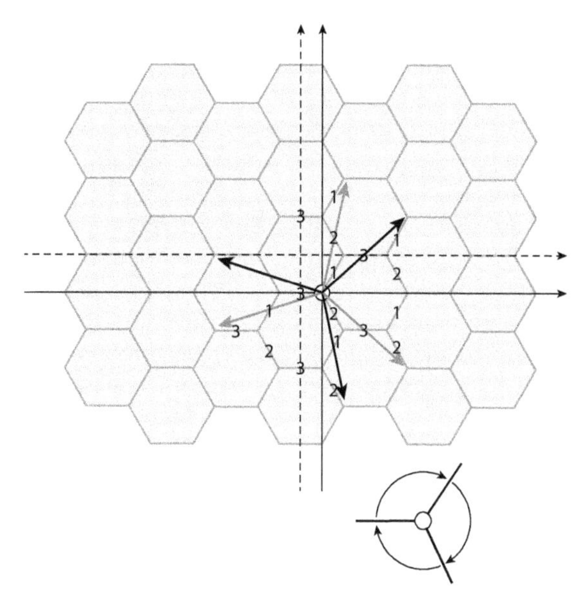

Figure 7.3. C_3 symmetry with respect to either the A-site rotation center or the center of the hexagonal lattice. If respected, the Hamiltonian should be invariant under cyclic $\boldsymbol{\delta}_1 \to \boldsymbol{\delta}_2 \to \boldsymbol{\delta}_3 \to \boldsymbol{\delta}_1$. The vector lines give the vectors to one type of third-next-nearest neighbors (the other type of third-next-nearest neighbor is the site B, exactly opposite to the site A in the same hexagonal unit cell). Notice that C_3 symmetry sends orange to orange and green to green, but the hopping of the green and orange can be different as in equation (7.25).

the six matrix elements can break up into two different matrix elements:

$$t_{1nnn}(e^{i\mathbf{k}\cdot(\delta_1-\delta_3+\delta_1)}+e^{i\mathbf{k}\cdot(\delta_3-\delta_2+\delta_3)}+e^{i\mathbf{k}\cdot(\delta_2-\delta_1+\delta_2)})$$

$$+t_{2nnn}(e^{i\mathbf{k}\cdot(\delta_1-\delta_2+\delta_1)}+e^{i\mathbf{k}\cdot(\delta_3-\delta_1+\delta_3)}+e^{i\mathbf{k}\cdot(\delta_2-\delta_3+\delta_2)}), \tag{7.25}$$

where $t_{1nnn} \neq t_{2nnn}$. In other words, C_3 symmetry does not require symmetry over all permutations of $\delta_{1,2,3}$, but over only the cyclic permutations. It is clear from figure 7.3 that the preceding term respects C_3 even though $t_{1nnn} \neq t_{2nnn}$.

To see the effect of the C_3 symmetry on the graphene band structure, we first analyze the nearest-neighbor Hamiltonian—where hopping occurs only through the nearest-neighbor bond. It is convenient to work in the basis in which the hopping occurs through $\boldsymbol{\delta}_1$, $\boldsymbol{\delta}_2$, $\boldsymbol{\delta}_3$. It is in this basis that the C_3 symmetry is easily imposed, and we can go to the basis 1, $\mathbf{a}_1$, $\mathbf{a}_2$ through a gauge transformation.

The off-diagonal matrix element of this Hamiltonian is

$$-t(e^{i\mathbf{k}\cdot\delta_1}+e^{i\mathbf{k}\cdot\delta_2}+e^{i\mathbf{k}\cdot\delta_3}) = -te^{i\mathbf{k}\cdot\delta_3}(1+e^{i\mathbf{k}\cdot\mathbf{a}_2}+e^{i\mathbf{k}\cdot\mathbf{a}_1}). \tag{7.26}$$

The gapless point happens at the k for which

$$1+e^{i\mathbf{k}\cdot\mathbf{a}_2}+e^{i\mathbf{k}\cdot\mathbf{a}_1}=0. \tag{7.27}$$

Because these are complex numbers of unit absolute value, they can be represented by a vector on the unit circle. The vector 1 is on the positive real axis, whereas the other two vectors, $e^{i\mathbf{k}\cdot\mathbf{a}_2}$, $e^{i\mathbf{k}\cdot\mathbf{a}_1}$, are above and below the real negative axis at the same angle so that their imaginary part cancels; however, their real parts add up to cancel 1. The only two solutions are $\mathbf{K}\cdot\mathbf{a}_2 = \frac{2\pi}{3}$, $\mathbf{K}\cdot\mathbf{a}_1 = \frac{4\pi}{3}$ and $\mathbf{K}'\cdot\mathbf{a}_1 = \frac{2\pi}{3}$, $\mathbf{K}'\cdot\mathbf{a}_2 = \frac{4\pi}{3}$, which are, of course, time reversals of each other (up to reciprocal lattice vectors). The gapless equality can then be written as

$$1+z+z^2=0; \qquad z=e^{i\mathbf{K}\cdot\mathbf{a}_2}=e^{i\frac{2\pi}{3}}, \qquad z^2=e^{i\mathbf{K}\cdot\mathbf{a}_1}=e^{i\frac{4\pi}{3}} \quad z^3=1, \tag{7.28}$$

where z is the third primitive root of unity. We now show that longer- range hopping not only cannot open a gap in perturbation theory (which was known from the local stability of Dirac points studied in section 7.3.3) but also cannot move the Dirac points $\mathbf{K}$, $\mathbf{K}'$, no matter how large the perturbation. However, we cannot guarantee the absence of other gapless points in the BZ upon the introduction of longer-range hopping. To prove that the gapless points are stable, we have to prove that any off-diagonal matrix element vanishes at $\mathbf{k} = \mathbf{K}$ when C_3 symmetry is present (if so, T guarantees that the matrix element will also vanish at $\mathbf{K}'$). Consider one A-site and the hoppings from it to the other sites. The hoppings are of two kinds:

1. There are hoppings that go through an even number of bonds (i.e., the vector that links a certain A-site with the site we want to hop onto can be expressed as a linear combination of a total even number of $\boldsymbol{\delta}_{1,2,3}$—the number of $\boldsymbol{\delta}_1$, $\boldsymbol{\delta}_2$, $\boldsymbol{\delta}_3$, when added, is even). These hoppings couple sites in the same sublattice, i.e., A to A and B to B. Because we cannot have any σ_z matrix by T and I combined, the term induced by these hoppings must be diagonal and proportional to the identity matrix (this is true even if C_3 symmetry is absent, just due to T and I). Hence, these terms are just an energy shift, which breaks the perfect particle-hole symmetry of graphene but cannot open a gap at the Dirac points.

2. More importantly, there are hoppings that go through an odd number of bonds and that couple A-sites with B-sites. These are important and must be treated with care. Its obvious that on the graphene lattice, $\boldsymbol{\delta}_{1,2,3}$ span the space, so any vector can be written in terms of them, but more importantly, any vector coupling an A-site and a B-site can be written in the form

$$\mathbf{v}_{AB} = n_1\boldsymbol{\delta}_1 + n_2\boldsymbol{\delta}_2 + n_3\boldsymbol{\delta}_3, \tag{7.29}$$

where $n_{1,2,3}$ are integers that crucially satisfy

$$n_1 + n_2 + n_3 = 1 \mod 3. \tag{7.30}$$

The preceding is true because of the chosen vector orientation of $\boldsymbol{\delta}_{1,2,3}$. We write each n as a multiple of 3 plus a remainder:

$$n_i = 3p_i + a_i, \qquad i = 1, 2, 3, \tag{7.31}$$

where, of course, $a_i \in [0, 1, 2]$ and, crucially, because of equation (7.30),

$$a_1 + a_2 + a_3 = -3(p_1 + p_2 + p_3) + 1 = 1 \mod 3. \tag{7.32}$$

Per C_3 symmetry, the off-diagonal matrix element coupling sites A and B is a cyclic permutation of $e^{i\mathbf{k}\cdot\mathbf{v}_{AB}}$ over the cyclic permutation of $\boldsymbol{\delta}$'s:

$$e^{i\mathbf{k}\cdot(n_1\boldsymbol{\delta}_1+n_2\boldsymbol{\delta}_2+n_3\boldsymbol{\delta}_3)} + e^{i\mathbf{k}\cdot(n_1\boldsymbol{\delta}_2+n_2\boldsymbol{\delta}_3+n_3\boldsymbol{\delta}_1)} + e^{i\mathbf{k}\cdot(n_1\boldsymbol{\delta}_3+n_2\boldsymbol{\delta}_1+n_3\boldsymbol{\delta}_2)} \tag{7.33}$$

We also have $\mathbf{a}_1 = \boldsymbol{\delta}_1 - \boldsymbol{\delta}_3$, $\quad \mathbf{a}_2 = \boldsymbol{\delta}_2 - \boldsymbol{\delta}_3$ and so the above matrix element becomes:

$$\begin{aligned}
&\exp(i\mathbf{k}\cdot(n_1\boldsymbol{\delta}_1 + n_2\boldsymbol{\delta}_2 + n_3\boldsymbol{\delta}_3)) + \exp(i\mathbf{k}\cdot(n_1\boldsymbol{\delta}_2 + n_2\boldsymbol{\delta}_3 + n_3\boldsymbol{\delta}_1)) + \exp(i\mathbf{k}\cdot(n_1\boldsymbol{\delta}_3 + n_2\boldsymbol{\delta}_1 + n_3\boldsymbol{\delta}_2)) \\
&= \exp(i\mathbf{k}\cdot\boldsymbol{\delta}_3)\big[\exp(i\mathbf{k}\cdot(3p_1\mathbf{a}_1 + 3p_2\mathbf{a}_2))\exp(i\mathbf{k}\cdot(a_1\mathbf{a}_1 + a_2\mathbf{a}_2)) \\
&\quad + \exp(i\mathbf{k}\cdot(3p_1\mathbf{a}_2 + 3p_3\mathbf{a}_1))\exp(i\mathbf{k}\cdot(a_1\mathbf{a}_2 + a_3\mathbf{a}_1)) \\
&\quad + \exp(i\mathbf{k}\cdot(3p_2\mathbf{a}_1 + 3p_3\mathbf{a}_2))\exp(i\mathbf{k}\cdot(a_2\mathbf{a}_1 + a_3\mathbf{a}_2))\big]
\end{aligned} \tag{7.34}$$

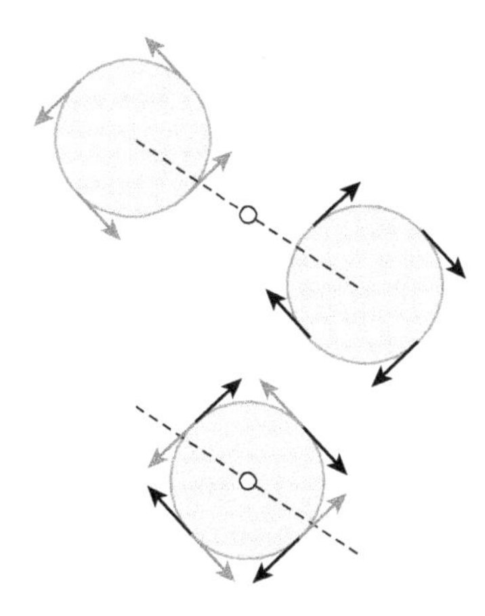

Figure 7.4. Dirac nodes carry vorticity (Berry phase). If a local gap is not allowed by T and I, the nodes can move around in the BZ without gapping only until they reach a T-invariant point, at which the two vortices annihilate and the Hamiltonian gaps.

We now particularize at $\mathbf{k} = \mathbf{K}$—the point for which the nearest-neighbor matrix element vanishes. Since

$$e^{i3\mathbf{K}\cdot\mathbf{a}_i} = (e^{i\mathbf{K}\cdot\mathbf{a}_i})^3 = 1 \tag{7.35}$$

for $i = 1, 2$, we see that the off-diagonal matrix element at $\mathbf{K}$ is

$$\begin{aligned} &e^{i\mathbf{k}\cdot\delta_3}(e^{i\mathbf{K}\cdot(a_1\mathbf{a}_1+a_2\mathbf{a}_2)} + e^{i\mathbf{K}\cdot(a_1\mathbf{a}_2+a_3\mathbf{a}_1)} + e^{i\mathbf{K}\cdot(a_2\mathbf{a}_1+a_3\mathbf{a}_2)}) \\ &\quad = e^{i\mathbf{k}\cdot\delta_3}(z^{2a_1+a_2} + z^{2a_3+a_1} + z^{2a_2+a_3}). \end{aligned} \tag{7.36}$$

Because we know $a_i \in 0, 1, 2$ and $a_1 + a_2 + a_3 = 1 \mod 3$, we can have only three distinct combinations of the numbers a_i. First, $a_1 = 0$, $a_2 = 0$, $a_3 = 1$: the matrix element is $z^0 + z^2 + z = 0$. Second, $a_1 = 1$, $a_2 = 1$, $a_3 = 2$: the matrix element is $z^3 + z^5 + z^4 = 1 + z^2 + z = 0$. Third, $a_1 = 0$, $a_2 = 2$, $a_3 = 2$: the matrix element is $z^2 + z^4 + z^6 = z^2 + z + 1 = 0$. Permutations of these three values of a obviously do not make a difference.

We hence proved that the matrix elements for any A to B hopping vanish at $\mathbf{k} = \mathbf{K}$ when C_3, T, and I are respected. We can say that the Dirac nodes are locally protected by T and I (we will see that two of them can annihilate and gap out, so they are not globally protected), whereas they are globally protected by C_3 symmetry.

7.4.2 Breaking of C_3 Symmetry

When C_3 symmetry is broken (we assume T and I are unbroken), there is nothing that stops the Dirac points from moving away from $\mathbf{K}$, $\mathbf{K}'$. It is easy to see that in the limit of high anisotropy, the graphene Hamiltonian is fully gapped. For example, for $t_3 \gg t_1, t_2$, we have $h(k) = -t_3\sigma_x$, with eigenvalues $E = \pm|t_3|$ and a gap equal to $2|t_3|$. It is hence clear that the absence of C_3 symmetry spoils the protection of Dirac modes. However, we know that for small anisotropy $t_c \approx t_a \approx t_b$, the Dirac nodes are stable because of the proof in section 7.4.1. How do the Dirac nodes gap?

In the anisotropic case, Dirac nodes come at points $\mathbf{K}_0$ ($\neq \mathbf{K}$) and $-\mathbf{K}_0$ related by TR invariance. As such, as long as $\mathbf{K}_0 \neq -\mathbf{K}_0 \pmod{\mathbf{b}_{1,2}}$, the Dirac nodes are stable and cannot gap because of the vorticity that they carry. However, when:

$$\mathbf{K}_0 = -\mathbf{K}_0 (\text{mod } \mathbf{b}_{1,2}), \tag{7.37}$$

the two Dirac points of vorticity $\pm\pi$ meet up and can annihilate, as in figure 7.4, at a TR-invariant point. The Hamiltonian at the point at which the Dirac nodes annihilate is quadratic (or flat) in the momentum multiplying one Pauli matrix and linear in the other. This has to be so for several reasons. First, the dispersion cannot be linear in both Pauli matrices because the wavefunction would then exhibit vorticity. We know this is not the case because the Dirac nodes annihilate, and the end result is a gapped Hamiltonian with zero vorticity. The second way of seeing this is the following: with T and I, the Hamiltonian for graphene takes the form

$$h(\mathbf{k}) = d_x(\mathbf{k})\sigma_x + d_y(\mathbf{k})\sigma_y + \epsilon(\mathbf{k}) \tag{7.38}$$

with no σ_z term. The $\epsilon(\mathbf{k})$ term is diagonal and, hence, does not influence the spectral gap, so we drop it in the further discussion. The time reversal ($h(-\mathbf{k}) = h^*(\mathbf{k})$) implies $d_x(-\mathbf{k}) = d_x(\mathbf{k})$, $d_y(-\mathbf{k}) = -d_y(\mathbf{k})$, and $\epsilon(\mathbf{k}) = \epsilon(-\mathbf{k})$. At a T-invariant point K_0, where the two Dirac nodes finally touch as anisotropy is increased, we have $\mathbf{K}_0 = -\mathbf{K}_0 \pmod{\mathbf{b}_{1,2}}$, which means

$$d_x(\mathbf{K}_0) = d_x(-\mathbf{K}_0); \quad d_y(\mathbf{K}_0) = d_y(-\mathbf{K}_0) = -d_y(\mathbf{K}_0) = 0. \tag{7.39}$$

We now see that only d_x is present in the Hamiltonian at $\mathbf{k} = \mathbf{K}_0$. Infinite-simally away from a T-invariant point $\mathbf{k} = \mathbf{K}_0 + \delta\mathbf{k}$, we have

$$d_x(\mathbf{K}_0 + \delta\mathbf{k}) = d_x(-\mathbf{K}_0 - \delta\mathbf{k}) = d_x(\mathbf{K}_0 - \delta\mathbf{k}). \tag{7.40}$$

gives Fourier-transforming to first order gives

$$d_x(\mathbf{K}_0) + \delta\mathbf{k}_i \frac{\partial d_x}{\partial k_i}\Big|_{k=K_0} = d_x(\mathbf{K}_0) - \delta\mathbf{k}_i \frac{\partial d_x}{\partial k}\Big|_{k=K_0}, \tag{7.41}$$

so the first order is missing: $\frac{\partial d_x}{\partial k_i}|_{k=K_0} = 0$ and

$$d_y(\mathbf{K}_0 + \delta\mathbf{k}) \approx \delta\mathbf{k}_i \frac{\partial d_y}{\partial k_i}\Big|_{k=K_0}. \tag{7.42}$$

The Hamiltonian is then

$$h(\mathbf{K}_0 + \delta\mathbf{k}) \approx \delta\mathbf{k}_i \frac{\partial d_y}{\partial k_i}\Big|_{k=K_0}\sigma_y + \delta\mathbf{k}_i\delta\mathbf{k}_j \frac{\partial^2 d_x}{\partial k_i \partial k_j}\Big|_{k=K_0}\sigma_x + d_x(\mathbf{K}_0), \tag{7.43}$$

with gap $2|d_x(\mathbf{K}_0)|$. When the Dirac nodes are touching, the gap is $2|d_x(\mathbf{K}_0)| = 0$. At the touching point, the Hamiltonian is linear in one direction but quadratic in the other. We want to stress that the preceding scenario for the gapping of Dirac nodes is absolutely necessary: for high anisotropy, the band structure is gapped,and for low anisotropy it is gapless, so there must be a place in between where the gap opens. Because the gapless Dirac nodes are locally stable (as proved earlier), to open a gap we need to bring them together first and annihilate them. This happens only at a T-invariant point.

7.5 Edge Modes of the Graphene Layer

So far in this chapter we have analyzed graphene with periodic boundary conditions and have focused on the symmetries of the problem and the stability of Dirac points. It turns out that graphene exhibits spectacular physics not only for the bulk (periodic boundary conditions), but also for the edge. We now focus on the new physics that arises in graphene once edges are

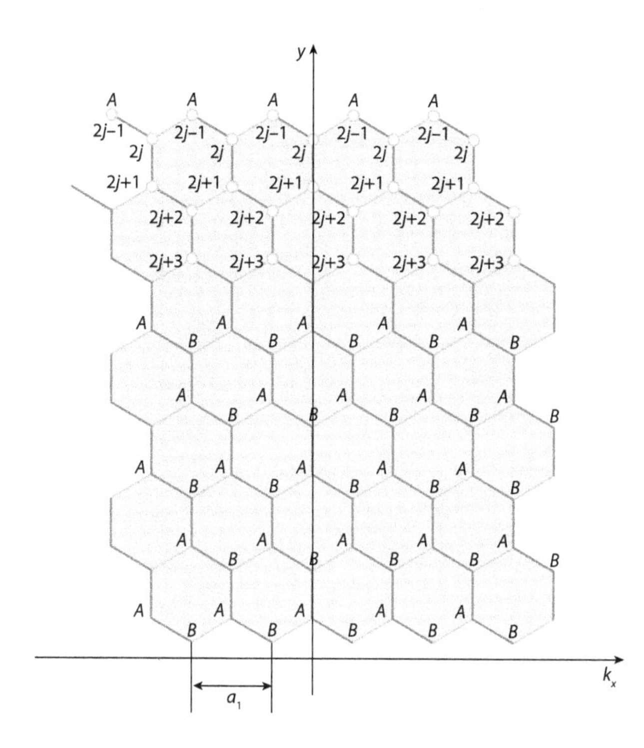

Figure 7.5. Graphene edge lattice with open-boundary conditions in the y- direction.

placed on the sample. We will see that a careful analysis of this situation strongly hints to the existence of the first topological insulator, the Chern insulator.

We analyze one graphene layer with edges on it. Instead of the transfer-matrix method, we will use the direct diagonalization of the Hamiltonian. As before, the isotropic Hamiltonian is

$$H = -t\sum_{i\neq j} c_i^\dagger c_j + \text{h.c.}, \tag{7.44}$$

with i, j on the hexagonal lattice and where we assume spinless fermions. For fermions with spin, we add an extra spin quantum number. We consider open-boundary conditions in the y-direction and periodic-boundary conditions in the x-direction as in figure 7.5, with "zigzag" chains (shown in the figure) oriented at an angle of $\pi/6$ angle with respect to the horizontal, represented by the color red in figure 7.5. Because we have distinct A- and B-sites in the unit cell, we can count them along one red chain (fig. 7.5) as even and odd—$2j-1$, $2j$, $2j+1$, $2j+2$, ...—sites. If we have periodic boundary conditions in the x-direction, we make the Fourier transform

$$c_{x,y}^\dagger = \frac{1}{N_x}\sum_{k_x} e^{ik_x x} c_{k_x,y}, \tag{7.45}$$

where N_x is the number of sites in the x-direction. The diagonalization of the Hamiltonian proceeds easily:

$$H = -t\sum_j \left(c^\dagger_{2j,k_x} c_{2j+1,k_x} + \text{h.c.}\right) - t\sum_j \left(c^\dagger_{2j} c_{2j-1}(e^{ik_x \frac{a_1}{2}} + e^{-ik_x \frac{a_1}{2}}) + \text{h.c.}\right), \tag{7.46}$$

where $a_1 = \frac{a\sqrt{3}}{2}$ and a is the graphene lattice constant. We make a gauge transformation only on the odd sites $c_{2j-1} \to c_{2j-1}e^{ik_x a_1 2}$ in the same way that we did before for the periodic

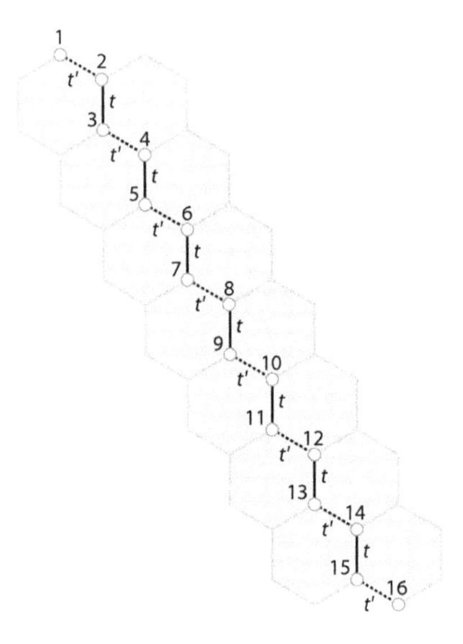

Figure 7.6. Graphene model with open boundary conditions in the y-direction. Depending on the value of k_x, the zig-zag red hoppings experience varying degrees of dimerization.

Hamiltonian, obtaining

$$H = -t \sum_j (c^\dagger_{2j,k_x} c_{2j+1,k_x} + c^\dagger_{2j,k_x} c_{2j-1,k_x}(1 + e^{-ik_x a_1}) + \text{h.c.}). \tag{7.47}$$

Hence the hopping, in ascending order, from odd to even sites ($2j-1 \to 2j$) is $t' = t(1 + e^{-ik_x a_1}) = t \cdot t_k$ with $t_k = 1 + e^{-ik_x a_1}$ as in figure 7.6. At $k_x = \pi/a_1$, the hopping is $t' = 0$, and hence we have two different situations, depending on whether the zigzag chain contains an even or odd number of sites. We analyze these cases separately next.

7.5.1 Chains with Even Number of Sites

For a chain with an even number of sites $2L_y$ (an integer) in the y-direction, where site 1 is on an A sublattice (these conditions guarantee that site $2L_y$ is on an B sublattice), we have that the two sites at the end of the chain do not couple with the dimerized bulk at $k_x = \pi/a_1$. In this case, we see that the states

$$|\psi_1\rangle = c^\dagger_{1,\pi/a_1}\,|0\rangle\,, \qquad |\psi_2\rangle = c^\dagger_{2L_y,\pi/a_1}\,|0\rangle \tag{7.48}$$

have $E = 0$ energy in the Hamiltonian and are, hence, zero edge modes. At $k_x = \pi/a_1$, these two states are exactly at energy 0; the bulk is completely dimerized, as in figure 7.7. Away from the point $k_x = \pi/a_1$, the edge modes have dispersion, but in the limit $L_y \to \infty$, the edge mode-energies go to zero. This can be seen by taking the single-particle wavefunction, $|\psi\rangle = \sum_i a_i\,|i\rangle$, and diagonalizing the Hamiltonian to obtain

$$t_k^* a_{2j-1} + a_{2j+1} = \frac{E}{-t} a_{2j}, \qquad t_k a_{2j} + a_{2j-2} = \frac{E}{-t} a_{2j-1}, \tag{7.49}$$

Two (quasi) zero-energy modes are present.

First, the mode localized close to the $j = 1$ edge: for $E \approx 0$ we obtain

$$\frac{a_{2j+1}}{a_{2j-1}} = -t_k^*; \qquad a_{2j} = 0 \text{ for all } j, \tag{7.50}$$

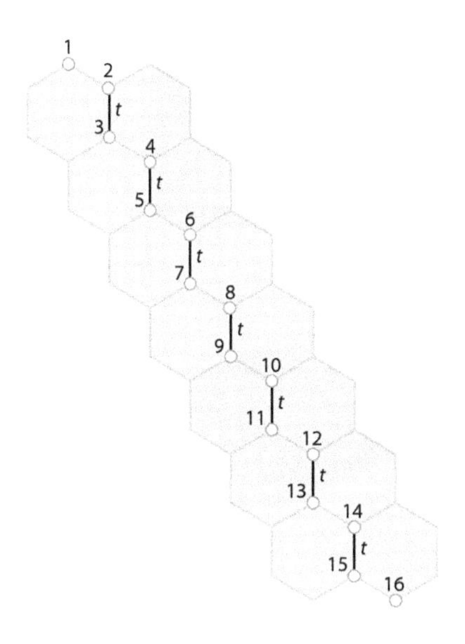

Figure 7.7. Graphene lattice and hoppings at $k_x = \pi$. The fully dimerized bulk is shown by the solid lines. There are no hoppings from site 1 to 2 or from site 15 to 16.

with the solution

$$a_{2j+1} = (-t_k^*)^j a_1. \tag{7.51}$$

It is rather clear that this is an edge mode as long as $t_k \neq 1$. Normally, we would say that both $t_k > 1$ and $t_k < 1$ are edge modes—localized on different edges—but this is not true. The reasoning is the following: we have assumed that the energy of the edge mode is $E = 0$ everywhere in k_x-space, even though, for modes away from $k_x = \pi$, the energy disperses on the order of $\exp(-L_y)$ (infinitesimal dispersion as we go to the thermodynamic limit but, nonetheless, still dispersion). The reason for this infinitesimal dispersion is the $2L_y$ site. The site $2L_y + 1$ does not exist, so $a_{2L_y+1} = 0$. As such, we have that, for $E = 0$ and $a_{2j} = 0$ for all j, we get

$$t_k^* a_{2L_y-1} + a_{2L_y+1} = \frac{E}{-t} a_{2L_y} = 0 = t_k^* a_{2L_y-1}. \tag{7.52}$$

Unless $t_k = 0$ (which happens only at $k_x = \pi$, equation (7.52) requires $a_{2L_y-1} \to 0$. This means that the solution $a_{2j+1} = (-t_k^*)^j a_1$ works only for $t_k < 1$, which is the same as $\frac{2\pi}{3} \leq k_x a_1 \leq \frac{4\pi}{3}$. These are exactly the Dirac points in graphene in terms of k_x. This represents an edge mode localized on the $j = 1$ edge of energy $E \to 0$ in the thermodynamic limit.

Second, the mode localized on the $j = 2L_y$ edge is obtained if

$$\frac{a_{2j-2}}{a_{2j}} = -t_k; \qquad a_{2j+1} = 0 \, \forall j, \tag{7.53}$$

which has as solution

$$a_{2L_y-2j} = (-t_k)^j a_{2L_y}. \tag{7.54}$$

Again, we have that $-t_k a_2 = 0$, so $a_2 \to 0$, which means $|t_k| < 1$, with identical conditions on k, as before. This is a mode localized at the $j = 2L_y$ edge of energy $E \to 0$ in the thermodynamic limit. The discussion of the edge modes just presented is reflected in the form

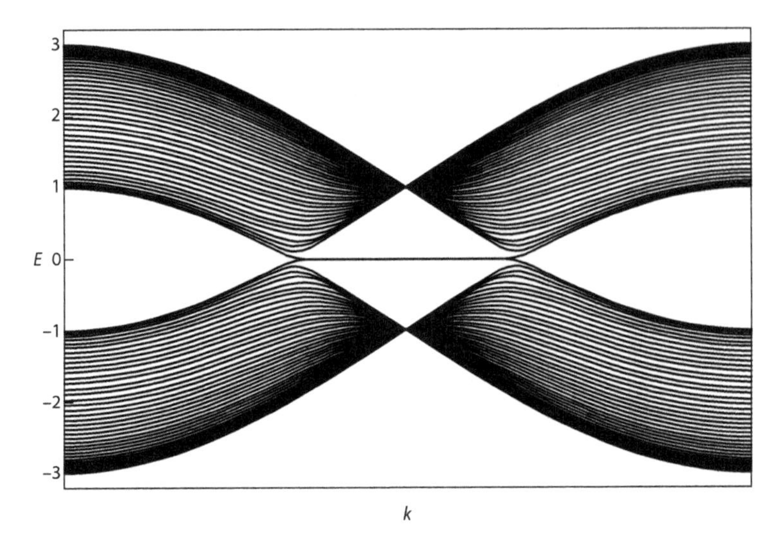

Figure 7.8. Graphene tight binding with edge modes. The boundaries are sites A and B (chains with even number of sites).

of the Hamiltonian:

$$h = -t \begin{bmatrix} 0 & t_k & 0 & 0 & 0 & 0 & 0 & 0 \\ t_k^* & 0 & 1 & 0 & 0 & 0 & 0 & 0 \\ 0 & 1 & 0 & t_k & 0 & 0 & 0 & 0 \\ 0 & 0 & t_k^* & 0 & 1 & 0 & 0 & 0 \\ 0 & 0 & 0 & 1 & 0 & t_k & 0 & 0 \\ 0 & 0 & 0 & 0 & t_k^* & 0 & 1 & 0 \\ 0 & 0 & 0 & 0 & 0 & 1 & 0 & t_k \\ 0 & 0 & 0 & 0 & 0 & 0 & t_k^* & 0 \end{bmatrix}, \tag{7.55}$$

with many sites in between, whose (almost) zero-energy eigenstates are

$$\psi_1 = (a_1, 0, a_3, 0, a_5, 0, \ldots, a_{2L_y-3}, 0, a_{2L_y-1}, 0), \tag{7.56}$$

where $t_k^* a_{2L_y-1} \to 0$ in order to have zero energy and

$$\psi_1 = (0, a_2, 0, a_4, 0, a_6, \ldots, 0, a_{2L_y-2}, 0, a_{2L_y}), \tag{7.57}$$

where $t_k a_2 \to 0$ to have zero energy.

We now have a picture for the edge modes in the thermodynamic limit. In the bulk, the bands will be the projection onto the k_x-axis of all the band dispersion on k_y, so the "bulk" bands will be fat. They will go to zero energy at the Dirac cones, or points. At every point in between the Dirac cones, there will be two, nondispersive edge modes at zero energy, which connect them as in figure 7.8. They are localized on one end or the other of the sample, and any local perturbation that could couple the modes on opposite edges cannot lift the degeneracy because it has to go through the whole sample to couple them.

For a chain with an even number of sites $2L_y$ in the y-direction, and with site 1 on the B sublattice, site $2L_y$ is guranteed to be on the A sublattice). Going through a similar calculation,

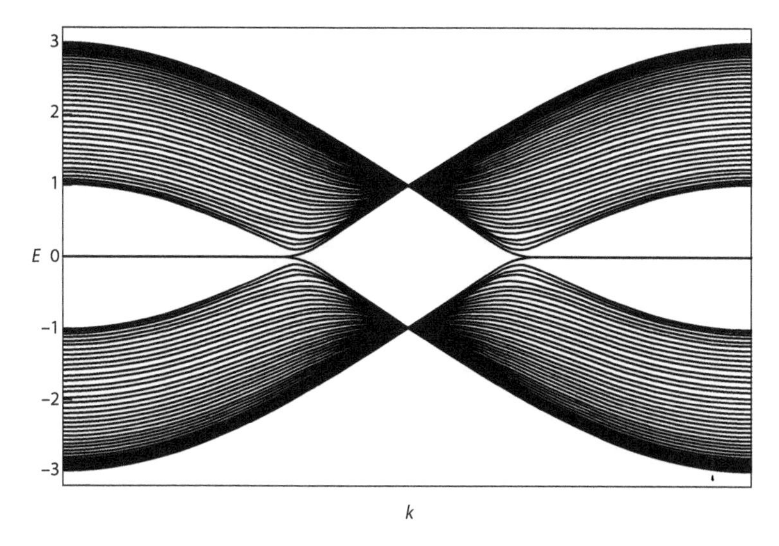

Figure 7.9. Graphene tight binding with edge modes. The boundaries are sites B and A (chains with even number of sites).

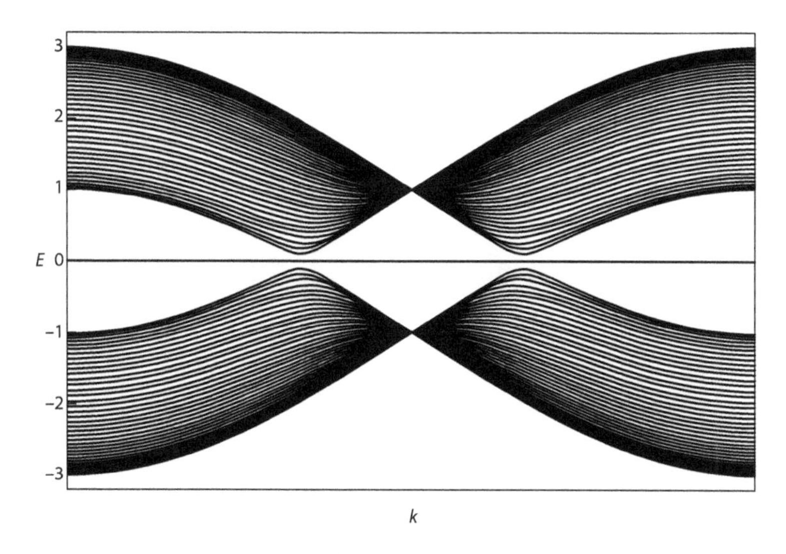

Figure 7.10. Graphene tight binding with edge modes. The boundaries are sites A and A (chains with odd number of sites).

we find edge modes from $\frac{4\pi}{3}$ to 2π and then continuing from 0 to $\frac{2\pi}{3}$. The edge modes will be at asymptotically zero energy and can be seen in figure 7.9.

7.5.2 Chains with Odd Number of Sites

For a chain with an odd number of sites, $2L_y - 1$ in the y-direction, if site 1 is on the A sublattice, then site $2L_y - 1$ is also on an A sublattice. This case, is of course, identical to the reflected case in which site 1 is site B and the site at $2L_y - 1$ is also B. In this case, it is easiest to use the matrix form of the Hamiltonian to immediately see the edge modes. As opposed to the case of an even

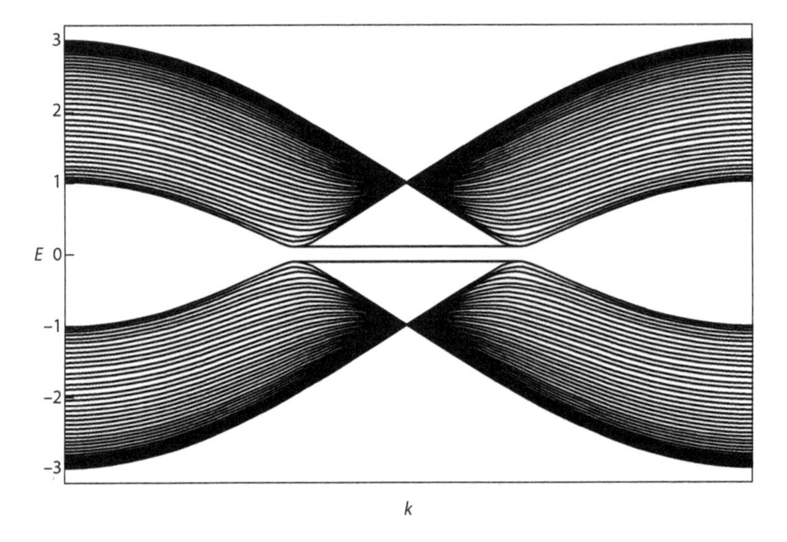

Figure 7.11. Graphene with Semenoff mass $m = 0.3t$. The boundaries are sites A and B (chains with even number of sites).

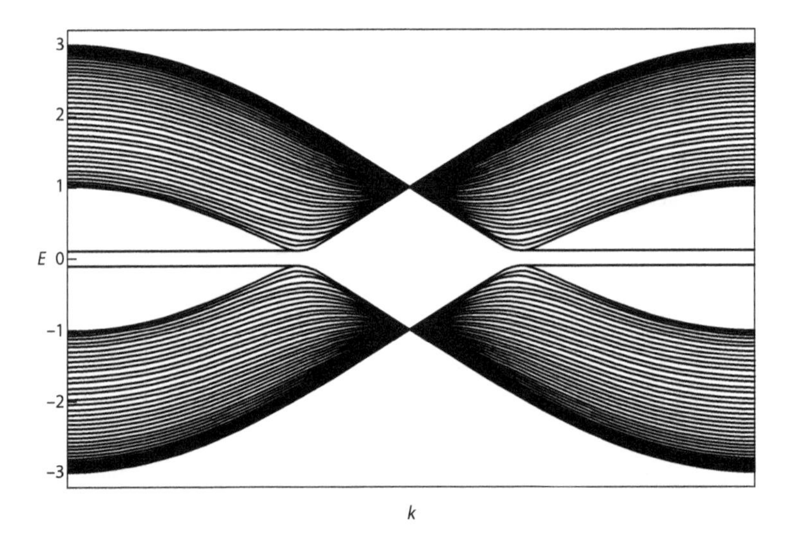

Figure 7.12. Graphene with Semenoff mass $m = 0.3t$. The boundaries are sites B and A (chains with even number of sites).

number of y sites, the Hamiltonian is missing the last column and row:

$$h = -t \begin{bmatrix} 0 & t_k & 0 & 0 & 0 & 0 & 0 \\ t_k^* & 0 & 1 & 0 & 0 & 0 & 0 \\ 0 & 1 & 0 & t_k & 0 & 0 & 0 \\ 0 & 0 & t_k^* & 0 & 1 & 0 & 0 \\ 0 & 0 & 0 & 1 & 0 & t_k & 0 \\ 0 & 0 & 0 & 0 & t_k^* & 0 & 1 \\ 0 & 0 & 0 & 0 & 0 & 1 & 0 \end{bmatrix}, \tag{7.58}$$

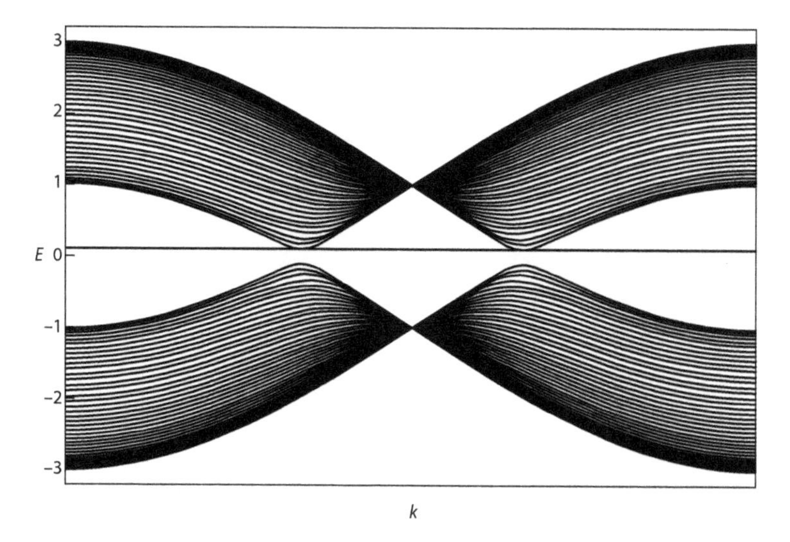

Figure 7.13. Graphene with Semenoff mass $m = 0.3t$. The boundaries are sites A and A (chains with odd number of sites).

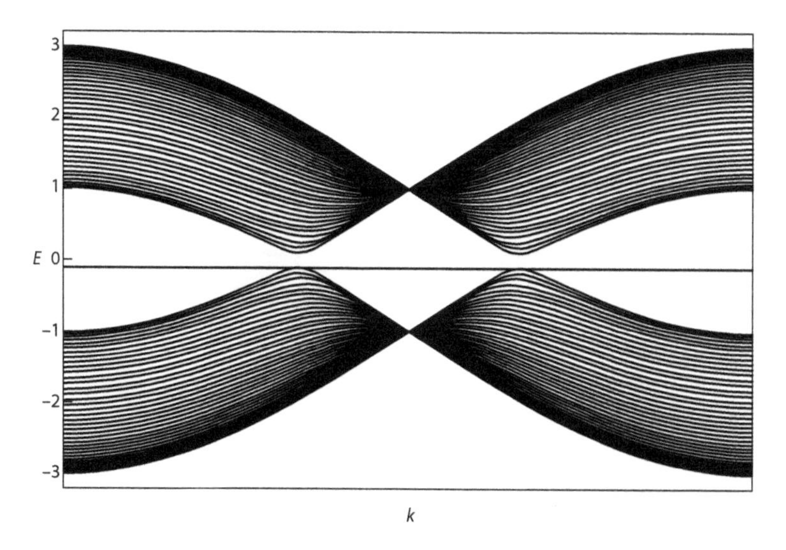

Figure 7.14. Graphene with Semenoff mass $m = 0.3t$. The boundaries are sites B and B (chains with even number of sites).

with many sites in between. Notice that in this case, the wavefunction

$$\psi_1 = (a_1, 0, a_3, 0, a_5, 0, \ldots, a_{2L_y-3}, 0, a_{2L_y-1}) \tag{7.59}$$

with

$$a_{2j+1} = (-t_k^*)^j a_1 \tag{7.60}$$

is always an *exact* energy solution of the Hamiltonian at any k. For $|t_k| < 1$—i.e., for k between $\frac{2\pi}{3}$ and $\frac{4\pi}{3}$, the edge state is localized on the 1-site, whereas for $|t_k| > 1$—i.e., in the remainder

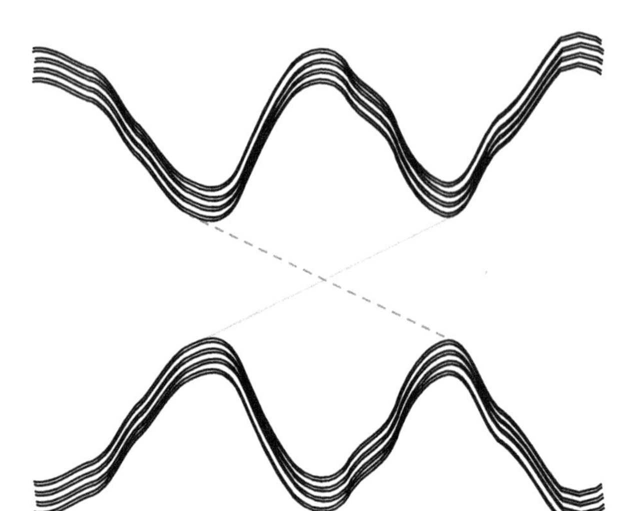

Figure 7.15. Sketch of graphene with Haldane mass. The boundaries are sites A and B (chains with even number of sites).

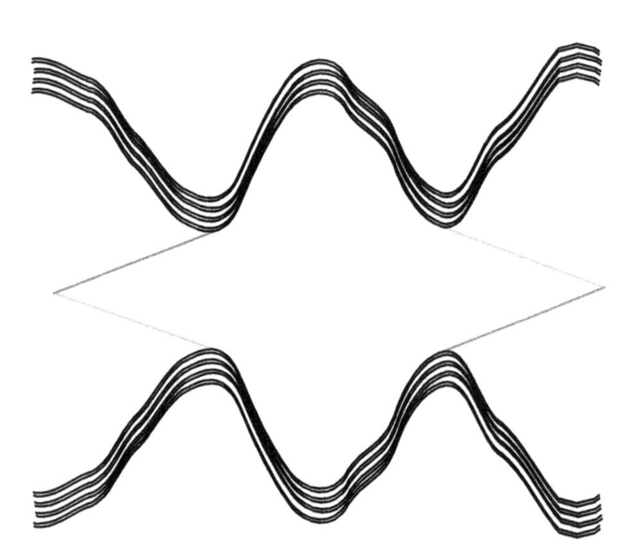

Figure 7.16. Sketch of graphene with Haldane mass. The boundaries are sites B and A (chains with even number of sites).

of the BZ—the state is localized on the $(2L_y - 1)$-site. This is clearly exemplified in figure 7.10. The edge mode in this case is *exactly* at zero energy, unlike in the previous case, in which it was split from zero energy by thermodynamically exponential (in L_y) spitting.

Notice that the other state that was available in the $2L_y$ chain,

$$\psi_1 = (0, a_2, 0, a_4, 0, a_6, ..., a_{2L_y-4}, 0, a_{2L_y-2}, 0), \tag{7.61}$$

cannot be a zero-energy state of the system because it would have to satisfy, at the same time, the conditions

$$t_k a_2 \to 0; \quad a_{2j} = t_k a_{2j+2}; \quad a_{2L_y-2} \to 0, \tag{7.62}$$

whose unique solution is the zero wavevector. The degeneracy of the zero mode at each point, including the special T-invariant points, is 1. This clearly shows that we are dealing with systems with spinless fermions for which time reversal does not induce Kramers' degeneracy.

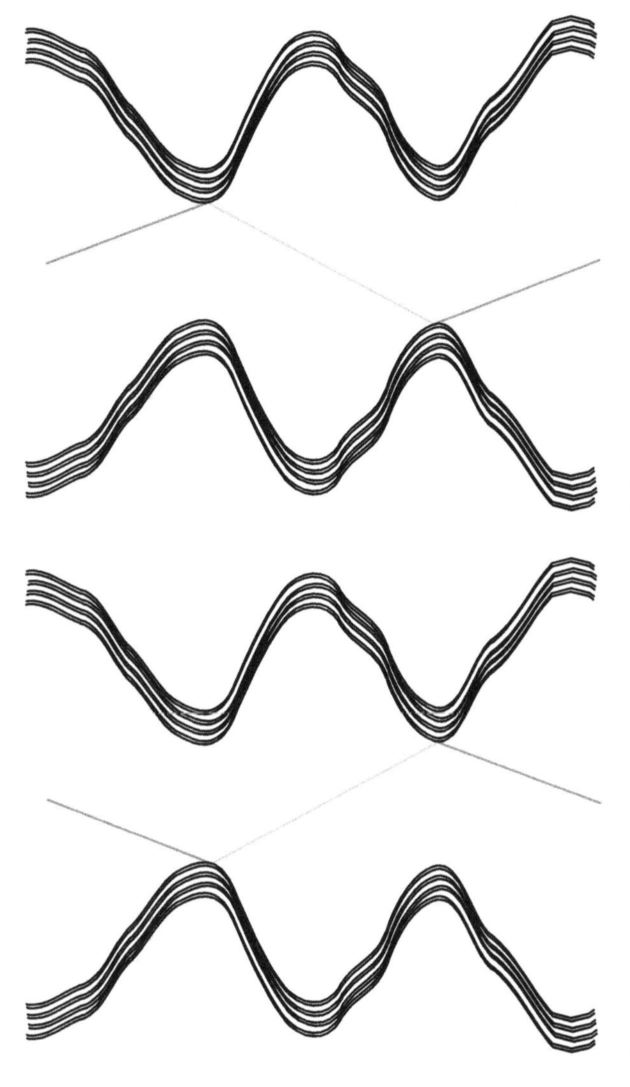

Figure 7.17. Sketch of graphene with Haldane mass. The boundaries are sites A and A (chains with odd number of sites).

Figure 7.18. Sketch of graphene with Haldane mass. The boundaries are sites B and B (chains with even number of sites).

7.5.3 Influence of Different Mass Terms on the Graphene Edge Modes

We now ask what happens to the edge modes upon opening a gap at the Dirac nodes. In the spinless case, several possibilities appear.

Case 1: First, we can add a term that gives different energies to lattice sites A and B, as in the case of BN. As before, this term is

$$h(k) = m\sigma_z, \tag{7.63}$$

with $m > 0$; m is called an inversion symmetry breaking mass, or Semenoff mass [39]. We ask how the gap opens. The edge modes that terminate (have high amplitude) on sites A will go up in energy, whereas the edge modes that terminate on sites B will go down in energy. We hence have the situations in figures 7.11, 7.12, 7.13 and 7.14, depending on whether the terminations are $A - B$, $B - A$, $A - A$, or $B - B$.

Case 2: We now add a term that breaks time reversal and gaps the system. This is called a Haldane mass; because we do not yet know how to add it on the lattice, we add it at the level of the Dirac fermion mass, which, per sections 7.3.2 and 7.3.1, means we need to add a mass of opposite sign to the Dirac fermions at **K** and **K**′. This mass term is a **k**-space-dependent term, which cannot be looked at as a term that adds energy on localized sites. It is not an on-site term. As such, the edge modes will do completely different things than in the case of a BN-type (inversion-breaking) mass. We have seen that in the inversion-breaking case, the edge modes remain linked to the Dirac mass; i.e., the mass is identical at both cones, and the edge modes still connect the cones on the same side of the gap. In the T-breaking case, a similar thing happens, but because the mass is negative at one cone and positive at the other cone, the edge modes will connect one cone with another *by crossing the gap*, as in figures 7.15–7.18. The Hall conductance can be inferred if we know where the edge modes in the figure are situated (i.e., on which side of the Laughlin cylinder), but it can be only ± 1. Thus, it is clear that a topological insulator with nonzero Hall conductance and that maintains lattice translational symmetry exists. We will find one in the next chapter.

7.6 Problems

1. *Berry Phase of Bilayer Graphene:* Show that two layers of graphene on top of each other in the Bernal cofiguration (with the sublattice A straight above the B one, and vice versa), and with just nearest-neighbor couplings, have a double Dirac vortex (quadratic vortex — with Berry phase 2π as we go around any equal-energy contour) at each of the points **K**, **K**′. Then show that by introducing next-neighbor couplings (both in and out of plane), the double Dirac vortex splits into four single Dirac nodes, three of vorticity π and one of vorticity $-\pi$. Show that the Dirac node of vorticity $-\pi$ stays exactly at the point, **K** (or **K**′) whereas the three Dirac nodes of vorticity π form an equilateral triangle around the point **K** (or **K**′).

2. *Inversion Eigenvalues in Graphene:* Graphene has inversion symmetry. Find the inversion eigenvalues of the bands in graphene at the four inversion symmetric points in the BZ.

3. *Breaking of C_3 Symmetry and Annihilation of Dirac Points:* Break the C_3 symmetry of the graphene lattice while still keeping inversion symmetry. As shown in this chapter, by increasing the term that breaks the C_3 symmetry, the two Dirac points initially (when C_3 symmetry is preserved) located at the points **K**, **K**′ begin moving toward each other in the BZ and eventually annihilate each other at a high-symmetry point, creating a fully gapped insulator. Show explicitly that this phase transition is accompanied by a change in the inversion eigenvalue at the high-symmetry point.

4. *Edge Modes of Gapless Graphene with Longer-Range Hopping*: Add next-nearest-neighbor hopping and solve for the edge modes of graphene with the zigzag boundary condition. Does the system still have an edge mode connecting the two Dirac points at momenta **K**, **K**′?

8

Simple Models for the Chern Insulator

We are now well prepared to discuss the simplest model of a topological insulator, the Chern insulator on the square lattice. Even though, historically, the first model of a Chern insulator was introduced by Haldane on the hexagonal lattice (which we study after the square-lattice call), a Chern insulator model on the square lattice is easier to introduce and understand. Before we do that, however, we must first understand the behavior of Dirac fermions upon opening a gap in their spectrum. The previous chapter focused on how to keep the Dirac fermions gapless. This chapter focuses on how to make them massive and the implications of making them massive. We will learn that massive Dirac fermions in the continuum exhibit a half-integer quantum Hall effect, and we will tie up some loose ends with regard to prior cryptic statements about smooth gauge, Stokes' theorem, and others. We will then show that Dirac fermions are very useful in obtaining the Hall conductance of systems that undergo a gap-closing-and-reopening transition.

8.1 Dirac Fermions and the Breaking of Time-Reversal Symmetry

In two dimensions, the model for a *continuum* Dirac fermion is a variant of the general two-level Hamiltonian:

$$h(\mathbf{k}) = \sum_i d_i(\mathbf{k})\sigma_i, \tag{8.1}$$

where $d_1 = k_x$, $d_2 = k_y$ up to a rotation. If $d_3 = 0$, then the Hamiltonian is gapless, whereas if $d_3 = M$ is a constant, then the Dirac fermion is gapped. We now show that when the σ matrices correspond to spin, the first case maintains time reversal, whereas the latter breaks it. Notice that here the momentum $\mathbf{k}$ is a continuum variable lying in the infinite Euclidean plane.

8.1.1 When the Matrices σ Correspond to Real Spin

If the Dirac equation matrix structure comes from real spin, then we have $T = -i\sigma_y K$, and we know that all three matrices σ_i are odd under time reversal. Then

$$Th(\mathbf{k})T^{-1} = -d_i(\mathbf{k})\sigma_i = -k_x\sigma_x - k_y\sigma_y - M\sigma_z. \tag{8.2}$$

If T is preserved, then the preceding should equal $h(-\mathbf{k}) = -k_x\sigma_x - k_y\sigma_y + M\sigma_z$. This is possible only when $M = 0$, i.e., when the system is gapless. Generalization to any $d_i(\mathbf{k})$ shows that if time reversal is to be preserved, then

$$Th(\mathbf{k})T^{-1} = -d_i(\mathbf{k})\sigma_i = h(-\mathbf{k}) = d_i(-\mathbf{k})\sigma_i; \quad \rightarrow \quad d_i(-\mathbf{k}) = -d_i(\mathbf{k}). \tag{8.3}$$

Notice that the conditions on the vector $d_i(\mathbf{k})$ are different from the graphene case, in which the matrices σ were not meant to represent the spin of the particle. At special points that are invariant (modulo a reciprocal lattice vector) under time reversal $\mathbf{k}_{\text{special}} = (0, 0), (0, \pi), (\pi, 0), (\pi, \pi) = -\mathbf{k}_{\text{special}} = \mathbf{G}/2$, we find that $d_i(\mathbf{G}/2) = -d_i(-\mathbf{G}/2) = -d_i(\mathbf{G}/2) = 0$ for each i; hence the system *must* be gapless. This can, of course, be more simply understood by just drawing the picture of a gapped Dirac fermion and seeing that the state $k = 0$ does not have a Kramers' pair. Because we are working in a spin-$\frac{1}{2}$ system, this necessarily implies the breaking of time reversal.

8.1.2 When the Matrices σ Correspond to Isospin

In this case, $T^2 = 1$ and the preceding condition for TR symmetry changes to $h^*(\mathbf{k}) = h(-\mathbf{k})$. This implies $d_1(\mathbf{k}) = d_1(-\mathbf{k})$, $d_2(\mathbf{k}) = -d_2(-\mathbf{k})$, and $d_3(\mathbf{k}) = d_3(-\mathbf{k})$. It then turns out that to maintain time reversal, we need d_1, d_3 to be even in $\mathbf{k}$ but d_2 to be odd. Of course, in this case the Chern number has to vanish.

8.2 Explicit Berry Potential of a Two-Level System

Let us now try to explicitly obtain the Berry potential and Berry field strength for a two-level system. The eigenstates of the Hamiltonian in equation (8.1), of energies $E_\pm = \pm\sqrt{d_1^2 + d_2^2 + d_3^2} \equiv \pm d$ are

$$\psi_{E_+} = \frac{1}{\sqrt{2d(d+d_3)}}\begin{pmatrix} d_3 + d \\ d_1 - id_2 \end{pmatrix}, \quad \psi_{E_-} = \frac{1}{\sqrt{2d(d-d_3)}}\begin{pmatrix} d_3 - d \\ d_1 - id_2 \end{pmatrix}, \tag{8.4}$$

which are properly normalized and orthogonal. We try to compute the Berry connection:

$$A_i(\mathbf{k}) = i\langle \psi_{E_-}|\partial_{k_i}|\psi_{E_-}\rangle = \frac{-1}{2d(d+d_3)}(d_2\partial_i d_1 - d_1\partial_i d_2). \tag{8.5}$$

We can then obtain the Berry curvature [40]:

$$F_{ij} = \frac{1}{2d^3}\epsilon_{abc} d_a \partial_i d_b \partial_j d_c = \frac{1}{2}\epsilon_{abc}\hat{d}_a\partial_i\hat{d}_b\partial_j\hat{d}_c, \tag{8.6}$$

where $\hat{d}_a = d_a/d$. We notice that the field strength is the Jacobian of the map $\mathbf{k} \to \mathbf{d}/d$, which is a map between the base manifold—a 2-D momentum space—and the target manifold—a 2-D sphere S^2—given by the equation $d^2 = 1$. When integrated over the base manifold, the Jacobian counts the winding number of the map, which is always an integer, the Chern number.

8.2.1 Berry Phase of a Continuum Dirac Hamiltonian

The 2-D Dirac Hamiltonian, in the continuum, corresponds to the case $d_3 = m$, $d_1 = k_x$, $d_2 = k_y$. Let us now compute the vector potential by plugging, values into this formula:

$$A_x = \frac{-k_y}{2\sqrt{k^2+m^2}(\sqrt{k^2+m^2}+m)}, \quad A_y = \frac{k_x}{2\sqrt{k^2+m^2}(\sqrt{k^2+m^2}+m)}. \tag{8.7}$$

Let us assume we are at finite chemical potential μ larger than the mass gap m, which endows us with a Fermi surface. Note that although we did not implicitly put the chemical potential into the Hamiltonian, the Berry vector potential remains unchanged because the chemical potential is a diagonal identity matrix term and the eigenstates are not changed by diagonal terms. If the eigenstates are unchanged, the Berry potential is unchanged as well. We then would like to integrate the Berry curvature over the Fermi surface, which is an azimuthal integration over the angle at a momentum $k_F = \sqrt{\mu^2 - m^2}$ (note that we are using $\hbar = 1$). Hence, for the integral of the Berry potential over the Fermi surface, we obtain

$$\int_{\text{Fermi surface}} d\mathbf{k} \cdot \mathbf{A} = \int_0^{2\pi} k_F \, d\theta A_\theta = \int_0^{2\pi} k_F d\theta \left(A_y \frac{k_x}{k_F} - A_x \frac{k_y}{k_F} \right)$$

$$= \int_0^{2\pi} \frac{k_F^2}{2\sqrt{k_F^2 + m^2}(\sqrt{k_F^2 + m^2} + m)} d\theta = \pi \frac{k_F^2}{\sqrt{k_F^2 + m^2}(\sqrt{k_F^2 + m^2} + m)} \approx |_{m/k_F << 1}$$

$$= \pi - \pi \frac{m}{k_F} \tag{8.8}$$

Notice that the Berry phase at the TR-invariant (gapless) $m = 0$ point is π; this be can easily understood because π is the only value (besides 0) that is TR invariant (modulo 2π). When time reversal is broken, the Berry phase around the Fermi surface picks up a contribution proportional to the ratio between the gap and the Fermi momentum—so if the Fermi level is really high, corresponding to large doping, then the Berry phase can be again very close to (but not equal) π, even though the time reversal is broken.

8.2.2 The Berry Phase for a Generic Dirac Hamiltonian in Two Dimensions

In the previous derivation of the Berry phase, we have assumed rotational invariance. However, around a degeneracy point or band, crossing in two-dimensions, we have the following general form for the Hamiltonian:

$$h(\mathbf{k}) = k_i \mathcal{A}_{ij} \sigma_j, \tag{8.9}$$

with $\mathcal{A}_{ij}$ a 2×2 matrix. We have neglected a possible diagonal term proportional to the identity matrix because it does not influence the eigenstates and, hence, does not enter the Berry phase.

The eigenvalue of the Hamiltonian $d = \sqrt{(\mathcal{A}_{1i} k_i)^2 + (\mathcal{A}_{2i} k_i)^2}$ and the Berry potential is

$$A_i(\mathbf{k}) = -\frac{1}{2d^2}(\mathcal{A}_{2j} k_j \mathcal{A}_{1i} - \mathcal{A}_{1j} k_j \mathcal{A}_{2i}). \tag{8.10}$$

We want to compute the Berry phase over the Fermi surface (see fig. 8.1), but, in general, the Fermi surface can be anisotropic. However, we have $A_i dk_i = A_\phi k(\phi) \, d\phi$ because an infinitesimal variation $d\mathbf{k} = k(\phi) d\phi \hat{1}_\phi$ on the Fermi surface. Here, $\hat{1}_\phi$ is the unit vector tangential to a surface of equal energy in k-space. We have that

$$A_\phi = A_y \frac{k_x}{k} - A_x \frac{k_y}{k}, \tag{8.11}$$

so the Berry phase is

$$\int_0^{2\pi} d\phi (A_y k_x - A_x k_y) = \int_0^{2\pi} d\phi \frac{1}{2d^2} k^2 \text{Det} \mathcal{A}, \tag{8.12}$$

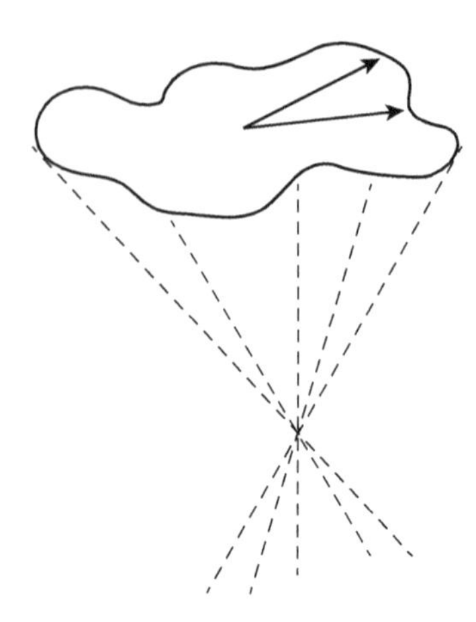

Figure 8.1. Fermi surface for generic, noncircular Dirac Hamiltonian.

which, upon substituting for d and using the integral

$$\int_0^{2\pi} d\phi \frac{1}{a\cos(\phi)^2 + b\sin(\phi)^2 + c\cos(\phi)\sin(\phi)} = \frac{4\pi}{\sqrt{4ab - c^2}}, \tag{8.13}$$

valid as long as $c^2 < 4ab$, which is the condition of not having nodal lines, brings us to a nice expression for the Berry phase:

$$\int_{\text{Fermi surface}} d\mathbf{k} \cdot \mathbf{A} = \pi \cdot \text{sign}(\text{Det}\,\mathcal{A}). \tag{8.14}$$

8.2.3 Hall Conductivity of a Dirac Fermion in the Continuum

The Berry curvature, necessary for the Chern number, vanishes on a lattice if you have time reversal. The Dirac fermion in the continuum does not have a lattice by definition, but the same theorem applies. This can be clearly seen from the form of the Berry curvature, which needs all three Pauli matrices to enter. For the continuum Dirac Hamiltonian, using equation (8.7) we have

$$F_{xy} = \frac{1}{2(m^2 + k^2)^{3/2}} m. \tag{8.15}$$

In contrast to the computation of the Berry phase, we now put the chemical potential in the gap ($|\mu| < m$) and integrate the Berry curvature over the occupied states (this is the Berry curvature of the occupied band):

$$\sigma_{xy} = \frac{e^2}{h}\frac{1}{2\pi}\int d^2k F_{xy} = \frac{e^2}{h}\frac{m}{4}\int_0^\infty dx \frac{1}{(m^2 + x)^{3/2}} = \frac{e^2}{h}\frac{\text{sign}(m)}{2}. \tag{8.16}$$

We have arrived at the notable result that massive 2-D Dirac fermions have a Hall conductance equal to one-half times the sign of their mass. This statement is puzzling because the Hall conductance, equal to the Chern (winding) number of the filled band, has to be an integer. This is true, but only if the problem is on a lattice—in this case, the reason we get one-half is because the Dirac fermion is in the continuum and has not been regularized properly. The bands of the continuum Dirac fermion do not have a bandwidth. The Hall conductance equals the Chern number and is an integer only if the base manifold (the BZ) is compact. In the continuum, the momentum runs over a noncompact manifold (the infinite Euclidean plane), and this does not apply.

How does having a lattice fix things? If we are on a lattice, the bands, which in the case of the Dirac fermion have infinite bandwidth (because the dispersion is $\pm\sqrt{k^2 + m^2}$), must bend down due to the fact that the problem has a finite bandwidth. At the points where the bands bend down (roughly), we will get another half a quantum of Hall conductance. These high-energy fermions, which contribute to conductances, are called *spectator fermions*. The moral of the story is that we *cannot* determine the full Hall conductance of a filled band by analyzing the physics of only a small part of the band. For example, in an insulator, the vicinity of the k-space point where the gap is the smallest can be modeled by a Dirac Hamiltonian. However, due to the presence of the spectator fermions just mentioned, we cannot calculate the Hall conductance of the filled lattice band just by looking at the points where the gap is the smallest and then performing the Dirac computation. The reason is that we might have some higher-energy modes that can add or subtract half-values. However, the change in Hall conductance upon closing or reopening the gap *can* be determined by focusing only on the vicinity of the point where the transition happens. Generically, a gap-closing-and-opening transition has to happen by varying a parameter, the mass of the Dirac fermion m. If we go from m negative to m positive (and, in the process, close the gap and reopen it), then we get the Hall conductance to change by 1. If we knew the value of the Hall conductance before the gap-closing transition, we would be able to find the value of the Hall conductance *after* closing and reopening the gap.

8.3 Skyrmion Number and the Lattice Chern Insulator

In the previous subsection, we saw that the Dirac fermions are nontrivial physical systems. So far we have looked at the Dirac Hamiltonian in the continuum, but it seems reasonable to assume that their lattice generalization will also yield nontrivial physics. We now would like to make the simplest lattice generalization of the continuum Dirac Hamiltonian: instead of k_x, k_y, we generalize this to $\sin(k_x), \sin(k_y)$, whereas the mass of the Dirac Hamiltonian can be generalized to $2 + M - \cos(k_x) - \cos(k_y)$. This lattice generalization becomes the continuum Dirac Hamiltonian for momentum close to $k = 0$. As we have already shown, the Hamiltonian for an insulator (spinless) must contain two bands. The lattice generalization of the Dirac Hamiltonian is, hence, an orbital model with two orbitals per site, one of s-type and one of p-type (or, more generally, orbitals of different parity). The coupling between the s and p orbitals thus must necessarily be an angular-momentum $L = 1$ coupling; hence, the lowest-order coupling is linear in k ($\sin(k_x) + i\sin(k_y)$). We also add intraorbital dispersions $2 - \cos(k_x) - \cos(k_y)$ because they are allowed by symmetry. Hence, to the lowest-order in Fourier modes, the Hamiltonian necessarily must look like

$$H = \sin(k_x)\sigma_x + \sin(k_y)\sigma_y + B(2 + M - \cos(k_x) - \cos(k_y))\sigma_z. \tag{8.17}$$

This Hamiltonian is fully gapped except at several values of M: at $M = 0$, it is gapless at $(k_x, k_y) = (0, 0)$; at $M = -2$, it is gapless at $(k_x, k_y) = (\pi, 0), (0, \pi)$ and at $M = -4$, it is gapless at $(k_x, k_y) = (\pi, \pi)$. By adiabatic continuity, as long as the Hamiltonian is gapped and the gap does not close, it remains in the same topological phase. By varying the parameter M, the model has phase transitions and different topological properties, which we now investigate.

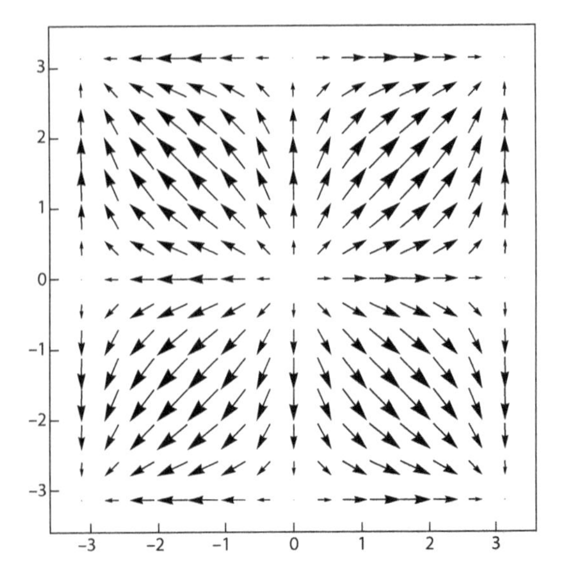

Figure 8.2. Vector plot for the components $d_1(k) = \sin(k_x)$, $d_2(k) = \sin(k_y)$ in the Chern insulator, $e_s = M/2B$.

8.3.1 $M > 0$ Phase and $M < -4$ Phase

In these two regimes, the Hamiltonian is fully gapped and the Chern number is zero, per direct computation. This can be explained in different ways. We can do a direct numerical computation using the Jacobian curvature, but this would not be very physically revealing. The most physically revealing is to realize that both $M > 0$ and $M < -4$ represent Hamiltonians topologically equivalent to the atomic limit (the limit in which all hoppings are set to zero, as if the lattice constant were infinity). The $M > 0$ regime is topologically the same phase as the phase $M \to \infty$, which has trivial (momentum-independent) eigenstates and zero Hall conductance. This is, in fact, an atomic limit, which does not show anything interesting because the energy bands are flat and only on-site energies are important so the wavefunctions are completely localized on the atomic sites. Here, $M < -4$ is topologically the same as $M \to -\infty$, which is also an atomic limit but with the on-site energies reversed from the $M \to \infty$ phase.

Another way of understanding that the Chern number is zero is by seeing that the d_3 component of the Hamiltonian does not change sign in the BZ. The Hall conductance is a Jacobian integrated over the whole BZs, which is identical to a skyrmion number of the vector $\hat{d}_i$. Because $\sin(k_x)$ and $\sin(k_y)$ form a vector that looks like figure 8.2, we see that to realize a skyrmion, we must look at the component of the d_3. The configuration of d_3 is ferromagnetic (see fig. 8.3) for the values of M in this section, and so there is no skyrmion in $\hat{d}_i$ and, hence, no nonzero Hall conductance.

8.3.2 The $-2 < M < 0$ Phase

As we decrease M from the atomic limit ($M = \infty$), at $M = 0$ the energy gap collapses at the $\Gamma = (0, 0)$ point. All other points remain gapped. Hence, we need to look only for the physics around that particular point if all we care about is the low-energy structure. To find out more information,we would like to obtain the continuum Hamiltonian near that point. We expand to obtain

$$H_{\Gamma+k} = k_x\sigma_x + k_y\sigma_y + M\sigma_z \tag{8.18}$$

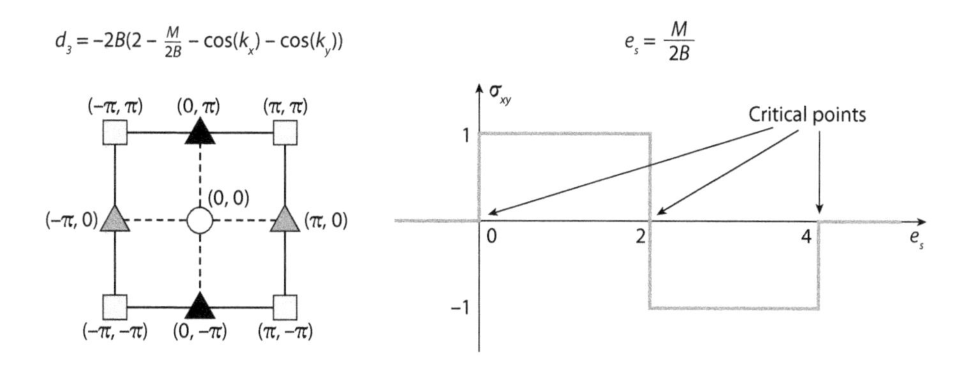

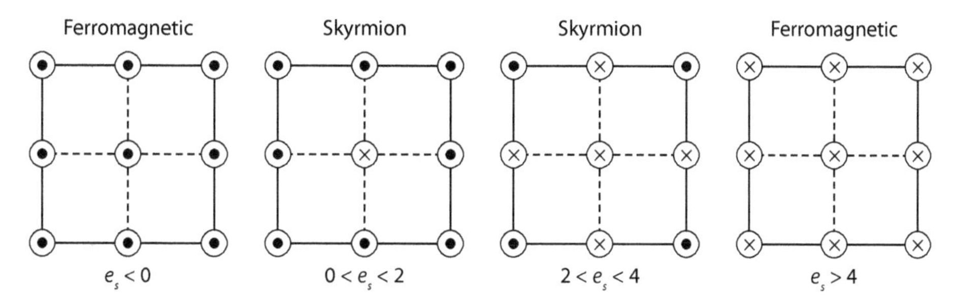

Figure 8.3. The values of $d_2(\mathbf{k})$ at high-symmetry points in the Brillouin zone and Chern numbers in the Chern insulator. Depending on whether d_3 dose or does not change sign in the BZ, we have or do not have skyrmion configurations.

up to linear terms in k. At the gap-closing-and-reopening transition, M goes from positive to negative, so the change in Hall conductance is

$$\Delta\sigma_{xy} = \frac{1}{2}\text{sign}(M_{<0}) - \frac{1}{2}\text{sign}(M_{>0}) = -1 \tag{8.19}$$

Because we know that the initial state had zero Hall conductance, by the preceding arguments, the new state has

$$\sigma_{xy} = -1. \tag{8.20}$$

The Hall conductance corresponds to the skyrmion number in the BZ. For $-2 < M < 0$, the configuration of d_3 looks like that in the figure 8.3 (which was plotted in terms of $e_s = M/B$). As such, $d_3 < 0$ around the Γ point (at exactly that point it is equal to M) and then becomes positive at $(\pi, 0)$, $(0, \pi)$, where it becomes equal to $2 + M$. This corresponds to a skyrmion number of -1. This skyrmion number can be understood as the difference between the two meron numbers of the continuum Dirac fermions with $M > 0$ and $M < 0$, which are $\frac{1}{2}$ and $-\frac{1}{2}$, respectively (fig. 8.4). The continuum Dirac equation has a meron in it because $d_3 = M$ is always in the same direction, but the k_x, k_y point away from the origin and, at large values, become dominant over d_3, and the vector lies in the plane. However, there is a nontrivial difference between $M > 0$ and $M < 0$. The difference between the two Hall conductances of the Dirac fermions (meron numbers) is the skyrmion number equal to -1.

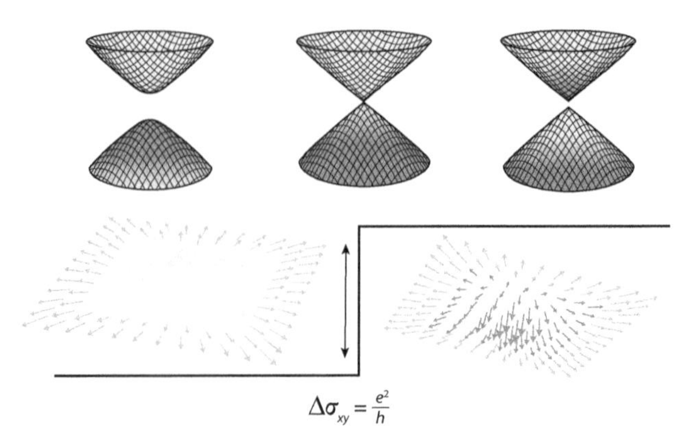

Figure 8.4. The Hall conductance equal to the skyrmion number as the difference between the Dirac equation meron numbers.

8.3.3 The $-4 < M < -2$ Phase

When we decrease M even further, we reach another phase transition. At $M = -2$, the gap closes at two points in the BZ: $(\pi, 0)$ and $(0, \pi)$. We can analyze what happens at this phase transition by looking only around those two points. We linearize the Hamiltonian around these points:

$$H_{(\pi,0)+\mathbf{k}} = -k_x\sigma_x + k_y\sigma_y + (2+M)\sigma_z, \tag{8.21}$$

$$H_{(0,\pi)+\mathbf{k}} = k_x\sigma_x - k_y\sigma_y + (2+M)\sigma_z, \tag{8.22}$$

and we can see that the Hall conductivity of each of these Dirac Hamiltonians is

$$\sigma_{xy} = -\text{sign}\frac{1}{2}(2+M). \tag{8.23}$$

The reason for the minus sign in front of the Hall conductance is the fact that there is now a minus sign in front of either σ_x or σ_y, respectively, for the two Hamiltonians in vicinity of $(\pi, 0)$ or $(0, \pi)$. For a generic Hamiltonian $H = k_\mu A_{\mu\nu}\sigma_\nu$, where in this case $\mu, \nu = 1, 2, 3$ and $k_\mu = (k_x, k_y, M)$—the Dirac mass—the Hall conductance is given by half the sign of the determinant of A: $\sigma_{xy} = \frac{1}{2}\text{sign}(\text{Det}(A))$. We hence have, for the transition between $M > -2$ and $M < -2$, a change in Hall conductance of

$$\Delta\sigma_{xy} = 2\left(-\frac{1}{2}\text{sign}(M+2)_{M<-2} - \left(-\frac{1}{2}\text{sign}(M+2)_{M>-2}\right)\right) = 2. \tag{8.24}$$

Because the phase before the transition ($M > -2$) had Hall conductance -1, the new Hall conductance is

$$\sigma_{xy} = -1 + 2 = 1. \tag{8.25}$$

8.3.4 Back to the Trivial State for $M < -4$

At $M = -4$, yet another point in the BZ becomes gapless: (π, π). Anywhere in between the values of $M = 0, -2, -4$, the model is gapped, as it is easily seen from the dispersion. Around the point (π, π), the Hamiltonian is

$$H_{(\pi,\pi)+\mathbf{k}} = -k_x\sigma_x - k_y\sigma_y + (4+M)\sigma_z, \tag{8.26}$$

and the change in Hall conductance between $M > -4$ and $M < -4$ is

$$\Delta\sigma_{xy} = \frac{1}{2}\text{sign}(4+M)_{M<-4} - \frac{1}{2}\text{sign}(4+M)_{M>-4} = -1, \tag{8.27}$$

which, when added to the previous Hall conductance, gives $\sigma_{xy} = 0$ for $M < -4$. We have now obtained the full-phase diagram of our model. For the Hall conductance, see the upper right panel in fig 8.3.

8.4 Determinant Formula for the Hall Conductance of a Generic Dirac Hamiltonian

So far, we have obtained the Hall conductance for rotationally invariant Dirac Hamiltonians. We now obtain the Chern number for the generic Dirac Hamiltonian:

$$h(\mathbf{k}) = k_a \mathcal{A}_{ab}\sigma_b + M\sigma_3, \tag{8.28}$$

where $\mathcal{A}_{ab}$ are numbers with $a, b = 1, 2$ (in a suitable σ basis) and M is the Dirac gap. The field strength reads

$$F_{12} = \frac{1}{2d^3}M(\mathcal{A}_{22}\mathcal{A}_{11} - \mathcal{A}_{21}\mathcal{A}_{12}) = \frac{1}{2d^3}M\text{Det}(\mathcal{A}) \tag{8.29}$$

where d is $d^2 = k_m k_i \mathcal{A}_{ij}\mathcal{A}_{mj} + M^2$. Going to polar coordinates $k_x = k\sin(\theta)$, $k_y = k\cos(\theta)$, we write the Hall conductance as the integral over the occupied states of the Berry curvature; using the identity

$$\int_0^{2\pi} \frac{d\theta}{(\cos(\theta)^2\mathcal{A}_{1j}\mathcal{A}_{1j} + \sin(\theta)^2\mathcal{A}_{2j}\mathcal{A}_{2j} + 2\sin(\theta)\cos(\theta)\mathcal{A}_{1j}\mathcal{A}_{2j})} = \frac{2\pi}{|\text{Det}(\mathcal{A})|}$$

we have that the Hall conductance equals

$$\sigma_{xy} = \frac{1}{2}\text{sign}(M)\text{sign}(\text{Det}(\mathcal{A})). \tag{8.30}$$

Notice that the Hamiltonian in equation (8.28) is not the most general Dirac Hamiltonian: this would be $k_\mu \mathcal{A}_{\mu\nu}\sigma_\nu$, with $\mu, \nu = 1, 2, 3$ and $k_3 = M$. The Hall conductance of such a Hamiltonian is tedious but straightforward to obtain and equals $\text{sign}(\text{Det}(\mathcal{A}))/2$.

8.5 Behavior of the Vector Potential on the Lattice

Let us further consider the behavior of the Berry potential in the Brillouin zone. We now want to show that the vector potential for the lattice Dirac Hamiltonian, as computed from the eigenstates, is perfectly well defined in the full Brillouin zone only if the system is topologically trivial (equivalently, no Hall conductance). The A_x Berry potential for our lattice

Hamiltonian is

$$A_x(\mathbf{k}) = \frac{\cos(k_x)}{2\sqrt{(2+M-\cos(k_x)-\cos(k_y))^2+(\sin(k_x))^2+(\sin(k_y))^2}} \times \frac{\sin(k_y)}{2+M-\cos(k_x)-\cos(k_y)+\sqrt{(2+M-\cos(k_x)-\cos(k_y))^2+(\sin(k_x))^2+(\sin(k_y))^2}}, \tag{8.31}$$

where we have separated the well-behaved part in the first line from the part that will eventually diverge on the topologically nontrivial side (second line). As we have seen, the first transition to a nontrivial topological insulator with Hall conductance $\sigma_{xy} = \frac{e^2}{h}$ takes place when the mass changes sign from positive to negative and the gap closes and then reopens at the point Γ. Other phase transitions of the Chern number take place at different places in the BZ, at different values of M, but we focus instead on the transition when M goes from $M = 0^+$ to $M = 0^-$ (for example, pick an arbitrary small value $M = -0.1$—adiabatic continuity guarantees that the physics will be the same in any phase where the gap does not close). Expand the second line of equation (8.31) around the Γ point for $k_x^2 + k_y^2 << \frac{M^2}{M+1}$:

$$\frac{k_y}{(M+|M|) + \frac{M+|M|+1}{2|M|}(k_x^2+k_y^2)}. \tag{8.32}$$

Now observe that if $M > 0$, the vector potential is well defined everywhere. However, for $M < 0$ (and small in absolute value so that we do not close the gap at other points in the BZ and our Γ point approximation is valid), we see that $M + |M| = 0$. Hence the vector potential $A_x(k)$ is proportional to

$$\frac{k_y}{\frac{1}{2|M|}(k_x^2+k_y^2)}, \tag{8.33}$$

and we see that if we look on the $k_x = 0$ axis, the vector potential has a $k_y/k_y^2 = 1/k_y$ singularity ($A_y(\mathbf{k})$ has a similar singularity but with k_y and k_x switched). Hence, for the topologically nontrivial side ($M < 0$), the gauge potential has singularities in the Brillouin zone.

8.6 The Problem of Choosing a Consistent Gauge in the Chern Insulator

The example of the lattice Dirac Hamiltonian is well suited for understanding the issue of choosing a smooth gauge in a Chern insulator. We can show that a smooth gauge can be chosen if the system has no Hall conductance—but not if the Hall conductance is nonzero. Let us look at the Bloch wavefunction of the occupied band,

$$\psi_{E_-} = \frac{1}{\sqrt{2d(d-d_3)}}\begin{pmatrix} d_3 - d \\ d_1 - id_2 \end{pmatrix}, \tag{8.34}$$

and let us use the simple Hamiltonian we have used so far: $d_1 = \sin k_x$, $d_2 = \sin k_y$, $d_3 = 2+M-\cos(k_x)-\cos(k_y)$. As we know from the skyrmion-number analysis, the Chern number equals 1 for $-2 < M < 0$ and equals -1 for $-4 < M < -2$, with zero Chern number for all other values of M. We now look at the region $-2 < M < 0$ and try to physically understand the cryptic comment made early on in the book that we are unable to choose a smooth gauge over the whole BZ in this case (because the Chern number is nonzero in this case) versus the case $M > 0$, where the Chern number is zero and where we are able to find a smooth gauge. We first define the S^2 manifold $\hat{d}_i = d_i/d$, where $d = \sqrt{d_1^2 + d_2^2 + d_3^2}$, and define the angles on the sphere:

$$\cos\theta = \hat{d}_3, \quad \sin\theta\cos\phi = \hat{d}_1, \quad \sin\theta\sin\phi = \hat{d}_2. \tag{8.35}$$

In terms of these angles, the Bloch function of the occupied band is

$$\psi^I_{E_-} = \begin{pmatrix} -\sqrt{\frac{1}{2}(1-\cos\theta)} \\ \dfrac{\sin\theta e^{-i\phi}}{\sqrt{2(1-\cos\theta)}} \end{pmatrix} \tag{8.36}$$

We placed an extra label I on the eigenstate of energy $E = -d$ to differentiate it from the gauge-transformed eigenstate II that we will soon introduce. Notice that when $\theta \to 0$, the eigenstate becomes

$$\psi^I_{E_-} \approx \begin{pmatrix} -\frac{\theta}{2} \\ e^{-i\phi} \end{pmatrix} \to \begin{pmatrix} 0 \\ e^{-i\phi} \end{pmatrix}, \tag{8.37}$$

where we have safely replaced $\theta/\sqrt{\theta^2}$ with 1 because $\theta \in [0, \pi]$ is positive. Notice that for $\theta = 0$, the wavefunction depends on the azimuthal angle in an ill-defined way because the azimuthal angle is not well defined. For example, at $\theta = 0$, two values of ϕ, 0 and π, represent the same point in space, but the Bloch state has a jump from (0,1) to (0,–1); hence, it is multivalued. This is one example of the fact that the wavefunction in the limit that $\theta = 0$ depends on the angle ϕ, and hence the gauge chosen is not smooth over the full S^2. However, we can define a gauge transformation of $\psi^I_{E_-}$,

$$\psi^{II}_{E_-} = e^{i\phi}\psi^I_{E_-}, \tag{8.38}$$

which gives the new Bloch state for the occupied band:

$$\psi^{II}_{E_-} = \begin{pmatrix} -\sqrt{\frac{1}{2}(1-\cos\theta)}e^{i\phi} \\ \frac{\sin\theta}{\sqrt{2(1-\cos\theta)}} \end{pmatrix}, \tag{8.39}$$

which is now obviously well defined at $\theta = 0$. Unfortunately, it is not well defined at $\theta = \pi$, for which it becomes

$$\psi^{II}_{E_-} = \begin{pmatrix} -e^{i\phi} \\ 0 \end{pmatrix}. \tag{8.40}$$

Hence, it has the same problem as $\psi^I_{E_-}$ had around $\theta = 0$, except now the problem is at $\theta = \pi$: the wavefunction is multivalued, depending on the ϕ-direction from which we reach $\theta = \pi$. There is no gauge in which we can make the wavefunction be well defined at *both* $\theta = 0$ and $\theta = \pi$.

Now comes the crucial part: for $M > 0$ (the phase with zero Hall conductance), we can see that $d_3(\mathbf{k}) = 2+M-\cos(k_x)-\cos(k_y)$ is *always* positive; at any point in the BZ, $d_3 > 0$. This means $\cos\theta = d_3/d > 0$ which means $\theta \in [0, \pi/2]$. Hence, θ never reaches the south pole π, so we can pick as a globally well-defined eigenstate the Bloch band $\psi^{II}_{E_-}$. However, if $-2 < M < 0$ (the phase of nonzero Hall conductance), then, at the point Γ, we have $d_3(0, 0) = M < 0$ and hence

$\frac{d_3(0,0)}{d(0,0)} = \text{sign}(M) = -1$, or $\theta = \pi$. Moving away from the point Γ, at the points $(\pi, 0)$ or $(0, \pi)$, we have $d_3(\pi, 0) = d_3(0, \pi) = 2 + M > 0$. At the π, π point, we also find $d_3(\pi, \pi) = 4 + M > 0$. Hence, at those points (where $d_{1,2} = 0$), we have $d_3/d = 1$ and, hence, $\theta = 0$. Thus, see that in this topologically nontrivial case, the map from the BZ torus to S^2 covers *both* the north and the south poles; hence, neither of the two gauges found before is going to be smooth in the full BZ. If we pick $\psi^I_{E_-}$ as an eigenstate, we have a problem at $\theta = 0$ (Γ point in the BZ), whereas if we pick $\psi^{II}_{E_-}$ as an eigenstate, we encounter a problem at $\theta = \pi$. Because $d_3(k)$ changes sign in the BZ, it will cover the full S^2 sphere, and we cannot pick a global, continuous, smooth gauge. This is generic for systems with nonzero Chern numbers.

The Chern number of the phase $-2 < M < 0$ is the winding number of the gauge transformation relating the two wavefunctions $\psi^I_{E_-}$, $\psi^{II}_{E_-}$. We pick a small circle around the south pole. In the region bounded by the circle that includes the south pole, $\psi^I_{E_-}$ is well defined, whereas in the region that includes the north pole, $\psi^{II}_{E_-}$ is well defined. We hence have defined a patching of the target manifold and smooth eigenstates on the two patches. On the circle, the two eigenstates are related by the gauge transformation equation (8.38): $e^{i\phi}$. The winding of this phase around the circle is equal to 2π, and the Chern number is $\frac{2\pi}{2\pi} = 1$.

8.7 Chern Insulator in a Magnetic Field

We would now like to add an external magnetic field to the Chern insulator and look at the structure of the Landau levels when the system is initially in a topologically nontrivial insulator state. The generic Hamiltonian for a Chern insulator with $M = M_0 \approx 0$ expanded around the minimum gap near the Γ point is

$$h(\mathbf{k}) = \begin{pmatrix} M(\mathbf{k}) & A_2 k_- \\ A_2 k_+ & -M(\mathbf{k}) \end{pmatrix}; \quad M(\mathbf{k}) = M_0 - B_0(k_x^2 + k_y^2); \quad k_\pm = k_x \pm k_y. \tag{8.41}$$

The case $M_0/B_0 > 0$ is a nontrivial Chern insulator because $M(\mathbf{k})$ changes sign at $k = \sqrt{M_0/B_0}$. Also, $M_0/B_0 < 0$ is a trivial insulator. We can add extra terms to the Hamiltonian, such as a full kinetic terms, etc., but this does not qualitatively change the physics presented next. Let us now apply a magnetic field $B > 0$ (so $eB < 0$, with magnetic length $l = 1/\sqrt{-eB}$). We define a pair of creation and annihilation operators satisfying the commutation relation $[a, a^\dagger] = 1$. For the momentum in terms of harmonic oscillator operators, we have

$$k_- = \frac{\sqrt{2}}{l} a, \quad k_+ = \frac{\sqrt{2}}{l} a^\dagger, \quad k_x^2 + k_y^2 = \frac{2}{l^2}\left(a^\dagger a + \frac{1}{2}\right). \tag{8.42}$$

The eigenstates of this Hamiltonian can be determined from

$$\begin{pmatrix} M_0 - B_0 \frac{2}{l^2}\left(a^\dagger a + \frac{1}{2}\right) & A_2 \frac{\sqrt{2}}{l} a \\ A_2 \frac{\sqrt{2}}{l} a^\dagger & -M_0 + B_0 \frac{2}{l^2}\left(a^\dagger a + \frac{1}{2}\right) \end{pmatrix} \begin{pmatrix} \psi_1 \\ \psi_2 \end{pmatrix} = E \begin{pmatrix} \psi_1 \\ \psi_2 \end{pmatrix} \tag{8.43}$$

We can solve the eigenvalue equation to find $(\psi_1, \psi_2) = (\theta_1 |n-1\rangle, \theta_2 |n\rangle)$, where $|n\rangle$ are the nth Landau-level wavefunctions. Notice that, because of the presence of $|n-1\rangle$, the preceding eigenstate is valid only for $n \geq 1$. The energy E_n can be obtained as a solution to the quadratic

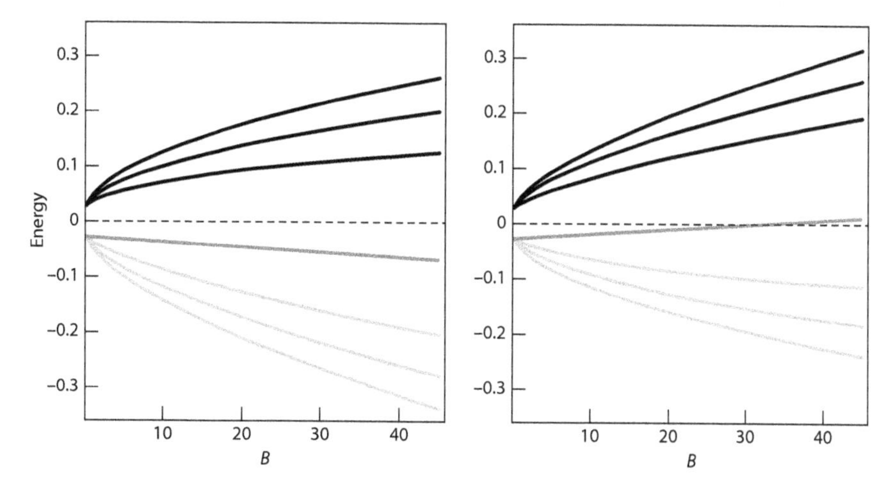

Figure 8.5. Landau-level behavior versus magnetic field B (only a few Landau levels are plotted) of a Chern Insulator (half of a quantum spin Hall state). Left: Topologically trivial Chern insulator. Right: Topologically nontrivial Chern insulator.

equation

$$E_{n\geq 1} = \frac{B_0}{l^2} \pm \sqrt{\left(M_0 - \frac{B_0}{l^2}n\right)^2 + \frac{2}{l^2}A_2^2 n}. \tag{8.44}$$

Energies come in pairs shifted by B_0/l^2. We are left with determining the dispersion of the $n = 0$ mode. Notice that if we were to use the preceding energy equation to obtain the $n = 0$ energy, this would give two $n = 0$ modes; we know this to be wrong because a single massive Dirac fermion has only one $n = 0$ mode—for the other one, the wavefunction is identically zero. The case $n = 0$ must be treated separately because $|n-1\rangle \rightarrow_{n=0} 0$. The wavefunction is then $(\psi_1, \psi_2) = (0, |0\rangle)$ (note that it is already normalized to unity). The energy then becomes

$$E_{n=0} = -M_0 + \frac{B_0}{l^2}. \tag{8.45}$$

Notice that if $M_0/B_0 > 0$ (topological nontrivial), then E_{n_0} traverses the bulk gap, as seen in figure 8.5. This is very different from the case of a topologically trivial insulator (such as undoped GaAs), in which the energy gap between Landau levels increases as we increase the magnetic field. This nonmonotonic dependence of the magnetic field in which the zero mode behaves differently from all other Landau levels is the hallmark of the nontrivial Chern insulator.

8.8 Edge Modes and the Dirac Equation

The Dirac Hamiltonian also explicitly shows the existence of edge modes at the boundary of two insulators with *different* Hall conductance. For $y > 0$, we consider our lattice Dirac model with $M > 0$ (say, $M = 0.1$), whereas for $y < 0$, we consider our lattice Dirac model with $-2 < M < 0$ (say $M = -0.1$). The only difference between the two lattice Hamiltonians is close to the point Γ, where the two continuum Hamiltonians are of Dirac form with negative and positive mass. Hence, in the continuum, the boundary between regions of Hall conductance 0 and Hall conductance 1 is modeled as a boundary between a Dirac fermion with Hall conductance $-\frac{1}{2}$ and a Dirac fermion with hall conductance $+\frac{1}{2}$. The moral of the story is that *differences* in Hall conductance not absolute values are important for edge states.

Let us now formalize the problem. We pick a boundary between a Dirac Hamiltonian with positive mass and one with negative mass in two dimensions—we could have picked any arbitrary dimension because the problem becomes one dimensional after Fourier-transforming in the directions parallel to the translationally invariant interface:

$$H(y) = -i\partial_x\sigma_x - i\partial_y\sigma_y + m(y)\sigma_z, \tag{8.46}$$

where $m(y) > 0$ for $y > 0$ and $m(y) < 0$ for $y < 0$. In fact, we need this to be true only at infinity, i.e., $m(y) > 0$ for $y \to +\infty$ and $m(y) < 0$ for $y \to -\infty$. Different interpolations of $m(y)$ would physically correspond to different boundary conditions. We look for a wavefunction solution:

$$\psi(x, y) = \phi_1(x)\phi_2(y), \tag{8.47}$$

which is separable, as the Hamiltonian indicates. We also look for a solution localized on the interface, because our previous experience indicates that boundaries between states of different Hall conductance carry gapless modes. This solution must be dependent on the only length scale in the problem, which is the mass gap:

$$\phi_2(y) = e^{-\int_0^y m(y')dy'}. \tag{8.48}$$

This solution has the desirable properties that it is localized on the edge and it is well behaved (finite) when y goes to $\pm$ infinity. Also $\phi_1(x)$ is a two-component spinor. We look for a chiral mode of energy $E = k_x$. At $k_x = 0$, we will find a zero-energy solution. Plugging in $\psi(x, y)$ into the Schrodinger equation, we get

$$-i\partial_x\sigma_x\phi_1(x) + m(y)(i\sigma_y\phi_1(x) + \sigma_z\phi_1(x)) = E\phi_1(x). \tag{8.49}$$

Obviously, for a zero-energy solution (such as the one at $k_x = 0$), we must cancel the term that depends on the mass, which is y-dependent and can be made very large by hand. So, we have

$$i\sigma_y\phi_1(x) + \sigma_z\phi_1(x) = 0 \longrightarrow \sigma_x\phi_1(x) = -\phi_1(x), \tag{8.50}$$

with the solution $\phi_1(x) = \chi(x)\frac{1}{\sqrt{2}}(1, -1)^T$ and $\chi(x)$ a scalar function of x. This scalar function satisfies the equation

$$i\partial_x\chi(x) = E\chi(x). \tag{8.51}$$

The solution $\chi(x) = e^{ik_x x}$ is a chiral propagating mode in the x-direction of energy $E = k_x$.

8.9 Haldane's Graphene Model

We finish this chapter by introducing what has historically been the first example of a topological insulator, the Chern insulator on the honeycomb lattice. In the late 1980s Haldane wanted to mimic the integer quantum Hall effect seen in the Landau-level problem while keeping the (full, not magnetic) translational symmetry of the lattice. This is actually rather

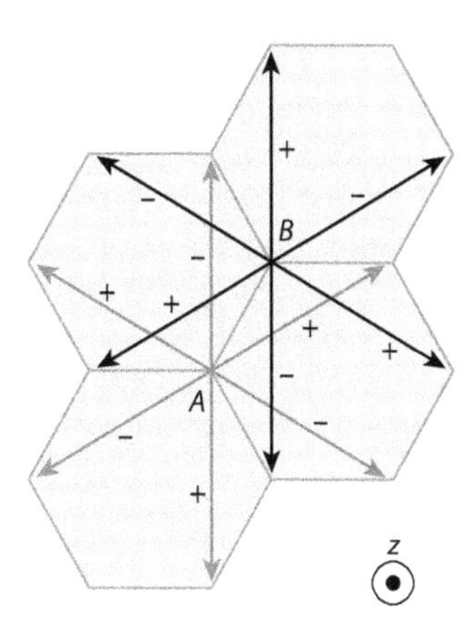

Figure 8.6. Phase convention for the Haldane model for a Chern insulator.

hard to do because, as we saw, magnetic fields with nonzero flux per plaquette enlarge the unit cell. The unbroken translational symmetry of the lattice is equivalent to having zero (mod 2π) flux per plaquette, which leads us to the rather paradoxical situation of trying to obtain Landau levels without a nonzero magnetic field. Haldane realized that time-reversal breaking, rather than overall nonzero flux per unit cell, is the essential ingredient of a nonzero Hall conductance. To keep the translational symmetry of the lattice, we need to break time reversal without a net flux per plaquette. The easiest way to do this is to put the magnetic phases on the next-nearest neighbors because going around a plaquette will not induce any nearest-neighbor phases. The Haldane model is on the honeycomb lattice, and the Hamiltonian is

$$H = t_1 \sum_{\langle i,j \rangle} c_i^\dagger c_j + t_2 \sum_{\langle\langle i,j \rangle\rangle} e^{-i\nu_{ij}\phi} c_i^\dagger c_j + M \sum_i \epsilon_i c_i^\dagger c_i, \tag{8.52}$$

where ϵ_i is ± 1, depending on whether i is on the A or B sublattice, and t is the nearest-neighbor-hopping energy, M is an on-site inversion symmetry-breaking term, and the new term is the next-nearest neighbor:

$$\nu_{ij} = \text{sign}(\hat{d}_1 \times \hat{d}_2)_z = \pm 1 \tag{8.53}$$

if $\hat{d}_{1,2}$ are the vectors along the two bonds constituting the next-nearest neighbors. The signs of the phases $i\phi$ are shown in figure 8.6. After Fourier transforming, the Haldane Hamiltonian for graphene can be expressed in compact form as

$$h(\mathbf{k}) = \epsilon(\mathbf{k}) + d_i(\mathbf{k})\sigma_i, \tag{8.54}$$

where

$$\epsilon(\mathbf{k}) = 2t_2(\cos(\phi)(\cos(\mathbf{k}\cdot\mathbf{a}_1) + \cos(\mathbf{k}\cdot\mathbf{a}_1) + \cos(\mathbf{k}\cdot(\mathbf{a}_1 - \mathbf{a}_2))), \tag{8.55}$$

$$d_1(\mathbf{k}) = \cos(\mathbf{k}\cdot\mathbf{a}_1) + \cos(\mathbf{k}\cdot\mathbf{a}_2) + 1, \tag{8.56}$$

$$d_2(\mathbf{k}) = \sin(\mathbf{k}\cdot\mathbf{a}_1) + \sin(\mathbf{k}\cdot\mathbf{a}_2), \tag{8.57}$$

$$d_3(\mathbf{k}) = M + 2t_2\sin(\phi)(\sin(\mathbf{k}\cdot\mathbf{a}_1) - \sin(\mathbf{k}\cdot\mathbf{a}_2) - \sin(\mathbf{k}\cdot(\mathbf{a}_1 - \mathbf{a}_2))). \tag{8.58}$$

8.9.1 Symmetry Properties of the Haldane Hamiltonian

We now want to analyze the symmetry properties of the Haldane model. First, we analyze the behavior under time reversal. For time reversal to be satisfied, it would mean that $\epsilon(\mathbf{k}) = \epsilon(-\mathbf{k})$, $d_1(\mathbf{k}) = d_1(-\mathbf{k})$, $d_2(\mathbf{k}) = -d_2(-\mathbf{k})$ and $d_3(\mathbf{k}) = d_3(-\mathbf{k})$. The first three are obviously satisfied by the Haldane Hamiltonian. The last condition is satisfied only for $\phi = 0, \pi$. We now analyze the behavior under inversion. For inversion around the middle of the unit cell (matrix representation σ_x) to be satisfied, it would mean that $\epsilon(\mathbf{k}) = \epsilon(-\mathbf{k})$, $d_1(\mathbf{k}) = d_1(-\mathbf{k})$, and $d_2(\mathbf{k}) = -d_2(-\mathbf{k})$, $d_3(\mathbf{k}) = -d_3(-\mathbf{k})$. The first three are obviously satisfied by the Haldane Hamiltonian, but the last one is satisfied only for $M = 0$. For both T and inversion to be satisfied, we must have both $\epsilon = 0$ and $\phi = 0$ or $\phi = \pi$.

8.9.2 Phase Diagram of the Haldane Hamiltonian

The interplay between the T-breaking mass term and the inversion-breaking mass gives the Haldane Hamiltonian, a nontrivial structure with three phases of Hall conductance, 0, 1, −1, very much like in our square-lattice example. Let us see how. We could directly compute the Hall conductance of the system using our Jacobian formula, but that would not be very physically illuminating. Instead, let us again use the Dirac argument.

We first notice that the system still has C_3 symmetry: we have proved before that the Hamiltonian without the new Haldane term has C_3 symmetry, and it is trivial to check that the Haldane term is invariant upon the permutation $\boldsymbol{\delta}_1 \to \boldsymbol{\delta}_2 \to \boldsymbol{\delta}_3 \to \boldsymbol{\delta}_1$. The inversion- and T-breaking terms induce gap in the spectrum, and we can understand the behavior of the model by looking at gap-closing-and-opening transitions. At which point in the BZ can these transitions happen? Because the model has C_3 symmetry, the gap-opening-and-closing transitions (when the system goes gapless) can happen only at the $\mathbf{K}$, $\mathbf{K}'$ points; thus, we focus on the Hamiltonian around those points. We expand the Haldane Hamiltonian around the point K to linear order:

$$h(\mathbf{K} + \boldsymbol{\kappa}) = \epsilon(\mathbf{K}) + \kappa_i \partial_i \epsilon(\mathbf{k} = \mathbf{K}) + d_a(\mathbf{K})\sigma_a + \kappa_i \partial_i d_a(\mathbf{k} = \mathbf{K})\sigma_a \tag{8.59}$$

for $\boldsymbol{\kappa} << \mathbf{K}$. After tedious algebraic simplification, we obtain

$$h(\mathbf{K} + \boldsymbol{\kappa}) = -3t_2 \cos(\phi) + \frac{3}{2} t_1 (\kappa_y \sigma_x - \kappa_x \sigma_y) + (M - 3\sqrt{3} t_2 \sin(\phi))\sigma_z. \tag{8.60}$$

Around $\mathbf{K}'$ we have the Hamiltonian

$$h(\mathbf{K}' + \boldsymbol{\kappa}) = -3t_2 \cos(\phi) - \frac{3}{2} t_1 (\kappa_y \sigma_x + \kappa_x \sigma_y) + (M + 3\sqrt{3} t_2 \sin(\phi))\sigma_z. \tag{8.61}$$

To obtain the value of the Hall conductance, as we did before, we have to start from a phase where we know the its value and trace the changes in the Chern number as we undergo gap-closing and reopening transitions. Such a phase is easy to find: we start from the case $M \to \pm\infty$. In that case the wavefunction is localized on the site A or B and the system is trivial, with eigenstates constant in momentum space. As such, the system has no Hall conductance. We pick $M \to +\infty$ as a reference point.

The Haldane Hamiltonian is similar to that in equation (8.28) and its Hall conductance can be readily obtained from equation (8.30). Without loss of generality, we now assume $\phi > 0$ (with $-\pi < \phi < \pi$), $t_2 > 0$ and start lowering M from $+\infty$. At

$$M = 3\sqrt{3} t_2 \sin \phi, \tag{8.62}$$

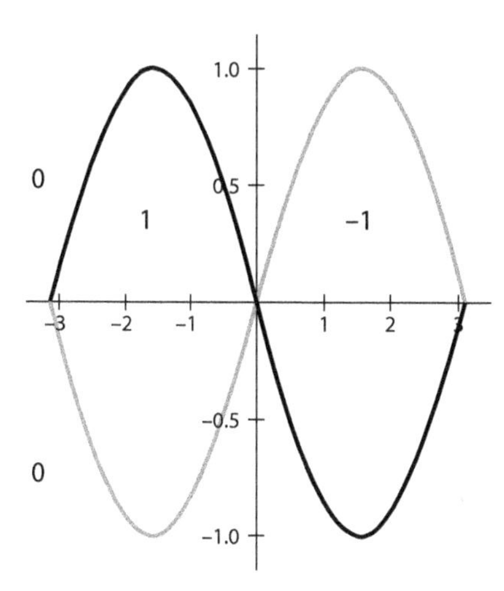

Figure 8.7. Phases of the Haldane model. The x-axis is ϕ and the y-axis is M. The numbers represent the Hall conductances in the phase. The vertical scale is in units of $3\sqrt{3}t_2$.

the K Dirac fermion goes through a gap-closing-and-opening transition and the Hall conductance changes from (equation (8.30)) $\frac{1}{2}\text{sign}(M - 3\sqrt{3}t_2 \sin(\phi)) = \frac{1}{2}$ to $-\frac{1}{2}$, with a change of Hall conductance equal to -1. The gap at K' stays open for $M = 3\sqrt{3}t_2 \sin(\phi)$, and only K goes through gap-closing-and-reopening transition to give a phase with

$$\sigma_{xy} = -1. \tag{8.63}$$

As we lower M even more, we reach the value where

$$M = -3\sqrt{3}t_2 \sin \phi, \tag{8.64}$$

at which point the K' Dirac fermion goes through a gap-closing and re-opening transition, which changes the Hall conductance by $\Delta\sigma_{xy} = 1$ (the point $\mathbf{K}$ remains gapped). The change in Hall conductance means that the phase $\sigma_{xy} = -1$ before the gap closing goes to a phase with $\sigma_{xy} = 0$ after the gap closing and reopening. This is also necessary because the phase $M < -3\sqrt{3}t_2 \sin \phi$ is adiabatically continuable to the state $M \to \infty$, which must have zero Hall conductance. The case $\phi < 0$ is treated in a similar fashion to give the phase diagram in figure 8.7.

8.10 Problems

1. *Berry Phase of a Generic Dirac Hamiltonian in Two Dimensions:* The most general form of a Hamiltonian in two dimensions is $h(\mathbf{k}) = k_i A_{ia}\sigma_a$, where $i = 1, 2$ and $a = 1, 2, 3$. Find the Berry phase of the Dirac fermion as it circles the gapless point $k_1 = k_2 = 0$ in terms of quantities dependent the matrix A.

2. *Berry Phase of a Gapless Hamiltonian with Cubic Crossing:* Find the Berry phase of a fermion circling the degeneracy point of a Hamiltonian that is cubic in the power of the momentum $k_+^3\sigma_- + k_-^3\sigma_+$, where $k_\pm = k_x \pm ik_y$, $\sigma_\pm = \sigma_x \pm i\sigma_y$. Generalize it to any power of k.

3. *Chern Number of a Gapped Cubic Crossing:* Now gap the cubic Hamiltonian $k_+^3\sigma_- + k_-^3\sigma_+$ by adding a term $m\sigma_z$. What is the Hall conductance of the resulting gapped systems. Generalize to any $k_\pm^n$ and prove a relation between the Chern number of gapped systems and the Berry phase of the gapless system.

4. *Lattice Hamiltonian for a Chern Number Higher than Unity:* In section 8.3 we saw that the Chern number of the lattice Dirac Hamiltonian $h(\mathbf{k}) = \sin k_x\sigma_x + \sin k_y\sigma_y + (M - \cos k_x - \cos k_y)\sigma_z$ equals unity (in absolute value). Show that the Hamiltonian $h(\mathbf{k}) = \sin 2k_x\sigma_x + \sin 2k_y\sigma_y + (M - \cos k_x - \cos k_y)\sigma_z$ has Chern numbers of absolute value both 1 and 3. (For which masses M does this occur?) Then generalize to find a model Hamiltonian for any odd Chern number system. How can we obtain even Chern numbers?

5. *Hall Conductance of a Generic Gapped Dirac Hamiltonian:* In the text, we obtained the Hall conductance for a Dirac Hamiltonian of the form $k_i\mathcal{A}_{ab}\sigma_b + M\sigma_3$, where $a, b = 1, 2$, to be as $\sigma_{xy} = \frac{1}{2}\text{sign}(M)\text{sign}(\text{Det}(\mathcal{A}))$. Generalize this formula to any gapped Hamiltonian $k_a\mathcal{A}_{aa}\sigma_a + M\sigma_3$.

6. *Phase Diagram of the Haldane Model without C_3 Symmetry:* Find the phase diagram of a modified Haldane model where C_3 symmetry has been broken by a trimerization term in the hopping along the $\boldsymbol{\delta}_1$ direction, for example. In the pure graphene system (without TR breaking), this term moves the Dirac nodes off their C_3 symmetric positions in the BZ. If large enough, it can even cause the Dirac nodes to meet and annihilate. Show the interplay between this term, the presence of Dirac nodes in the TR-invariant system, and the appearance of the Chern insulator once the Dirac nodes are gapped by the TR-breaking terms (phases).

9

Time-Reversal-Invariant Topological Insulators

The Haldane model of a Chern insulator shows that a nontrivial insulator with nonzero Hall conductance can exist when TR symmetry is broken. More than 15 years after the Haldane model was published, it was realized that keeping symmetries *intact* gives rise to systems as interesting as the ones where symmetries are broken. Kane and Mele [4, 5] first realized that by doubling the Haldane model of the Chern insulator by introducing spin in the problem, we can obtain an insulator that maintains TR symmetry but has a robust, gapless pair of helical (not chiral) edge states. In this chapter we introduce the Kane and Mele model first and then introduce the HgTe model [7] for a topological insulator. Such an insulator was the first experimentally realizable topological insulator, and its experimental discovery blew the field of topological insulators wide open.

9.1 The Kane and Mele Model: Continuum Version

In this chapter, we chose the same gauge choices and lattice orientation as in the original Kane and Mele papers. As such, we need to rotate the graphene lattice by $\pi/2$ compared with the one we used for the Haldane model: $k_x \to k_y$ and $k_y \to -k_x$. The K, K' Haldane Hamiltonians become

$$h(\mathbf{K}+\boldsymbol{\kappa}) = -3t_2\cos(\phi) - \frac{3}{2}t_1(\kappa_x\sigma_x + \kappa_y\sigma_y) + (M - 3\sqrt{3}t_2\sin(\phi))\sigma_z. \tag{9.1}$$

For $\boldsymbol{\kappa} \ll \mathbf{K}$ and around $\mathbf{K}'$, we have

$$h(\mathbf{K}'+\boldsymbol{\kappa}) = -3t_2\cos(\phi) - \frac{3}{2}t_1(-\kappa_x\sigma_x + \kappa_y\sigma_y) + (M + 3\sqrt{3}t_2\sin(\phi))\sigma_z. \tag{9.2}$$

We begin with a low-energy description of graphene by expanding the wavefunction around the points $\mathbf{K}$, $\mathbf{K}'$. We can mix the wavefunction around $\mathbf{K}$ and $\mathbf{K}'$ in a four-component spinor:

$$\Psi(r) = ((u_A(\mathbf{K}), u_B(\mathbf{K})), (u_A(\mathbf{K}'), u_B(\mathbf{K}')))\psi(r). \tag{9.3}$$

For the moment, assume no Haldane phase ($\phi = 0$), and denote the Dirac velocity $v_F = -\frac{3at_1}{2}$ with $t_1 < 0$. The low-energy description of graphene around each Dirac node is then given by the Hamiltonian

$$H_0 = -i\hbar v_f\psi^\dagger(\sigma_x\tau_z\partial_x + \sigma_y\partial_y)\psi. \tag{9.4}$$

where $\tau_z = \begin{pmatrix} 1 & 0 \\ 0 & -1 \end{pmatrix}$ acts in the $\mathbf{K}$, $\mathbf{K}'$-point space. In matrix notation,

$$H_0 = -i\hbar v_F \psi^\dagger \begin{bmatrix} \sigma_x \partial_x + \sigma_y \partial_y & 0 \\ 0 & -\sigma_x \partial_x + \sigma_y \partial_y \end{bmatrix} \psi, \tag{9.5}$$

where the two Hamiltonian blocks correspond to low energies around the points $\mathbf{K}$ and $\mathbf{K}'$. In this low-energy language, we now want to look at the possible terms that can open a gap. As we know, σ_z at both $\mathbf{K}$ and $\mathbf{K}'$ can open a gap that is odd under inversion. The other term, the Haldane mass, is $\sigma_z \tau_z$; it provides for a mass that changes sign between the points $\mathbf{K}$ and $\mathbf{K}'$. The term is even under inversion (which changes both sublattices and $\mathbf{K}$ with $\mathbf{K}'$), but it is odd under T.

9.1.1 Adding Spin

The Hamiltonian in equation(9.5) is for spinless fermions. We now want to add spin to the system, which doubles it. For each spin $\uparrow, \downarrow$, the starting Hamiltonian is in equation(9.5), and the Hamiltonian becomes an 8×8 matrix if we include all spin, points $\mathbf{K}$, $\mathbf{K}'$, and sublattice A, B structure in a single low-energy matrix. This is the low-energy gapless graphene Hamiltonian with spin. As per our previous analyses, we suspect something interesting can happen if we open gaps in the system. Several gaps are possible: the boring ones are just doubled versions of the gaps that we encountered in the spinless version. If the gaps do not make use of the new spin structure, we are bound to obtain just doubled versions of the Haldane model. However, with added spin, another gap-opening term is possible:

$$H_{SO} = \lambda_{SO} \psi^\dagger \sigma_z \tau_z s_z \psi, \tag{9.6}$$

where s_z is the electron spin. The s_i matrices act on spin space, the σ_i act on the A, B sublattice space, and the τ_i act on the low-energy $\mathbf{K}$, $\mathbf{K}'$ space. This term respects TR symmetry, as we show shortly. It is easy to prove that this is the only $q = 0$ constant term allowed by mirror symmetry. For broken mirror symmetry — which can happen either by an applied electric field in the z-direction or by the substrate — we can have a term of the Rashba type [41], $(\mathbf{s} \times \mathbf{p}) \cdot \hat{z}$:

$$H_R = \lambda_R \psi^\dagger (\sigma_x \tau_z s_y - \sigma_y s_x) \psi \tag{9.7}$$

Let us now analyze the TR properties of these terms. Because we have added spin to the system, the TR operator has changed to $T = -i s_y K$ instead of just K for spinless fermions (where this K is complex conjugation, not to be confused with the momentum K in graphene). The Hamiltonian at K without spin-orbit coupling can be written as

$$H_0(\mathbf{K} + \mathbf{k}) = \psi^\dagger \begin{bmatrix} h_\uparrow(\mathbf{k}) & 0 \\ 0 & h_\downarrow(\mathbf{k}) \end{bmatrix} \psi, \tag{9.8}$$

where $h_\uparrow = h_\downarrow = k_x \sigma_x + k_y \sigma_y$ are the spinless graphene Hamiltonians around $\mathbf{K}$. We now want to add a *constant* in terms of $\mathbf{K} + \mathbf{k}$ that can open a gap. To open a gap, any extra added Hamiltonian ΔH must anticommute with H_0. If the added term commutes with any of the matrices in the Hamiltonian, then it will not open a gap: for example, $k_x \sigma_x$ and $a\sigma_x$ commute — an $a\sigma_x$-term does not open a gap but rather shifts the gapless point $k_x \to k_x - a$. Hence, the terms allowed that would open a gap must anticommute with σ_i because $H_0(\mathbf{K} + \mathbf{k}) = I_s \times k_i \sigma_i$, where I_s is the identity matrix in spin space. Hence, the only term allowed must contain σ_z in A, B sublattice space:

$$\Delta H(\mathbf{K}+\mathbf{k}) = \mathcal{D}\sigma_z, \tag{9.9}$$

where $\mathcal{D}$ is a spin-space matrix. Of course, the simplest such term is the inversion-breaking mass (σ_z) for spin $\uparrow$, $\downarrow$ at both $\mathbf{K}$, $\mathbf{K}'$: this corresponds to $\mathcal{D}$ being the identity in spin space. However, this is just two copies of the inversion symmetry-breaking mass term, one for spin up and one for spin down. It does not bring us anything exciting. This is, of course, because the term does not involve spin. The next simple term is to give the $\uparrow$, $\downarrow$ Dirac fermions at $\mathbf{K}$ opposite masses:

$$\Delta H(\mathbf{K}+\mathbf{k}) = s_z\sigma_z = \begin{bmatrix} \sigma_z & 0 \\ 0 & -\sigma_z \end{bmatrix}. \tag{9.10}$$

We now figure out the form of the Hamiltonian at $\mathbf{K}'$ such that the system is time-reversal invariant. In matrix form, the T operator is

$$T = \begin{bmatrix} 0 & -I \\ I & 0 \end{bmatrix} K, \tag{9.11}$$

with K complex conjugation and I the identity 2×2 matrix. For the Bloch Hamiltonian to be T invariant, we need

$$H(-\mathbf{k}) = TH(\mathbf{k})T^{-1} = \begin{bmatrix} h_\downarrow(\mathbf{k})^* & 0 \\ 0 & h_\uparrow(\mathbf{k})^* \end{bmatrix}. \tag{9.12}$$

The initial gapless Hamiltonian, $H_0(k)$, defined over the whole BZ is obviously TR invariant, as amply discussed in previous chapters. Time reversal relates the low-energy Hamiltonian at $\mathbf{K}$ with that at $\mathbf{K}'$ by flipping the spin:

$$\Delta H(-\mathbf{K}-\mathbf{k}) = \Delta H(\mathbf{K}'-\mathbf{k}) = -s_z\sigma_z = \begin{bmatrix} -\sigma_z & 0 \\ 0 & \sigma_z \end{bmatrix}. \tag{9.13}$$

Thus, we have obtained the form of the low-energy gap-opening Hamiltonian around the points $\mathbf{K}$ and $\mathbf{K}'$ that preserves T. Around $\mathbf{K}$, it is $s_z\sigma_z$. Around $\mathbf{K}'$, it is $-s_z\sigma_z$. If we consider the eight-component spinor that contains both spin, lattice, and $\mathbf{K}$, $\mathbf{K}'$ quantum numbers, then a TR invariant mass is indeed $\lambda_{SO}s_z\sigma_z\tau_z$, as stated before. This term can obviously arise only from spin-orbit coupling because it couples the spin with the momentum quantum numbers. Such a term opens a gap in the spectrum:

$$E_k = \pm\sqrt{(\hbar v_f k)^2 + \lambda_{SO}^2}. \tag{9.14}$$

Each band is doubly degenerate, due to the fact that we can write the full tight-binding model (not only the expansion around $\mathbf{K}$ and $\mathbf{K}'$) as

$$H = t_1 \sum_{\langle ij\rangle} c_{i\sigma}^\dagger c_{j\sigma} + i\lambda_{SO} \sum_{\langle\langle ij\rangle\rangle} v_{ij} c_{i\sigma}^\dagger s_{\sigma\sigma'}^z c_{j\sigma'}, \tag{9.15}$$

which is just two copies of the Haldane model. The v_{ij}'s are identical bond signs, as in the Haldane model.

9.1.2 Spin ↑ and Spin ↓

We now delve further into the physical properties of the doubled Haldane model by first separately analyzing the physics of the model for spin ↑ and ↓. For spin ↑, the Hamiltonians at **K** and **K**′ are

$$h(\mathbf{K}+\mathbf{k}) = k_x\sigma_x + k_y\sigma_y + \lambda_{SO}\sigma_z, \qquad h(\mathbf{K}'+\mathbf{k}) = -k_x\sigma_x + k_y\sigma_y - \lambda_{SO}\sigma_z. \tag{9.16}$$

This is the $M = 0$ limit of the Haldane model studied in chapter 8. The Haldane mass term is λ_{SO}. As such, in this regime, from our analysis of the Haldane model in the previous chapter, we know that the spin ↑ has a Hall conductance equal to 1.

For spin ↓, the Hamiltonians at **K** and **K**′ are

$$h(\mathbf{K}+\mathbf{k}) = k_x\sigma_x + k_y\sigma_y - \lambda_{SO}\sigma_z, \qquad h(-\mathbf{K}+\mathbf{k}) = -k_x\sigma_x + k_y\sigma_y + \lambda_{SO}\sigma_z. \tag{9.17}$$

This is the $M = 0$ limit of the Haldane model studied previously, but with an opposite Haldane term from the spin ↑ Hamiltonian. As such, from our analysis of the Haldane model, we know that the spin ↑ has a Hall conductance equal to -1.

We have hence uncovered the physics of the Kane and Mele model. We have a quantum Hall effect for spin ↑ with Hall conductance 1 and a quantum Hall effect for spin ↓ with Hall conductance -1. If we chose to place edges on our graphene sample then, on each edge separately, we will have a chiral (from the Hall conductance 1) and an antichiral (from the Hall conductance -1) edge mode, crossing at a time-reversal invariant degeneracy point, either $k_\parallel = 0$ or $k_\parallel = \pi$, where $k_\parallel$ is the momentum along the edge. As we have both chiral and antichiral modes on the same edge (i.e., states traveling in both directions in close proximity), a gap would usually open due to backscattering. However, in this case, backscattering single-particle terms are forbidden due to TR invariance. We proved in chapter 4 that the scattering matrix elements between TR-invariant pairs are zero for an odd fermion number. This means that if we have an *odd* number of fermion pairs of edge modes (odd number of Kramers' doublets), we cannot open a gap by a single-particle backscattering term—no such TR-invariant terms can be written (however TR-invariant multiparticle interaction terms can be written). If we have an even number of Kramers' pairs on an edge, we can write single-particle backscattering terms. This suggests the existence of two separate classes and, thus, a Z_2 classification of noninteracting topological insulators, which we further develop later.

9.1.3 Rashba Term

The Hamiltonian in equation(9.15) is just two decoupled Haldane Hamiltonians, one for spin ↑ and one for spin ↓ with opposite Hall conductances. To couple the two Hamiltonians, we can also add a Rashba term at the point **K**, $(\mathbf{s} \times \boldsymbol{\sigma}) \cdot \hat{1}_z$:

$$h(\mathbf{K}+\mathbf{k}) = h(\mathbf{K}) = \sigma_x s_y - \sigma_y s_x. \tag{9.18}$$

By TR invariance, we obtain the Hamiltonian at the point $\mathbf{K}' = -\mathbf{K}$:

$$h(\mathbf{K}') = s_y h^*(\mathbf{K}) s_y = -\sigma_x s_y^3 + \sigma_y s_y s_x s_y = -\sigma_x s_y - \sigma_y s_z. \tag{9.19}$$

The Hamiltonians at **K** and **K**′ can be written in compact form for both points **K** and **K**′:

$$H_R = \sigma_x \tau_z s_y - \sigma_y s_x. \tag{9.20}$$

The Rashba term by itself (without the Δ_{SO} term) does not open a full gap in the spectrum. This is due to the fact that this term does not anticommute with the Hamiltonian. At the point $\mathbf{K}$, we can add $\lambda_R(s_y\sigma_x - s_x\sigma_y)$ to the $H_0(\mathbf{K}+\mathbf{k})$ Hamiltonian and get the four eigenvalues

$$-\lambda_R \pm \sqrt{k^2+\lambda_R^2}, \qquad \lambda_R \pm \sqrt{k^2+\lambda_R^2}. \tag{9.21}$$

This shows that the Hamiltonian is gapless at the point K regardless of the value of the Rashba coupling. Out of four bands (two from the sublattice and two from spin), two are fully gapped and two are gapless. Even though it does not open a full gap by itself, the Rashba term can be used to influence (reduce or enhance) the Kane and Mele gap ($\sigma_z s_z \tau_z$) because that gap term anticommutes with the Rashba term.

9.2 The Kane and Mele Model: Lattice Version

For completeness, we now provide the form of the Kane and Mele model over the whole hexagonal lattice BZ and not only the continuum version around the points $\mathbf{K}$ and $\mathbf{K}'$. As previously mentioned, we rotate the graphene lattice we used in the chapter 8 to keep consistency with the Kane and Mele model. We have the lattice vectors $\boldsymbol{\delta}_3 = a(0,-1)$, $\boldsymbol{\delta}_1 = a(-\sqrt{3}/2, 1/2)$, $\boldsymbol{\delta}_2 = a(\sqrt{3}/2, 1/2)$, and the lattice translation vectors $\mathbf{a}_1 = \boldsymbol{\delta}_1 - \boldsymbol{\delta}_3$, $\mathbf{a}_2 = \boldsymbol{\delta}_2 - \boldsymbol{\delta}_3$. The primitive BZ wavevectors are $\mathbf{b}_1 = \frac{2\pi}{3a}[-\sqrt{3}, 1]$, $\mathbf{b}_2 = \frac{2\pi}{3a}[\sqrt{3}, 1]$, and the corners of the BZ (graphene gapless points) are $\mathbf{K} = \frac{2\pi}{3}(-1/\sqrt{3}, 1)$, and $\mathbf{K}' = \frac{2\pi}{3}(1/\sqrt{3}, 1)$. We measure lengths in units of the minimum distance between two A sites (or two B sites): $l = 2a\sin(\pi/3) = \sqrt{3}a$. The lattice Kane and Mele Hamiltonian whose continuum form we explored in section 9.1.2 is

$$H = t\sum_{\langle ij\rangle} c_i^\dagger c_j + i\lambda_{SO}\sum_{\langle\langle ij\rangle\rangle} v_{ij} c_i^\dagger s_z c_j + i\lambda_R \sum_{\langle ij\rangle} c_i^\dagger(\mathbf{s}\times\mathbf{d}_{ij})_z c_j + \lambda_v \sum_i \epsilon_i c_i^\dagger c_i, \tag{9.22}$$

where $c_i = (c_{i\uparrow}, c_{i\downarrow})$ and

$$v_{ij} = \frac{2}{\sqrt{3}}(\mathbf{d}_1 \times \mathbf{d}_2)_z \tag{9.23}$$

are the Haldane phases. The λ_{SO} term does not violate mirror symmetry with respect to the lattice plane: if $z \to -z$, then the v_{ij} changes sign, but so does the spin s_z to keep the Hamiltonian unchanged. On the other hand, the Rashba term explicitly violates mirror symmetry. The λ_v term is the inversion-symmetry-breaking term—$\epsilon_i = \pm 1$ —depending on whether i is the A or B site. To maintain consistency with the notation in the the original Kane and Mele paper, we define the following momentum notation:

$$\mathbf{k}\cdot\mathbf{a}_1 = y - x, \qquad \mathbf{k}\cdot\mathbf{a}_2 = y + x. \tag{9.24}$$

We now label our momenta in terms of x, y. Because we now have two sites per unit cell and two spins, the Bloch Hamiltonian can be most conveniently written in terms of the $SU(4)$ 4×4 matrices. We introduce several of these matrices: $\Gamma_1 = \sigma^x \times I$, $\Gamma_2 = \sigma_z \times I$, and, hence, $\Gamma_{12} = \frac{1}{2i}[\Gamma_1, \Gamma_2] = -\sigma_y \times I$. In momentum space, we diagonalize the nearest-neighbor hopping of the preceding Hamiltonian and keep the Bloch part (which acts on the spinor $\psi = (\psi_{A\uparrow}\ \psi_{B\uparrow}\ \psi_{A\downarrow}\ \psi_{B\downarrow})$) to obtain

$$t\sum_{\langle ij\rangle >} c_i^\dagger c_j \to t(1+\cos(y-x)+\cos(y+x))\Gamma_1 - 2t\cos(x)\cos(y)\Gamma_{12}. \tag{9.25}$$

The inversion-symmetry-breaking sublattice mass term is, in the Kane and Mele notation, $\lambda_v \sum_i c_i^\dagger c_i \to \Gamma_2$. For the spin-orbit coupling term, we have, based on the signs in the Haldane model (but rotated by $\pi/2$ for our configuration),

$$i\lambda_{SO} \sum_{\langle\langle ij \rangle\rangle} v_{ij} c_i^\dagger s^z c_j \to 2\lambda_{SO} - (2\sin(2x) - 4\sin(x)\cos(y))\Gamma_{15}, \tag{9.26}$$

where we have defined $\Gamma_5 = \sigma_y s_z$; hence, $\Gamma_{15} = \sigma_z s_z$. This differs from the equivalent term in the original Kane and Mele paper by an overall minus sign. Up to now the Hamiltonian is

$$h(\mathbf{k}) = d_1(\mathbf{k})\Gamma_1 + d_{12}(\mathbf{k})\Gamma_{12} + d_2(\mathbf{k})\Gamma_2 + d_{15}(\mathbf{k})\Gamma_{15}, \tag{9.27}$$

where the terms $d_1(\mathbf{k})$, $d_{12}(\mathbf{k})$, $d_2(\mathbf{k})$, and $d_{15}(\mathbf{k})$ can be read from equations (9.25) and (9.26). We will come back and explain the purpose and choice of the Γ matrices in a little while. As it stands, the Hamiltonian still has one symmetry, the mirror symmetry under reflection around the xy-plane, $z \to -z$. Under this symmetry, $s_z \to -s_z$ and also $v_{ij} \to -v_{ij}$ because the term is the z-component of a cross product. To break this symmetry, we must add a term with a defined z-direction, the easiest of which is a Rashba-type term that corresponds to the application of an electric field in the plane:

$$i\lambda_R \sum_{\langle ij \rangle} c_i^\dagger (\mathbf{s} \times \mathbf{d}_{ij})_z c_j = i\lambda_R \sum_{\langle ij \rangle} c_i^\dagger \epsilon_{zmn} s^m d_{ij}^n c_j, \tag{9.28}$$

where $\mathbf{d}_{ij}$ is the distance between nearest-neighbor sites i, j. Due to its intricate form (it couples all s_x, s_y components of spin), this term is actually the most tedious to write explicitly. We find it has the following form:

$$h_R(\mathbf{k}) = d_3(\mathbf{k})\Gamma_3 + d_4(\mathbf{k})\Gamma_4 + d_{23}(\mathbf{k})\Gamma_{23} + d_{24}(\mathbf{k})\Gamma_{24}, \tag{9.29}$$

where the tight-binding expressions have the form

$$d_3(\mathbf{k}) = \lambda_R(1 - \cos(x)\cos(y)), \quad d_4(\mathbf{k}) = -\sqrt{3}\lambda_R \sin(x)\sin(y), \tag{9.30}$$

$$d_{23}(\mathbf{k}) = -\lambda_R \cos(x)\sin(y), \quad d_{24}(\mathbf{k}) = \sqrt{3}\lambda_R \sin(x)\sin(y). \tag{9.31}$$

It is now time to talk in more detail about the matrices used in defining the Hamiltonian. These types of matrices will be used extensively in the future because as the minimal models of 2-D and 3-D T-invariant topological insulators are four-band models. The Hamiltonian is a 4×4 Hermitian matrix (sublattice and spin), so it will be the sum of an identity matrix and the 15 matrix generators of the $SU(4)$ group. These matrices can be classified in the usual way: $15 = 5 + 10$, where we have 5 generators of a Clifford algebra (also sometimes referred to as Dirac algebra):

$$\{\Gamma_a, \Gamma_b\} = 2\delta_{ab}, \quad a, b = 1, 2, 3, 4, 5; \tag{9.32}$$

the 10 remaining generators are

$$\Gamma_{ab} = \frac{1}{2i}[\Gamma_a, \Gamma_b]. \tag{9.33}$$

We have already chosen several of these matrices, so let us complete the full basis:

$$\Gamma_{1,2,3,4,5} = (\sigma_x I,\ \sigma_z I,\ \sigma_y s_x,\ \sigma_y s_y,\ \sigma_y s_z). \tag{9.34}$$

The Γ_{ab} can be easily obtained by commuting the Clifford generators but will not be written here. equation (9.34) is but one of the representations of the Clifford algebra. We will use another representation soon in connection to the HgTe model for a quantum-spin Hall state, but the general theory of the algebra is the same. The Clifford matrices chosen by Kane and Mele are invariant under time reversal $T\Gamma_i T^{-1} = \Gamma_i$ (with $T = i(I \otimes s_y)K$), although other representations where this is not true are possible.

A generic 4×4 Hamiltonian (without any specific symmetries) has the form

$$h(\mathbf{k}) = \sum_{a=1}^{5} d_a(\mathbf{k})\Gamma_a + \sum_{a<b=1}^{5} d_{ab}(\mathbf{k})\Gamma_{ab}. \tag{9.35}$$

The T invariance of the Γ_a implies the T oddness of the Γ_{ab}'s: $T\frac{1}{2i}[\Gamma_a, \Gamma_b]T^{-1} = -\frac{1}{2i}[\Gamma_a, \Gamma_b]$. From these properties of the Γ_a, Γ_{ab} matrices, a TR Hamiltonian $Th(\mathbf{k})T^{-1} = h(-\mathbf{k})$ must have the following properties:

$$d_a(-\mathbf{k}) = d_a(\mathbf{k}), \qquad d_{ab}(-\mathbf{k}) = -d_{ab}(\mathbf{k}). \tag{9.36}$$

Although the matrices used in the Kane and Mele paper [32,33] are representation dependent and will be changed in future chapters, the representation-independent statement is that out of the 15 $SU(4)$ generator matrices of a four-band Hamiltonian that includes spin, we can generically chose only 5 of them to be TR even, whereas the rest have to be TR odd.

For $\lambda_R = 0$, we saw this model is easy to diagonalize and gives two copies of the Haldane model and a pair of counter-propagating edge modes on *each edge* if diagonalized in a cylinder geometry that has open-boundary conditions on two edges. The gap is of order $|6\sqrt{3}\lambda_{SO} - 2\lambda_v|$. For $\lambda_v > 3\sqrt{3}\lambda_{SO}$, the system is an inversion-symmetry-breaking dominated phase and, if diagonalized so that the edge is of zig-zag type (see chap. 7), it will have edge modes connecting the cones (similar to the ones studied in the gapless graphene case of the previous chapters, but in this case they will be dispersive), but they will not cross the bulk gap (see fig. 9.1(b)). In contrast, for $\lambda_v < 2\sqrt{3}\lambda_{SO}$, we see that the system has edge modes crossing the bulk gap, which were easily understood from the continuum argument previously given. With $\lambda_R = 0$, this model has S_z conservation and we can easily (as we have) talk about spin$\uparrow$ and spin $\downarrow$ sectors. The symmetry then is $U(1) \times U(1)$, one $U(1)$ being the charge $U(1)$ and the other being the $U(1)$ of S_z conservation. With $\lambda_R \neq 0$, the S_z symmetry is broken, and we can no longer talk about two copies of fermions, one with spin $\uparrow$ and one with spin $\downarrow$, having separate Chern numbers. However, with $\lambda_R \neq 0$ but small so that we do not close the bulk gap, we observe that the edge modes do not disappear. When diagonalized in a cylindrical geometry (open boundary conditions in one direction, closed in the other), we find that edge modes traverse the bulk energy gap as long as λ_R has not closed the bulk gap.

Each edge has a *pair* of counterpropagating edge modes, which cross at some T-invariant point. This crossing is protected by T symmetry. As long as time reversal is preserved, every $\mathbf{k}$-point in the system must have a T orthogonal counterpart at $-\mathbf{k}$. The T-invariant points must each have two states, by Kramers' theorem. That means the gap can never open for a *single* pair of counterpropagating modes. For every edge in the system, we will observe a pair

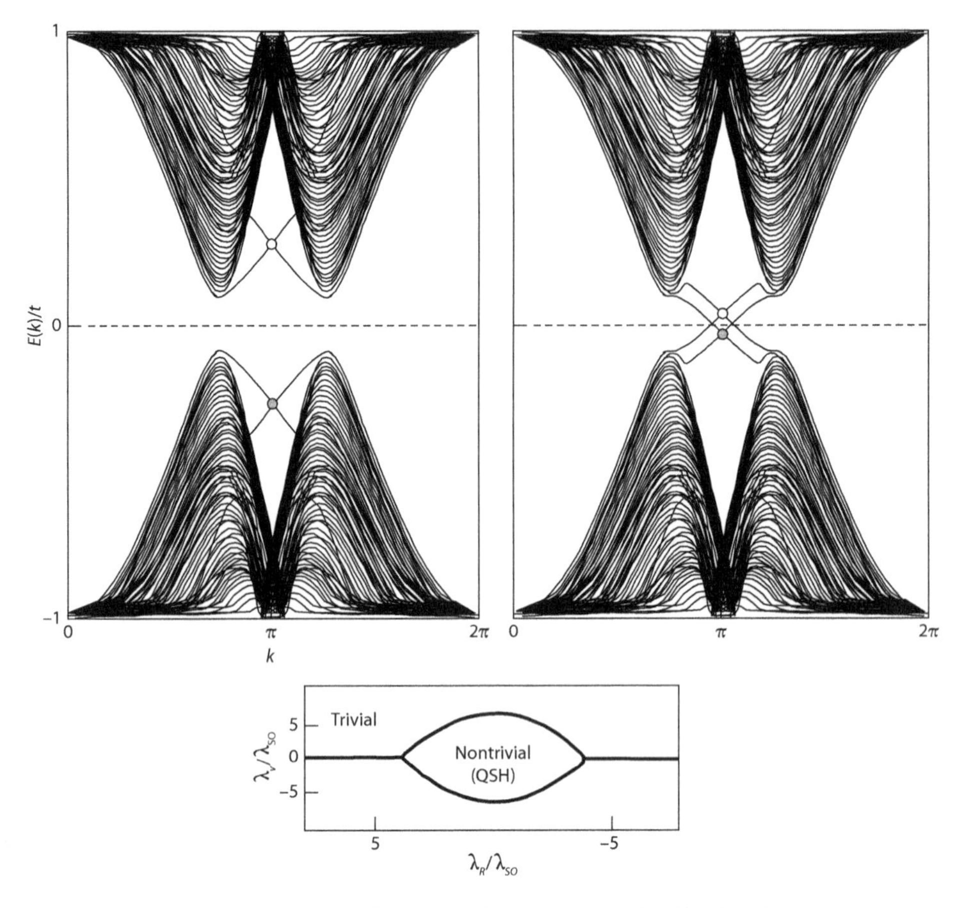

Figure 9.1. Energy bands for a one-dimensional zigzag strip in the (a)QSH phase $\lambda_v = 0.1$ t and (b) the insulating phase $\lambda_v = 0.4$ t. In both cases $\lambda_{SO} = 0.06$ t and $\lambda_R = 0.05$ t. The edge states on a given edge cross at $ka = \pi$.The inset show the phase diagram as a function of λ_v and λ_R for $0 < \lambda_{SO} \ll t$.

of edge modes. The dispersions of different edges can cross at some non-T-invariant point, as in fig. 9.1(a), but the matrix elements coupling these points are zero because they have to fully traverse the system length. Hence there is no one-body, T-invariant local perturbation term that can couple the two branches of a single pair of edge modes. The situation could not be more different for an even number of a pair of edge modes *on the same edge*. In this case, the modes can open gaps at the intersection points between the two pairs (which can be away from a T-invariant BZ point). T does not protect that gap opening because a gap opening at $k \neq 0$ can be accompanied by another gap opening at $-k \neq 0$, which can cancel the T breaking of the first. Different pairs of the two modes scatter between themselves and can open a gap because the matrix elements are on the same edge.

If S_z is a good quantum number, threading a flux through the system takes one spin $\uparrow$ from the left edge, A, to the right edge, B, and takes spin $\downarrow$ from edge B to edge A. Hence the system pumps quantized spin and has a quantized spin–Hall conductance of 2 (in units of e^2/h). When S_z is not a good quantum number, the system does not have quantized spin–Hall conductance, but spin still gets pumped. We will see in chapter 10 that the system has a quantized TR polarization.

9.3 First Topological Insulator: Mercury Telluride Quantum Wells

Unfortunately, the initial proposal [4] of a topological insulator in graphene was shown to be experimentally hard to achieve [42, 43]: the gap opened by the spin-orbit interaction turns out to be extremely small, of the order of 10^{-3} meV. From theoretical investigations of the type-III HgTe semiconductor quantum wells, it was then shown [7] that the topological insulating state should be realized in the "inverted" regime, where the quantum-well thickness d_{QW} is greater than a certain critical thickness d_c. The mechanism by which the topological insulator comes about, band inversion, was later shown to be generic [9]. Most topological insulators discovered today come about through band inversion.

9.3.1 Inverted Quantum Wells

We first try to give a general overview of the band structure for inverted quantum wells. The form of the Hamiltonian from which we will show the existence of a topological insulator in these quantum wells can be determined from symmetry requirements only, at least when expanded close to the Γ-point ($\mathbf{k} = 0$). The central feature of the type-III quantum wells is band inversion: the barrier material such as cadmium telluride (CdTe) has a normal semiconductor band progression (similar to, say, gallium arsenide (GaAs)), with the Γ_6 s-type band lying above the Γ_8 p-type band and the well material HgTe having an inverted band progression whereby the s-type Γ_6 band lies below the p-type Γ_8 band. The Γ_8 band is an angular momentum $\frac{3}{2}$ band (made out of the angular momentum 1 of the p-orbitals and electron spin $\frac{1}{2}$, which are coupled by spin-orbit coupling), and its projections on the S_z quantum number are $\pm\frac{1}{2}$ and $\pm\frac{3}{2}$. The Γ_6 band is a spin-$\frac{1}{2}$ band. In both these materials, the gap is the smallest near the Γ-point in the BZ (fig.9.2). Therefore, we shall restrict ourselves to a six-band model and start with the following six basic atomic states per unit cell combined into a six-component spinor: $\Psi = \left(|\Gamma_6, \frac{1}{2}\rangle, |\Gamma_6, -\frac{1}{2}\rangle, |\Gamma_8, \frac{3}{2}\rangle, |\Gamma_8, \frac{1}{2}\rangle, |\Gamma_8, -\frac{1}{2}\rangle|\Gamma_8, -\frac{3}{2}\rangle\right)$. This would be the band labeling in a three-dimensional material. In quantum wells grown in the [001] direction, the cubic, or spherical symmetry, is broken down to the axial rotation symmetry in the plane. These six bands combine to form the spin $\uparrow$ and spin-$\downarrow$ ($\pm$) states of three quantum well subbands that have been labeled $E1$, $H1$, $L1$ in the literature [44]. The $L1$ subband is separated from the other two [44], and we neglect it, leaving an effective four-band model for thin quantum wells. At the Γ-point with in-plane momentum $k_{\|} = 0$, m_J is still a good quantum number. At this point the $|E1, m_J\rangle$ quantum-well subband state is formed from the linear combination of the $|\Gamma_6, m_J = \pm\frac{1}{2}\rangle$ and the $|\Gamma_8, m_J = \pm\frac{1}{2}\rangle$ states, whereas the $|H1, m_J\rangle$ quantum-well subband state is formed from the $|\Gamma_8, m_J = \pm\frac{3}{2}\rangle$ states. Away from the Γ-point, the $E1$ and the $H1$ states can mix. Because the $|\Gamma_6, m_J = \pm\frac{1}{2}\rangle$ state has opposite parity from the $|\Gamma_8, m_J = \pm\frac{3}{2}\rangle$ state under 2-D spatial reflection, the coupling matrix element between these two states must be an odd function of the in-plane momentum k. This is also true by considering the fact that the coupling between an $m_J = \frac{3}{2}$ state and an $m_J = \frac{1}{2}$ state has to be an $\Delta L_z = 1$ coupling. The in-plane momentum $k_+ = k_x + ik_y$ provides that coupling. From these symmetry considerations, we deduce the general form of the effective Hamiltonian for the $E1$ and the $H1$ states, expressed in the basis of $|E1, m_J = \frac{1}{2}\rangle$, $|H1, m_J = \frac{3}{2}\rangle$ and $|E1, m_J = -\frac{1}{2}\rangle$, $|H1, m_J = -\frac{3}{2}\rangle$:

$$H_{\text{eff}}(k_x, k_y) = \begin{pmatrix} H(\mathbf{k}) & 0 \\ 0 & H^*(-\mathbf{k}) \end{pmatrix}, \quad H(\mathbf{k}) = \epsilon(\mathbf{k}) + d_i(\mathbf{k})\sigma_i, \tag{9.37}$$

where σ_i are the Pauli matrices. The form of $H^*(-\mathbf{k})$ in the lower block is determined from TR symmetry (as we will see in chap. 12) and $H^*(-\mathbf{k})$ is unitarily equivalent to $H^*(\mathbf{k})$ for this system. If inversion symmetry and axial symmetry around the growth axis are not broken,

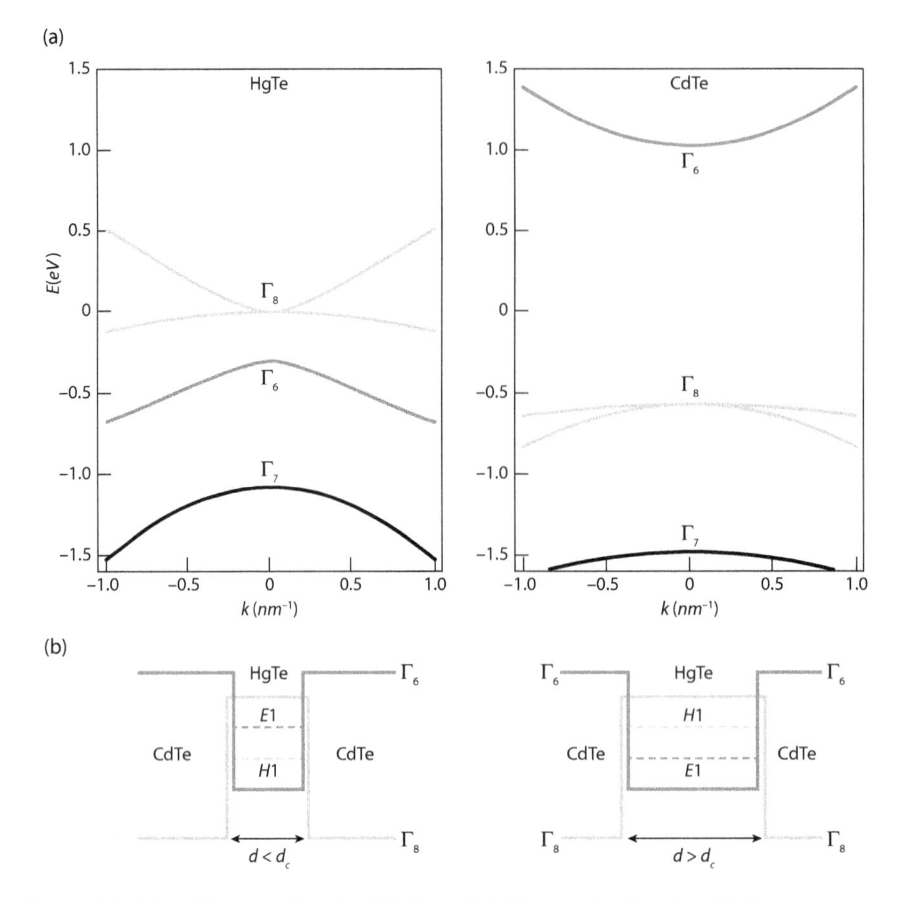

Figure 9.2. (A) Bulk energy bands of HgTe and CdTe near the Γ point. (B) The CdTe/HgTe/CdTe quantum well in the normal regime $E1 > H1$ with $d < d_c$ and in the inverted regime $H1 > E1$ with $d > d_c$. In this, and all subsequent figures, $\Gamma_8/H1$ ($\Gamma_6/E1$) symmetry is correlated with the color red (blue).

then the interblock matrix elements vanish, as presented. In real HgTe, however, inversion symmetry is softly broken, and off-diagonal terms are present. However, because they are small and do not close the bandgap, we can analyze the physics in the absence of these terms and argue by adiabatic continuity that the same topological physics remains valid when the inversion breaking-terms are added.

We see that, to the lowest order in k, the Hamiltonian matrix decomposes into 2×2 blocks, for each spin $\uparrow, \downarrow$. The preceding Hamiltonian has more symmetry than only inversion symmetry — in fact it has $U(1) \times U(1)$ symmetry, which are the two quantum Hall effects in the two spins separately. From the symmetry arguments given before, we deduce that $d_3(\mathbf{k})$ is an even function of k, whereas $d_1(\mathbf{k})$ and $d_2(\mathbf{k})$ are odd functions of k — they are $L_z = 1$ terms. Therefore, we can generally expand them in the following form:

$$d_1 + id_2 = A(k_x + ik_y) \equiv Ak_+,$$
$$d_3 = M - B(k_x^2 + k_y^2), \quad \epsilon_k = C - D(k_x^2 + k_y^2), \tag{9.38}$$

where the constants A, M, B, C, and D can be obtained from a first-principle (or perturbative) calculation.Their approximate values are given in Table 9.1, and the band dispersion is shown

TABLE 9.1
Parameters for $Hg_{0.32}Cd_{0.68}Te$/HgTe quantum wells.

d (nm)	A (eV)	B (eV)	C (eV)	D (eV)	M (eV)
5.8	−3.62	−18.0	−0.0180	−0.594	0.00922
7.0	−3.42	−16.9	−0.0263	0.514	−0.00686

in fig. 9.2 The Hamiltonian in the (2×2) subspace therefore takes the form of the $(2+1)$-D Dirac Hamiltonian, plus an $\epsilon(k)$ term, which drops out in the quantum Hall response. We discard this term from now on. Within each 2×2 subblock, the Hamiltonian has the Hall conductance

$$\sigma_{xy} = -\frac{1}{8\pi^2} \int \int dk_x dk_y \hat{\mathbf{d}} \cdot \left(\partial_{\mathbf{x}} \hat{\mathbf{d}} \times \partial_{\mathbf{y}} \hat{\mathbf{d}}, \right) \tag{9.39}$$

in units of e^2/h, where $\hat{\mathbf{d}}$ denotes the unit $\hat{d}_i(k)$ vector introduced in the Hamiltonian (eq. (9.37)). Although we are in the continuum, the Dirac Hamiltonian is regularized properly by the presence of the k^2 terms in the $d_3(\mathbf{k})$ term. However, if the k^2 terms were not present, the Dirac Hamiltonian would have Hall conductance $\frac{\text{sign}(M)}{2}$. The presence of the quadratic terms renders the Hall conductance an integer that is based on whether or not the vector $\hat{d}_3$ winds from $\mathbf{k} = 0$ to $\mathbf{k} = \infty$ from the north to the south pole of the S^2 sphere. If $M/B > 0$, then it is clear that d_3 points to one pole at $\mathbf{k} = 0$ and to another at $\mathbf{k} = \infty$. This is the continuum regularization of a Dirac Hamiltonian and will be, for example, used in the $p + ip$ model of a topological superconductor in chapter 16. The most important quantity in the Hamiltonian is the mass, or gap parameter, M, which is the energy difference between the $E1$ and $H1$ levels at the Γ-point. The overall constant C sets the zero of energy to be the top of the valence band of bulk HgTe. In a quantum-well geometry, the band inversion in HgTe necessarily leads to a level crossing at some critical thickness d_c of the HgTe layer. For thickness $d < d_c$, i.e., for a thin HgTe layer, the quantum well is in the "normal" regime, where the CdTe is predominant; hence, the band energies at the Γ-point satisfy $E(\Gamma_6) > E(\Gamma_8)$. For $d > d_c$ the HgTe layer is thick and the well is in the inverted regime where HgTe dominates and $E(\Gamma_6) < E(\Gamma_8)$. As we vary the thickness of the well, the $E1$ and $H1$ bands must, therefore, cross at some d_c, and the gap parameter M changes sign between the two sides of the transition (fig. 9.3). The form of the effective Dirac Hamiltonian and the sign change of M at $d = d_c$ for the HgTe/CdTe quantum wells deduced previously from general arguments is sufficient to conclude the existence of the quantum-spin Hall (QSH) state in this system. As we change the quantum-well thickness d across d_c, M changes sign, and the gap closes at the Γ point leading to a vanishing d_i $(\mathbf{k} = 0)$ vector at the transition point $d = d_c$. The sign change of M leads to a well-defined change of the Hall conductance $\Delta\sigma_{xy} = 1$ across the transition. So far, we have discussed only one 2×2 block of the effective Hamiltonian, H. General TR symmetry dictates that $\sigma_{xy}(H) = -\sigma_{xy}(H^*)$; therefore, the total charge Hall conductance vanishes, whereas the spin Hall conductance, given by the difference between the two blocks, is finite and is given by $\Delta\sigma_{xy}^{(s)} = 2$ (in units of e^2/h). From the general relationship between the quantized Hall conductance and the number of edge states, we conclude that the two sides of the phase transition at $d = d_c$ must differ in the number of pairs of helical edge states by 1. Thus, one side of the transition must be Z_2 odd and topologically distinct from a fully gapped conventional insulator. To find out which side must be topologically nontrivial, we have to compute the value of M/B, but the overwhelming evidence points to the inverted side being topologically nontrivial. Indeed, in the inverted gap regime where $\frac{M}{2B} = 2.02 \times 10^{-4}$ at 7.0 nm and not in the normal regime (where $\frac{M}{2B} < 0$), and the

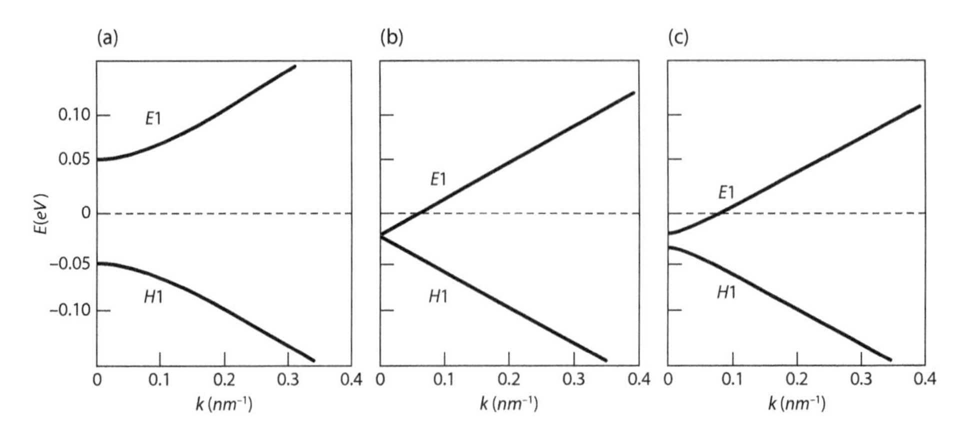

Figure 9.3. (A) Energy (eV) of $E1$ and $H1$ bands at $k_{\|} = 0$ versus. quantum-well thickness d (Å). (b) Energy dispersion relations $E(k_x, k_y)$ of the $E1$, $H1$ subbands at 4.0 nm,6.35 nm and 7.0 nm from left to right. Colored shading indicates the symmetry type of band at that k-point. Places where the cones are more red (blue) indicates that the dominant states are $H1(E1)$ states at that point. Purple shading is a region where the states are more evenly mixed in character. For 4.0 nm the lower (upper) band is dominantly $H1(E1)$. At 6.35 nm the bands are evenly mixed near the band crossing and retain their $d < d_c$ behavior moving further out in k-space. At $d = 7.0$ nm, the regions near $k_{\|} = 0$ have flipped their character but eventually revert back to the $d < d_c$ further out in k-space. Only this dispersion shows the meron structure. (c) Schematic meron configurations representing the $d_i(\mathbf{k})$ vector near the point Γ. The shading of the merons has the same meaning as the preceding dispersion relations. The change in meron number across the transition is exactly equal to 1, leading to a quantum jump of the spin Hall conductance $\Delta\sigma_{xy}^{(s)} = 2e^2/h$. All plots are for $Hg_{0.32}Cd_{0.68}Te/HgTe$ quantum wells.

inverted case is the topologically nontrivial regime supporting a QSH state. Figure 9.3 shows the energies of both the $E1$ and $H1$ bands at $k_{\|} = 0$ as a function of quantum-well thickness d obtained from our analytical solutions. At $d = d_c \sim 6.4$ nm, these bands cross.

9.4 Experimental Detection of the Quantum Spin Hall State

We now discuss the experimental detection of the QSH state. We want to focus only on purely electrical measurements because spin measurements are difficult to achieve. By sweeping the gate voltage, we can measure the two-terminal conductance G_{LR} from the p-doped to bulk-insulating to n-doped regime (figs. 9.4 and 9.5). In the bulk- insulating regime, G_{LR} vanishes at low temperatures for a normal insulator at $d < d_c$, whereas G_{LR} approaches a value close to $\frac{2e^2}{h}$ for $d > d_c$ and for samples shorter than the mean free inelastic scattering path. This has been observed in a remarkable experiment performed at the University of Wurzburg [8].

Strikingly, in a six-terminal measurement, the QSH state would exhibit vanishing electric voltage drop between the terminals μ_1 and μ_2 and between μ_3 and μ_4 in the zero-temperature limit and in the presence of a finite electric current between the L and R terminals. In other words, longitudinal resistance should vanish in the zero-temperature limit with a power-law dependence, over distances larger than the mean free path. Because of the absence of backscattering, in the absence of leads along the path between source and drain, the helical edge currents flow without dissipation, and the voltage drop occurs only at the drain side of the contact [45]. Finally, a spin filtered measurement can be used to determine the spin

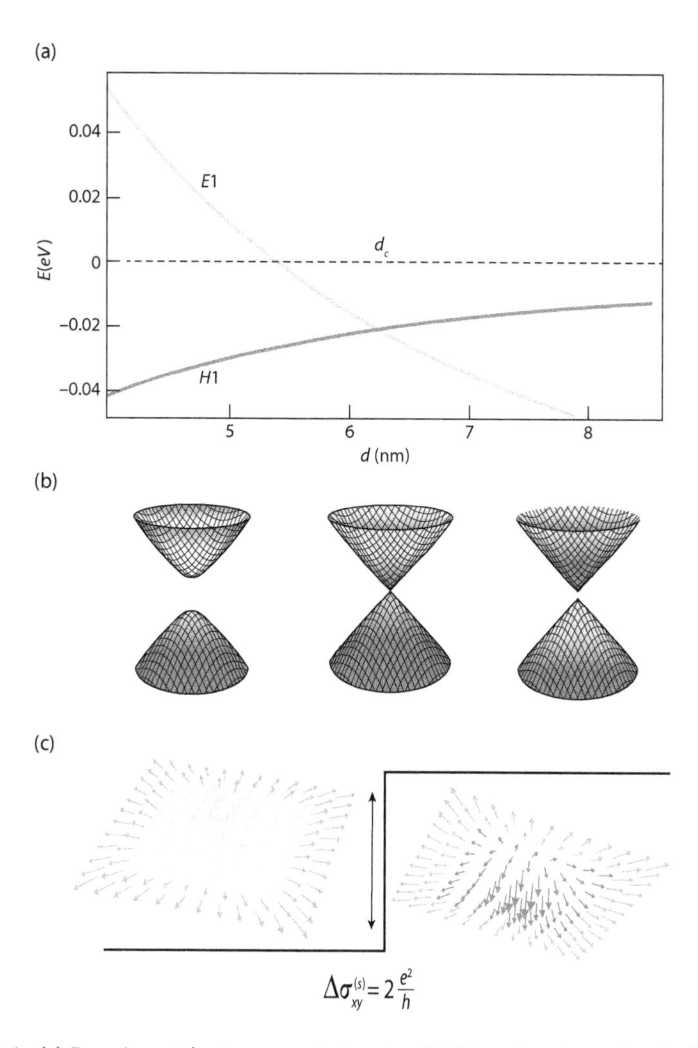

Figure 9.4. (a) Experimental setup on a six-terminal Hall bar showing pairs of edge states with spin-up (spin-down) states given in green (purple). (b) A two-terminal measurement on a Hall bar would give G_{LR} close to $2e^2/h$ contact conductance on the QSH side of the transition and zero on the insulating side. In a six-terminal measurement, the longitudinal voltage drops $\mu_2 - \mu_1$ and $\mu_4 - \mu_3$ vanish on the QSH side with a power law as the zero-temperature limit is approached. The spin Hall conductance $\sigma^{(s)}_{xy}$ has a plateau with the value close to $2(e^2/h)$.

Hall conductance $\sigma^{(s)}_{xy}$, although we expect this to be much more experimentally challenging. Numerical calculations [46] show that it should take a value close to $\sigma^{(s)}_{xy} = 2(e^2/h)$. This experiment has not yet been performed.

9.5 Problems

1. *Numerical Evaluation of the Kane-Mele Hamiltonian Spectrum:* Exactly diagonalize the Kane-Mele Hamiltonian for the topological insulator with a zigzag edge to see the evolution of the zero modes connecting the $\mathbf{K}$, $\mathbf{K}'$ Dirac cones upon opening the TR gap.

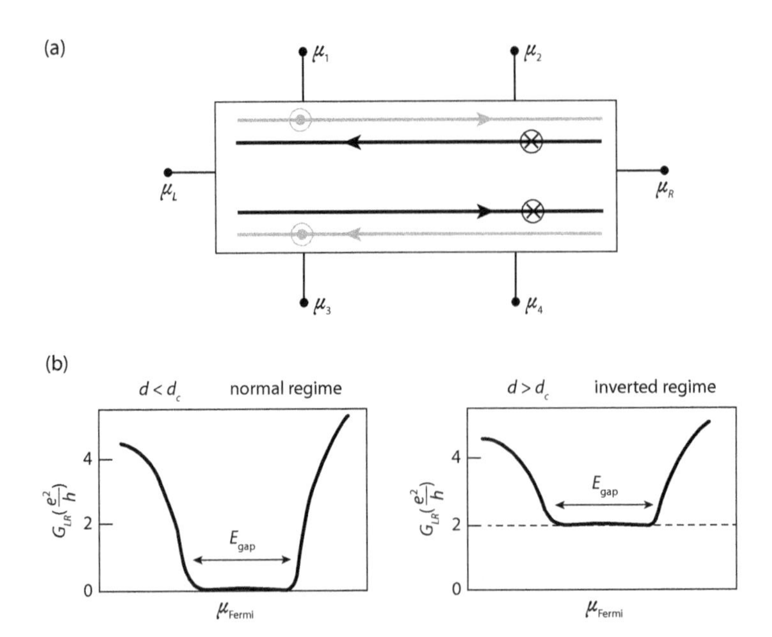

Figure 9.5. Dispersion relations for the $E1$ and $H1$ subbands for (a) $d = 4.0$ nm and (b) $d = 6.35$ nm, and (c) $d = 7.0$ nm.

2. *Clifford Algebra and Time Reversal:* Show that for a four-band system, with a Hamiltonian expandable in 4×4 matrices, we can choose only five of these matrices, even under time reversal $T^2 = -1$.

3. *HgTe Hamiltonian with Inversion Symmetry Breaking*: The HgTe system does not have inversion symmetry (even though the model used to exhibit a QSH effect in HgTe does). Add a small inversion-symmetry-breaking term (what is the most general form of such a term?) to the HgTe Hamiltonian and plot the phase diagram as a function of the inversion-symmetry-breaking term. What happens to the phase-transition point where the gap closes from the trivial side and reopens the topologically nontrivial side at $\mathbf{k} = 0$ upon the introduction of an inversion-symmetry-breaking term.

4. *HgTe in the Presence of a Magnetic Field*: What is the form of the HgTe Hamiltonian in an applied magnetic field perpendicular to the quantum well? In the presence of the edge states, find the form of the longitudinal conductance G_{xx} as a function of the temperature when the Fermi level is in the gap opened by the magnetic field in the edge modes at the Dirac point.

10

Z_2 Invariants

We have so far established that a new phase of matter exists in TR-invariant insulators, at least in two dimensions. The physical imprint of the state is the presence of gapless counterpropagating edge modes on each edge of a sample. These modes are protected from opening a gap by TR invariance if and only if there exist an odd number of pairs of them. That is, an even number of edge-state pairs are not protected from opening a gap. The edge-mode discussion suggests that there is a Z_2-type order in the QSH state. In this chapter we detail the original Kane and Mele [4, 5] calculation for a topological invariant. We look at their first definition of a Z_2 quantity, which differs from (but ends up being equivalent to) their more widely known second definition. We will prove that there exists a Z_2 topological invariant and that there is a topological classification of QSH insulators (2-D insulators) with two different phases, between which we cannot go without closing the bulk gap. Although this original invariant has been superceded by invariants based on sewing matrices, it is still useful to learn it because of its intuitive character. We consider only 4×4 Hamiltonians, which are the smallest generic Hamiltonians for T-invariant 2-D and 3-D topological insulators. Our discussion easily generalizes to Hamiltonians with more bands. We assume that the system has a gap, so we focus only on the occupied eigenstates. In the following, every time a band $|u_i(\mathbf{k})\rangle$ appears, we assume the i describes an *occupied* band.

10.1 Z_2 Invariant as Zeros of the Pfaffian

We would like to obtain an index whose behavior tells us about whether the system is a topological insulator or not. Because we know our index must contain the time-reversal operator, it is intuitive to consider the antisymmetric matrix of overlaps of the ith eigenstate with the time reversal of the jth eigenstate:

$$\begin{aligned}\langle u_i(\mathbf{k})| T |u_j(\mathbf{k})\rangle &= (u_i(\mathbf{k}))^\star_m U_{mn} (u_j(\mathbf{k}))^\star_n = -(u_j(\mathbf{k}))^\star_n U_{nm} K (u_i(\mathbf{k}))_m \\ &= - \langle u_j(\mathbf{k})| T |u_i(\mathbf{k})\rangle , \end{aligned} \tag{10.1}$$

where $T = UK$, with U a unitary, antisymmetric matrix (in order for $T^2 = -1$). For a 2×2 matrix, we hence have that

$$\langle u_i(\mathbf{k})| T |u_j(\mathbf{k})\rangle = \epsilon_{ij} P(\mathbf{k}), \tag{10.2}$$

where $P(\mathbf{k})$ is the Pfaffian of the matrix

$$P(\mathbf{k}) = \mathrm{Pf}[\langle u_i(\mathbf{k})| T |u_j(\mathbf{k})\rangle]. \tag{10.3}$$

For a 2×2 matrix, the Pfaffian picks the A_{12} component. The Pfaffian is not gauge invariant. Under a gauge transformation $|u_i'(\mathbf{k})\rangle = R_{ij}(\mathbf{k}) |u_j(\mathbf{k})\rangle$ (with R a unitary matrix), we have that

$$P'(\mathbf{k}) = \text{Pf}[\langle u'_i(\mathbf{k})| T |u'_j(\mathbf{k})\rangle] = \text{Pf}[R^\star_{im} \langle u_m(\mathbf{k})| T |u_n(\mathbf{k})\rangle (R^{\star T})_{nj}]$$
$$= \text{Det}[R^\star]P(\mathbf{k}), \tag{10.4}$$

where we have used the identity $\text{Pf}(R^\star AR^\dagger) = \text{Det}(R^\star)Pf(A)$. The Pfaffian is, hence, invariant under an $SU(2)$ rotation (which has determinant 1), but under a $U(1)$ transformation $U = e^{i\phi}$, it changes to $P'(\mathbf{k}) = e^{-2i\theta}P(\mathbf{k})$. Thus, the absolute value of the Pfaffian $|P(\mathbf{k})|$ is a gauge-invariant quantity.

We now want to define two different spaces of the Bloch wavefunctions $|u_i(\mathbf{k})\rangle$ on the BZ torus, which differ by their properties under time-reversal: an even subspace, for which the space spanned by $|u_i(\mathbf{k})\rangle$ is equivalent (up to a $U(2)$ transformation) to the space spanned by $T|u_i(\mathbf{k})\rangle$, and an odd subspace, for which they are orthogonal. In these subspaces, the Pfaffian index has absolute value either 1 or 0. The behavior of the Pfaffian in these subspaces will reveal the topological nature of the system.

10.1.1 Pfaffian in the Even Subspace

We first want to look at how the Hamiltonian $h(\mathbf{k})$ in equation (9.35) behaves at TR-invariant points $\mathbf{k} = \mathbf{G}/2$. Due to the antisymmetry of the $d_{ab}(\mathbf{k})$'s with respect to inverting $\mathbf{k}$, the Hamiltonian is

$$h\left(\frac{\mathbf{G}}{2}\right) = d_a\left(\frac{\mathbf{G}}{2}\right)\Gamma_a, \tag{10.5}$$

with the property $Th(\mathbf{G}/2)T^{-1} = h(-\mathbf{G}/2) = h(\mathbf{G}/2)$. Assume that at T invariant points, we have the eigenstate $|u_i(\mathbf{G}/2)\rangle$ of the Hamiltonian $h(\mathbf{G}/2)$ with energy $E_i(\mathbf{G}/2)$

$$h\left(\frac{\mathbf{G}}{2}\right)\left|u_i\left(\frac{\mathbf{G}}{2}\right)\right\rangle = E_i\left(\frac{\mathbf{G}}{2}\right)\left|u_i\left(\frac{\mathbf{G}}{2}\right)\right\rangle \tag{10.6}$$

At T-invariant points, if $|u_i(\mathbf{G}/2)\rangle$ is an eigenstate, then so is $T|u_i(\mathbf{G}/2)\rangle$, orthogonal to it. Consider the bands 1 and 2, which are necessarily degenerate at the T-invariant point, with energy $E_1 = E_2 = E$. Because we have only (in the generic case) two states and because we know that if one state, $|u_1(\mathbf{G}/2)\rangle$, is an eigenstate, then $T|u_1(\mathbf{G}/2)\rangle$ is also an eigenstate orthogonal to $|u_1(\mathbf{G}/2)\rangle$ at the same energy, it must be that $T|u_1(\mathbf{G}/2)\rangle \propto |u_2(\mathbf{G}/2)\rangle$. Hence, the T-invariant points belong to an "even" subspace, in which $T|u_i(\mathbf{G}/2)\rangle$ is equivalent to $|u_i(\mathbf{G}/2)\rangle$ up to a $U(2)$ rotation. Generically, in an even subspace we have

$$T|u_i(\mathbf{k})\rangle = M_{ij}|u_j(\mathbf{k})\rangle, \tag{10.7}$$

where M is a unitary matrix. The matrix $\langle u_i(\mathbf{k})| T |u_j(\mathbf{k})\rangle = M_{ij}$ contains only off-diagonal elements $i, j = 1, 2$. Because the matrix M is unitary, we have $M_{12}M^\star_{12} = 1$, so see that

$$|P(\mathbf{k})| = 1 \tag{10.8}$$

in the even subspace. Most importantly, $|P(\mathbf{k})| = 1$ at T-invariant points. In the Kane and Mele Hamiltonian, as presented, the even subspace has $d_{ab} = 0$.

10.1.2 The Odd Subspace

The odd subspace is defined by the fact that the subspace spanned by $T\left|u_i(\mathbf{k})\right\rangle$ is orthogonal to the subspace spanned by $\left|u_i(\mathbf{k})\right\rangle$. Shortly, we reveal some of its properties. The encompassing definition of the odd subspace involves the points $\mathbf{k}$ where the subspace spanned by $T\left|u_i(\mathbf{k})\right\rangle$ is orthogonal to the subspace spanned by $\left|u_i(\mathbf{k})\right\rangle$. In this subspace, we have that $\left\langle u_1(\mathbf{k})\right| T\left|u_2(\mathbf{k})\right\rangle = 0$, so that

$$P(\mathbf{k}) = 0. \tag{10.9}$$

If the Pfaffian has a zero at some $\mathbf{k}$, it will also have a zero at $-\mathbf{k}$. In fact, the Pfaffians at $\mathbf{k}$ and $-\mathbf{k}$ are related to each other. We have the following relation between the wavefunctions at $\mathbf{k}$ and the ones at $-\mathbf{k}$:

$$\left|u_j(-\mathbf{k})\right\rangle = B_{ji}(\mathbf{k})T\left|u_i(\mathbf{k})\right\rangle, \tag{10.10}$$

where $B_{ji}(\mathbf{k})$ is a unitary matrix at each k. B is called a sewing matrix. The preceding relation is valid for both inversion-symmetric and inversion-asymmetric cases. For the inversion-asymmetric cases, we can denote the generically spin-split bands by 1 and 2, and hence the matrix has only off-diagonal elements, but for inversion-symmetric cases, the matrix B can have diagonal elements as well because the bands are degenerate. After some algebra, we obtain

$$P(-\mathbf{k}) = Pf[\left\langle u_i(-\mathbf{k})\right| T\left|u_j(-\mathbf{k})\right\rangle] = \mathrm{Det}[B^*(\mathbf{k})]Pf[\left\langle u_{a_1}(\mathbf{k})\right| T\left|u_{a_2}(\mathbf{k})\right\rangle]^*. \tag{10.11}$$

Hence, if the Pfaffian at $\mathbf{k}$ vanishes, so does the one at $-\mathbf{k}$. Moreover, because $P(-\mathbf{k})$ is related to $P^*(\mathbf{k})$, the *phases* of the Pfaffian close to a degeneracy point $\mathbf{k}$ and $-\mathbf{k}$ advance in opposite order, which means that they have *opposite* vorticity:

$$v = \frac{1}{2\pi}\int_C d\mathbf{k} i\nabla \log(P(\mathbf{k})), \tag{10.12}$$

where the countour C surrounds the point $\mathbf{K}$ at which $P(\mathbf{K}) = 0$.

10.1.3 Example of an Odd Subspace: $d_a = 0$ Subspace

We now give examples of the odd subspace for the Kane and Mele QSH Hamiltonian defined in equation (9.35). It turns out that the subspace of k points for which $d_a = 0$'is *included* in the odd subspace but does *not* saturate it. For each point $\mathbf{k}$ where $d_a = 0$, we have that

$$H(\mathbf{k}) = -H(-\mathbf{k}). \tag{10.13}$$

Because of TR symmetry, however, $TH(\mathbf{k})T^{-1} = H(-\mathbf{k}) = -H(\mathbf{k})$. If at momentum k the two occupied states have energies $E_1 \leq E_2 \leq 0$, we have

$$H(\mathbf{k})\left|u_1(\mathbf{k})\right\rangle = E_1(\mathbf{k})\left|u_1(\mathbf{k})\right\rangle, \qquad H(\mathbf{k})\left|u_2(\mathbf{k})\right\rangle = E_2(\mathbf{k})\left|u_2(\mathbf{k})\right\rangle. \tag{10.14}$$

Then the TR partner of these states will have energies

$$\begin{aligned} H(\mathbf{k})T\left|u_1(\mathbf{k})\right\rangle &= -TH(\mathbf{k})\left|u_1(\mathbf{k})\right\rangle = -E_1(\mathbf{k})T\left|u_1(\mathbf{k})\right\rangle = E_3(\mathbf{k})T\left|u_1(\mathbf{k})\right\rangle, \\ H(\mathbf{k})T\left|u_2(\mathbf{k})\right\rangle &= -TH(\mathbf{k})\left|u_2(\mathbf{k})\right\rangle = -E_2(\mathbf{k})T\left|u_2(\mathbf{k})\right\rangle = E_4(\mathbf{k})T\left|u_2(\mathbf{k})\right\rangle, \end{aligned} \tag{10.15}$$

where, obviously, $E_3(\mathbf{k}) \geq E_4(\mathbf{k}) \geq 0$. The fact that $E_4(\mathbf{k}) > 0$ and $E_2(\mathbf{k}) < 0$ assures that the system is an insulator except at the $\mathbf{k} = \mathbf{G}/2$ points, where the $d_a = 0$ subspace would give a gapless Hamiltonian. Hence, because they are at different energies, we have that $\langle u_1(\mathbf{k}) | T | u_2(\mathbf{k}) \rangle = 0$ and, hence, $P(\mathbf{k}) = 0$ for all the $\mathbf{k}$ points for which $d_a = 0$. However, the odd subspace contains more $\mathbf{k}$-points than just those at which $H(\mathbf{k}) = -H(-\mathbf{k})$.

10.1.4 Zeros of the Pfaffian

We now claim that the zeros of the Pfaffian in *half* of the BZ are a topological invariant. We cut the BZ in half, making sure that the points $\mathbf{k}$, $-\mathbf{k}$ belong to different halves of the BZ. Assume that in *half* the BZ, we have one zero of the Pfaffian. This zero has a vorticity and hence cannot disappear directly—similar to the situation of the graphene Dirac node. In this case, we have a vortex in the Pfaffian. In graphene, the Bloch wavefunction vortex that gave the Dirac node could usually disappear in several different ways. The first is by breaking time-reversal or inversion. In the Pfaffian case studied here, we keep T unbroken, but we have broken inversion. Thus, we have no other symmetry to break—breaking T will destroy the vortex, but the classification we aim to obtain is for topological insulators with T symmetry, so we want to keep this unbroken. The classification in which we are interested does not require a preserved inversion symmetry. In graphene, the Dirac nodes could move towards a T-invariant point and annihilate there because they have opposite vorticity. In contrast, for our Pfaffian case, the zeros of the Pfaffian can move toward a T-invariant point (and that is the only place they could possibly annihilate if there is one zero in half the BZ), but it turns out that the zeros *cannot* annihilate. If they could, it would mean that we can have a zero of the Pfaffian exactly at the T-invariant point. However, we have proved in section 10.1.1 that the T-invariant point belongs to the *even* subspace, which has Pfaffian of unit modulus. Hence, one Pfaffian zero in half the BZ is stable globally.

An even number of zeros, of any vorticity, can be annihilated without joining at a T-invariant point. If we have two vortices of different vorticity in half a BZ, they can always meet up and annihilate. If they have identical vorticity signs, they can meet up with the zeros in the other half of the BZ and annihilate because in this case they do not need to meet up at a T-invariant point. An odd number of zeros can again annihilate two by two until the very last one, which cannot annihilate because of the reasons stated in the previous paragraph. To count the number of zeros, we can define the index

$$I = \frac{1}{2\pi i} \int_C d\mathbf{k} \cdot \nabla \log(P(\mathbf{k})), \tag{10.16}$$

where the contour C is half the BZ (defined in such a way that both $\mathbf{k}$ and $-\mathbf{k}$ are never both included in the same half of the BZ). If T is broken, then the zeros are no longer prevented from annihilating at a T-invariant k-point, and the topological distinction is lost. Note that we still have not proved that if you have an odd number of zeros in half the BZ, then the state is nontrivial (has edge states). We have showed only that the insulating state with T for half-integer spin has a topological classification due to the fact that an even number of zeros of the Pfaffian can annihilate, but an odd number cannot. It is now easy to connect the QSH state with the presence of an odd number of zeros of the Pfaffian in half the BZ. A trivial insulator is connected to the atomic limit. In the atomic limit, Bloch wavefunctions do not depend on momentum because the hoppings are zero. The Bloch wavefunctions are constant. Hence, the Pfaffian is always the same and equals 1 (a simple way to see this is that at the time-reversal momenta, we proved that the pfaffian is unity—but in the atomic limit, this should be the same everywhere in the BZ). Zeros of the Pfaffian can appear and disappear through bulk-phase transitions in which the bulk gap collapses. A single zero in half the BZ is obtained if the system undergoes a phase transition to a QSH state.

10.1.5 Explicit Example for the Kane and Mele Model

We now show the explicit case of the Pfaffian invariant in the Kane and Mele model. For simplicity, we look at the mirror-symmetric case $\lambda_R = 0$ and $d_3 = d_4 = d_{23} = d_{24} = 0$. We have

$$H(\mathbf{k}) = d_1(\mathbf{k})\Gamma^1 + d_2(\mathbf{k})\Gamma^2 + d_{12}(\mathbf{k})\Gamma^{12} + d_{15}(\mathbf{k})\Gamma^{15}. \tag{10.17}$$

At the $\mathbf{K}$-point, we have $d_1(\mathbf{K}) = d_{12}(\mathbf{K}) = 0$ and the Hamiltonian has a $U(1) \times U(1)$ conservation law and, hence, decomposes into spin $\uparrow$ and spin $\downarrow$ sub-blocks. Hence the spin $\uparrow$ Hamiltonian is

$$h_\uparrow(\mathbf{K}) = \begin{bmatrix} \lambda_v + 3\sqrt{3}\lambda_{SO} & 0 \\ 0 & -(\lambda_v + 3\sqrt{3}\lambda_{SO}) \end{bmatrix}, \tag{10.18}$$

whereas for spin $\downarrow$ we have

$$h_\downarrow(\mathbf{K}) = \begin{bmatrix} \lambda_v - 3\sqrt{3}\lambda_{SO} & 0 \\ 0 & -(\lambda_v - 3\sqrt{3}\lambda_{SO}) \end{bmatrix}. \tag{10.19}$$

We again start looking at $\lambda_v > 0$ large, for which we know that the phase cannot have anything nontrivial because as the system has just a large on-site energy difference between A-sites and B-sites. The occupied bands are

$$E_1 = -(\lambda_v + 3\sqrt{3}\lambda_{SO}), \qquad |u_1(\mathbf{k})\rangle = (0, 1, 0, 0)^T, \tag{10.20}$$

$$E_2 = -\lambda_v + 3\sqrt{3}\lambda_{SO}, \qquad |u_2(\mathbf{k})\rangle = (0, 0, 0, 1)^T, \tag{10.21}$$

The T matrix is

$$T = \begin{bmatrix} 0 & 0 & 1 & 0 \\ 0 & 0 & 0 & 1 \\ -1 & 0 & 0 & 0 \\ 0 & -1 & 0 & 0 \end{bmatrix} K, \tag{10.22}$$

and hence by direct computation, $\langle u_1(\mathbf{K})| T |u_2(\mathbf{K})\rangle = 1$. This shows that the Pfaffian can equal 1 even if the point K is not part of the even subspace (i.e., not all d_{ab} are zero here). However, if the point K is part of the even subspace, then the Pfaffian must have unit modulus.

Now decrease λ_v so that it becomes smaller than $3\sqrt{3}\lambda_{SO}$. At this point, there is a level crossing, and the upper occupied band at $\mathbf{K}$ changes from energy $-\lambda_v + 3\sqrt{3}\lambda_{SO}$ to energy $\lambda_v - 3\sqrt{3}\lambda_{SO}$. The occupied bands now are

$$E_1 = -(\lambda_v + 3\sqrt{3}\lambda_{SO}), \qquad |u_1(\mathbf{k})\rangle = (0, 1, 0, 0)^T, \tag{10.23}$$

$$E_2 = \lambda_v - 3\sqrt{3}\lambda_{SO}, \qquad |u_2(\mathbf{k})\rangle = (0, 0, 1, 0)^T, \tag{10.24}$$

and hence

$$\langle u_1(\mathbf{K})| \, T \, |u_2(\mathbf{K})\rangle = 0. \tag{10.25}$$

The space spanned by $|u_i(K)\rangle$ is orthogonal to that spanned by $T\,|u_i(K)\rangle$. Of course, by time reversal, a level crossing at K is accompanied by a level crossing at K', but the key is to look at level crossings in only *half* of the BZ. We see that a level crossing in half the BZ changes the Pfaffian at point **K** from 1 to 0. The system is in a topological (QSH) phase until it goes through a bulk phase transition again. We have now just proved that a topological phase transition with change of Pfaffian from unity to zero happens through a Dirac mass gap changing sign in *half* the BZ. This behavior is generic. The appearance of a Pfaffian zero in half the BZ is directly related to the switching of a Dirac fermion mass. We can find zeros of $P(\mathbf{k})$ by tuning k_x, k_y; hence, if they exist they are generically isolated points in the BZ. If they do not exist, we are in a trivial phase. If C_3 symmetry is present, the zeros of the Pfaffian can be only at the only C_3 symmetric places in the BZ, which are the $\mathbf{K}$, $\mathbf{K}'$ points.

10.2 Theory of Charge Polarization in One Dimension

Although the Pfaffian Z_2 invariant is useful for distinguishing different phases in 2-D TR-invariant insulators, it lacks the physical content that the Hall conductance displays in understanding the physical response of the Chern insulator. In other words, the Hall conductance is a topological invariant related to physical transport (Hall effect of charge) in a Chern insulator. In a TR insulator, what is the quantity being transported and how is it related to a Z_2 invariant? For this, it turns out we first need to go back and analyze the theory of charge polarization in insulators. This will then give us both the Hall conductance and the Z_2 TR polarization—the quantity being transported in a topological insulator. Our discussion follows that of of papers by Resta and Vanderbilt [47].

Let us consider a one-dimensional system with unit lattice constant and number of sites L. Let us also assume TR invariance for spin-$\frac{1}{2}$ systems—an insulator will have an even number, $2N$, of occupied bands. The normalized, occupied eigenstates can be written in terms of periodic Bloch functions as

$$|\psi_{n,k}\rangle = \frac{1}{\sqrt{L}} e^{ikx} \, |u_{n,k}\rangle \, . \tag{10.26}$$

Define the Wannier functions associated with each unit cell:

$$|R, n\rangle = \frac{1}{2\pi} \int dk e^{-ik(R-r)} \, |u_{n,k}\rangle \, , \tag{10.27}$$

where R is a lattice vector. The Wannier functions are not unique because $|u_{n,k}\rangle$ can be gauge transformed, and our preceding definition is a particular gauge choice. This gauge transformation multiplies the $|u_{n,k}\rangle$ by a possibly k-dependent phase. In addition to this phase multiplication, the many-body wavefunction—which is a Slater determinant of occupied states and hence does not care about their energies—is invariant under mixing by a $U(2N)$ transformation, $|u_{n,k}\rangle \to \sum_m U_{nm}(k)\,|u_{m,k}\rangle$. This transformation mixes the eigenstates of the Hamiltonian, so the gauge-transformed u's are not individual eigenstates. However the Slater determinant of the occupied bands is unchanged up to a phase, as can easily be seen by remembering that U_{mn} is a unitary matrix times a $U(1)$ phase. The total charge polarization (in units of the electron charge) for a 1-D system is given by the sum over all the bands of the center of charge of the Wannier states:

$$P = \sum_n \int dr \, \langle 0, n| \, r \, |0, n\rangle \, ; \tag{10.28}$$

hence, we find the explicit form

$$\begin{aligned} P &= \sum_n \frac{1}{(2\pi)^2} \frac{1}{L} \int dr \int \int dk_1 dk_2 e^{i(k_2 - k_1)r} \left\langle u_{n,k_1} \right| r \left| u_{n,k_2} \right\rangle \\ &= \sum_n \frac{1}{2\pi} \int dk_x i \left\langle u_{n,k_x} \right| \nabla_{k_x} \left| u_{n,k_x} \right\rangle , \end{aligned} \tag{10.29}$$

where, because we are in one dimension, we took $r = x$. The preceding formula immediately shows the relationship between charge polarization and the Berry phase:

$$P = \frac{1}{2\pi} \int dk_x A(k_x), \qquad A(k_x) = i \sum_n \left\langle u_{n,k_x} \right| \nabla_{k_x} \left| u_{n,k_x} \right\rangle . \tag{10.30}$$

The integral in the polarization can be taken over a closed contour in the BZ. Under a gauge transformation $|u(k)\rangle \to e^{i\phi} |u(k)\rangle$, in which the $U(1)$ phase advances by $2\pi m$ when k_x advances around the BZ, the polarization varies by

$$P \to P + m. \tag{10.31}$$

This is identical to the fact that the polarization is defined only up to a lattice vector (in a translational invariant system), as per the equation (10.28). We will come back to this and talk more about polarization later. Although the polarization is not a gauge-invariant quantity, changes in the polarization induced by adiabatic and continuous changes in the Hamiltonian through a closed cycle are gauge invariant. For example, let us consider a Hamiltonian that depends on an extra parameter k_y, $H = H[k_y]$, and assume that the wavefunctions $|u_{k,n}(t)\rangle$ are defined continuously between two momenta K_{y1} and K_{y2} (for all k_x); then, we may write

$$P[K_{y2}] - P[K_{y1}] = \frac{1}{2\pi} \left(\int_{c_2} dk_x A(k_x, k_y) - \int_{c_1} dk A(k_x, k_y) \right) , \tag{10.32}$$

where $c_{1,2}$ are the loops $k_x = -\pi \to \pi$ for $k_y = K_{y1}, K_{y2}$, respectively. We can use Stokes' theorem to write this as $\int_{K_{y1}}^{K_{y2}} dk_y \int dk_x F(k_x, k_y)$, where

$$F(k_x, k_y) = i \sum_n (\langle \nabla_{k_y} u_{k_x,n}(k_y) | \nabla_{k_x} u_{k_x,n}(k_y) \rangle - cc). \tag{10.33}$$

The space of k_x, k_y is that of a cylinder—k_x is periodic, whereas k_y takes values between K_{y1} and K_{y2}. The integral is over the cylinder spanned by k_x and k_y, and the sum is over all the n occupied bands. We can relabel the argument as an index $u_{k_x,n}(k_y) = u_{k_x,k_y,n}$. We can now close the cylinder into a torus by making $K_{y1} = K_{y2} + 2\pi$. For a periodic cycle $H[k_y + 2\pi] = H[k_y]$, the change in polarization is given by the integral over the torus. This is nothing but the Chern number. It characterizes the charge *difference* that is pumped in each cycle. As proved before in this book, $F(-k_x, -k_y) = -F(k_x, k_y)$ for a TR-invariant system, and hence the Chern number of a T system is zero.

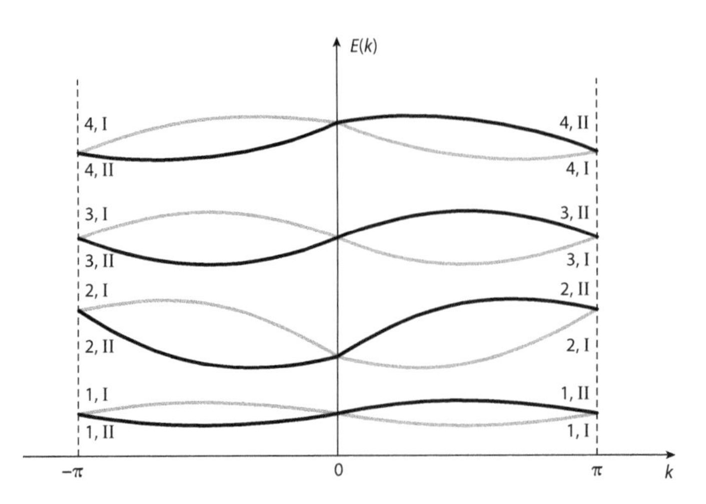

Figure 10.1. 1-D band structure with time reversal and no other symmetry. The bands are spin split.

10.3 Time-Reversal Polarization

We have so far defined the charge polarization, which is the sum of the Berry potential integrated over the BZ. The difference in the charge polarization over the full BZ is the Hall conductance. We would now like to understand what modification we should make to the system to introduce the information about TR invariance. We still consider a 1-D band structure, but we now add the information about the fact that we have a spin-$\frac{1}{2}$ system with TR invariance. Based on intuition, this means that we must not look at the full BZ when we compute properties of the system but at only half of the zone. Alternatively, because the energy bands come in pairs, we can basically obtain the full properties of a system if we look at only one of the bands in a Kramers' pair over the full BZ. Analyzing one of the bands in a Kramers' pair over the whole BZ allows us to infer information about the other band using time reversal. The bottom line is that the strategy of attack for the TR case is to look at *halves*: either all the bands in half the BZ or half of the bands in the full BZ. We now formalize such a strategy.

We assume for simplicity that no accidental degeneracies occur; if they do, we can always separate the bands at almost every k so that we remove them. The bands can then be numbered as in figure 10.1, where roman indices I, II denote the Kramers' pairs and $\alpha = 1, 2, \ldots, N$ denotes the non-Kramers' pair index. This is a specific numbering of the bands—we will soon replace it with a sewing matrix formalism, which is more versatile and easily extendable to other symmetries and also more mathematical in nature. Time reversal transforms eigenstates at k of bands I into eigenstates at $-k$ of bands II, and vice versa, but only up to a phase factor:

$$\left|u^{\mathrm{I}}_{-k,\alpha}\right\rangle = e^{i\chi_{k,\alpha}} T \left|u^{\mathrm{II}}_{k,\alpha}\right\rangle \tag{10.34}$$

The TR operator acting on this then gives

$$T\left|u^{\mathrm{I}}_{-k,\alpha}\right\rangle = Te^{i\chi_{k,\alpha}} T \left|u^{\mathrm{II}}_{k,\alpha}\right\rangle = e^{-i\chi_{k,\alpha}} T^2 \left|u^{\mathrm{II}}_{k,\alpha}\right\rangle = -e^{-i\chi_{k,\alpha}} \left|u^{\mathrm{II}}_{k,\alpha}\right\rangle ; \tag{10.35}$$

hence,

$$\left|u^{\mathrm{II}}_{-k,\alpha}\right\rangle = -e^{i\chi_{-k,\alpha}} T \left|u^{\mathrm{I}}_{k,\alpha}\right\rangle . \tag{10.36}$$

We notice that the preceding representation is not explicitly invariant under the gauge transformation $|u_{k,n}\rangle \to \sum_m U_{nm}(k)\,|u_{k,m}\rangle$. However, for the purpose of our calculation, we will soon transform it into such an invariant representation. As we said before in our general strategy, we would like to look at the polarization of *one* member of the TR Kramers' pair, and we define

$$P^s = \frac{1}{2\pi}\int_{-\pi}^{\pi} dk A^s(k), \qquad A^s(k) = i\sum_a \langle u^s_{k,a}|\,\nabla_k\,|u^s_{k,a}\rangle\,, \tag{10.37}$$

where $s = \mathrm{I}, \mathrm{II}$ is the index of the TR pair. This quantity is the polarization of band s. For one of the bands, it is the polarization of half of the Kramers' pair over the full BZ. It is defined up to a multiple of the lattice constant—under changes in the phases (gauge) of $|u^s_{k,a}\rangle$. However, if we look at one specific TR band, I ($s = \mathrm{I}$), the polarization seems to depend on the specific choice of the labels I, II—it seems to break the $U(2N)$ symmetry of the many-body wavefunction. We now show this is not so. The trick is to separate the integral of the P^s (eq. (10.37)) into two integrals each over half of the BZ and then relate one to the other by the TR equations. In other words,

$$P^{\mathrm{I}} = \frac{1}{2\pi}\int_0^{\pi} (A^I(k) + A^I(-k)). \tag{10.38}$$

We now relate $A^{\mathrm{I}}(-k)$ to $A^{\mathrm{II}}(k)$ through the TR equations

$$\begin{aligned} A^{\mathrm{I}}(-k) &= -i\sum_a \langle u^{\mathrm{I}}_{-k,a}|\,\nabla_k\,|u^{\mathrm{I}}_{-ka}\rangle \\ &= -i\sum_a \langle Tu^{\mathrm{II}}_{k,a}|\,\nabla_k\,|Tu^{\mathrm{II}}_{ka}\rangle - i\sum_a i\nabla_k\chi_{k,a}\langle Tu^{\mathrm{II}}_{k,a}|Tu^{\mathrm{II}}_{ka}\rangle \\ &= -i\sum_a \langle Tu^{\mathrm{II}}_{k,a}|\,\nabla_k\,|Tu^{\mathrm{II}}_{ka}\rangle + \sum_a \nabla_k\chi_{k,a} \end{aligned} \tag{10.39}$$

The sign of the last term in the preceding expression differs from that of the Fu and Kane (eq. (3.14) of [14]). We used the fact that $\langle Tu^{\mathrm{II}}_{k,a}|Tu^{\mathrm{II}}_{ka}\rangle = \langle Tu^{\mathrm{II}}_{k,a}|\, e^{-i\chi_{k,a}}e^{i\chi_{k,a}}\,|Tu^{\mathrm{II}}_{ka}\rangle = \langle u^I_{-k,a}|u^I_{-ka}\rangle = 1$. We also obtain (by letting $T = UK$, with U unitary)

$$\begin{aligned} \langle Tu^{\mathrm{II}}_{k,a}|\,\nabla_k\,|Tu^{\mathrm{II}}_{ka}\rangle &= (U_{mn}u^{\mathrm{II}*}_{k,a,n})^*\nabla_k U_{mp}u^{\mathrm{II}*}_{k,a,p} = (U^\dagger)_{nm}U_{mp}u^{\mathrm{II}}_{k,a,n}\nabla_k u^{\mathrm{II}*}_{k,a,p} \\ &= u^{\mathrm{II}}_{k,a,n}\nabla_k u^{\mathrm{II}*}_{k,a,n} = -u^{\mathrm{II}*}_{k,a,n}\nabla_k u^{\mathrm{II}}_{k,a,n} = -\,\langle u^{\mathrm{II}}_{k,a}|\,\nabla_k\,|u^{\mathrm{II}}_{k,a}\rangle \end{aligned} \tag{10.40}$$

Hence, we finally have

$$A^{\mathrm{I}}(-k) = i\sum_a \langle u^{\mathrm{II}}_{k,a}|\,\nabla_k\,|u^{\mathrm{II}}_{k,a}\rangle + \sum_a \nabla_k\chi_{k,a} = A^{\mathrm{II}}(k) + \sum_a \nabla_k\chi_{k,a}. \tag{10.41}$$

We can then use equation (10.41) to trade part of the polarization of one of the bands in the Kramers' pair for that of the other band. We find

$$\begin{aligned} P^{\mathrm{I}} &= \frac{1}{2\pi}\left(\int_0^{\pi} (A^{\mathrm{I}}(k) + A^{\mathrm{II}}(k)) + \int_0^{\pi}\sum_a \nabla_k\chi_{k,a}\right) \\ &= \frac{1}{2\pi}\left(\int_0^{\pi} A(k) + \sum_a(\chi_{\pi,a} - \chi_{0,a})\right), \end{aligned} \tag{10.42}$$

where $A(k) = A^{\mathrm{I}}(k) + A^{\mathrm{II}}(k)$ is the full Berry connection, the sum of the Berry potential of bands I and II. The last term can be written in a very nice way by introducing the $U(2N)$ *sewing matrix*:

$$B_{mn}(k) = \langle u_{-k,m} | \, T \, | u_{k,n} \rangle \, . \tag{10.43}$$

Notice that this is *not* equal to the matrix we gave earlier to obtain the Pfaffian Z_2 index. The current matrix is *not* antisymmetric at all points in the BZ, but it is unitary, as we show in section 10.3.2. The previous matrix, which was $\langle u_{k,m} | \, T \, | u_{k,n} \rangle$, is antisymmetric at all points in the BZ, but it is not unitary. The unitarity of the matrix B and the fact that it relates states at k with ones at $-k$ through time reversal make this sewing matrix one of the most important mathematical constructs in the theory of topological insulators. In the representation $|u^{\mathrm{I}}_{-k,a}\rangle = e^{i\chi_{k,a}} T \, |u^{\mathrm{II}}_{k,a}\rangle$, $|u^{\mathrm{II}}_{-k,a}\rangle = -e^{i\chi_{-k,a}} T \, |u^{\mathrm{I}}_{k,a}\rangle$ we have that

$$B^{\mathrm{II,I}}_{mn}(k) = \langle u^{\mathrm{II}}_{-k,m} | \, T \, | u^{\mathrm{I}}_{k,n} \rangle = -\delta_{mn} e^{-i\chi_{-k,n}}, \qquad B^{\mathrm{I,II}}_{mn}(k) = -B^{\mathrm{II,I}}_{mn}(-k) \tag{10.44}$$

The matrix $B^{\mathrm{I,II}}_{mn}$ is the sewing matrix between the Kramers' pair bands I and II. There could be multiple such Kramers' pairs, indexed by $m = 1, 2, \ldots, N$, but the sewing matrix couples the Kramers' pairs only two by two. Hence, the matrix B becomes a block-diagonal matrix of the subblock matrices:

$$\begin{bmatrix} 0 & -e^{-i\chi_{-k,n}} \\ e^{-i\chi_{k,n}} & 0 \end{bmatrix} . \tag{10.45}$$

We see that irrespective of $\chi_{k,n}$, at $k = 0, \pi$, the preceding matrix is antisymmetric (and only at $k = 0, \pi$). In section 10.3.1 we show this without going to a specific representation of the bands. We have that

$$\sum_a (\chi_{\pi,a} - \chi_{0,a}) = i \log \left[\frac{Pf[B(\pi)]}{Pf[B(0)]} \right] , \tag{10.46}$$

where we differ from [14] by another sign that cancels the previous sign difference in equation (10.39) to give matching final results. Hence, for the polarization of band I, we obtain

$$P^{\mathrm{I}} = \frac{1}{2\pi} \left(\int_0^{\pi} dk A(k) + i \log \left[\frac{Pf[B(\pi)]}{Pf[B(0)]} \right] \right) . \tag{10.47}$$

If we add the polarization of the bands I and II, we obtain the full-charge polarization:

$$P = \frac{1}{2\pi} \int_{-\pi}^{\pi} dk A(k) = P^{\mathrm{I}} + P^{\mathrm{II}} . \tag{10.48}$$

As mentioned before, the change in this polarization as we undertake a cycle of k_y from 0 to 2π is zero because the system has no Chern number. However, we want to define the TR polarization $P_T = P^{\mathrm{I}} - P^{\mathrm{II}} = 2P^{\mathrm{I}} - P$:

$$P_T = \frac{1}{2\pi} \left(\int_0^{\pi} dk \; A(k) - \int_{-\pi}^{0} dk A(k) + 2i \log \left[\frac{\mathrm{Pf}[B(\pi)]}{\mathrm{Pf}[B(0)]} \right] \right) . \tag{10.49}$$

We would now like to obtain the difference between $A(k)$ and $A(-k)$ in order to massage the first term in a better form. To do this, it is efficient to now move to a representation-independent formulation, in which we do not assume that the sewing matrix B is formed of the phases $\chi_{k,n}$ present so far.

10.3.1 Non-Abelian Berry Potentials at k, $-k$

If we apply the TR operator to a band $|u_{k,\beta}\rangle$, we will get another state of equal energy, but at momentum $-k$. If the system is also inversion symmetric, then we have a manifold of doubly degenerate states at each k, but in either case the symmetry gives

$$|u_{-ka}\rangle = \sum_{\beta} B^*_{\alpha\beta}(k) T \, |u_{k,\beta}\rangle \, , \tag{10.50}$$

where B is the k-dependent sewing matrix and α, β are band indices, which here run from 1 to $2N$. Each band eigenstate $|u_{k,\alpha}\rangle$ is a vector of components $(u_{k\alpha})_n$, where n now runs over all components of the vector $|u_{k,\alpha}\rangle$. Note that we have not made the $U(2N)$-breaking gauge choice of labeling Kramers' doublets as I, II. The matrix B is unitary, is explicitly equal to $\langle u_{-k,\alpha}| \, T \, |u_{k,\beta}\rangle = B_{\alpha\beta}(k)$, and has the property that

$$B_{\alpha\beta}(k) = -B_{\beta\alpha}(-k). \tag{10.51}$$

We would now like to relate the Berry phase potential at k with the one at $-k$. Because we are in a multiband system, we can write an identity for the *non-Abelian* Berry vector potential as

$$\begin{aligned} a_i^{\alpha\beta}(-k) &= i \, \langle u_{-k,\alpha}| \, \partial_{-k_i} \, |u_{-k,\beta}\rangle \\ &= -i \sum_{m\theta\gamma} B_{\alpha\theta}(k) B^*_{\beta\gamma}(k) (u_{k\theta})_n U^*_{mn} U_{mp} \partial_k (u_{k\gamma})^*_p \\ &\quad + B_{\alpha\theta}(k) (\partial_k B^*_{\beta\gamma}(k)) (u_{k\theta})_n U^*_{mn} U_{mp} (u_{k\gamma})^*_p \quad (\text{because } \; U^*_{mn} U_{mp} = \delta_{np}) \\ &= -i \sum_{n,\gamma,\theta} B_{\alpha\theta}(k) B_{\beta,\gamma}(k)^* (u_{k\theta})_n \partial_k (u_{k\gamma})^*_n - B_{\alpha\theta}(k) (\partial_k B_{\beta,\gamma}(k)^*) (u_{k\theta})_n (u_{k\gamma})^*_n \\ &= B_{\alpha\theta}(k) (-i (u_{k\theta})_n \partial_k (u_{k\gamma})^*_n) B^*_{\beta\gamma} - i B_{\alpha\theta} \partial_k B^*_{\beta\theta} \end{aligned} \tag{10.52}$$

because the non-Abelian vector potential at k reads $a_i^{\gamma\theta}(k) = i \, \langle u_{k\gamma}| \, \partial_k \, |u_{k\theta}\rangle = i(u_{k,\gamma})^*_n \partial_k (u_{k\theta})_n$, we see that the first term of the last line in equation (10.52) contains the complex conjugate of the Berry potential at k, namely, $a_i^{\theta\gamma *}(k) = -i(u_{k,\theta})_n \partial_k (u_{k\gamma})^*_n$. We finally obtain that the non-Abelian vector potential at $-k$ is a non-Abelian gauge transformation of the complex conjugate of the non-Abelian vector potential at k, with the gauge-transformation matrix being B, $a_i^{\alpha\beta}(-k) = B_{\alpha\theta}(k) a_i^{\theta\gamma *}(k) B^*_{\beta\gamma} - i B_{\alpha\theta} \partial_k B^*_{\beta\theta}$, or, in matrix form,

$$a_i(-k) = B(k) a_i^*(k) B^\dagger(k) - i B \partial_i B^\dagger. \tag{10.53}$$

The Abelian Berry phase is just a sum over the diagonal components of the non-Abelian potential; in other words,

$$\begin{aligned} A_i(-k) &= \text{Tr}[a_i(-k)] = \text{Tr}[B^\dagger(k) B(k) a_i^*(k)] - \text{Tr}[i B \partial_i B^\dagger] \\ &= \text{Tr}[a_i(k)] + i \text{Tr}[B^\dagger(k) \partial_i B] = A_i(k) + i \text{Tr}[B^\dagger(k) \partial_i B]. \end{aligned} \tag{10.54}$$

To identify the Abelian potential with its complex conjugate, we remember that the diagonal part of the Berry potential is real: $a_i^{\alpha\alpha *} = (i a^*_m \partial_k a_m)^* = i a^*_m \partial_k a_m = a_i^{\alpha\alpha}$ We hence find that the difference of the Abelian potential at k and $-k$ is just $\text{Tr}[B^\dagger(k) \partial_i B] = (A_i(-k) - A_i(k))/i$. We have thus found the relation between the Berry potentials at k and $-k$ and the sewing matrix.

10.3.2 Proof of the Unitarity of the Sewing Matrix B

We have repeatedly used the fact that the matrix B is unitary to obtain the relation between the vector potential at k and $-k$. It is now time to prove such a relation:

$$B^*_{\alpha\theta} B_{\alpha\beta} = \delta_{\theta\beta}, \tag{10.55}$$

where double index implies summation and the matrix B can be written by components as $B_{\alpha\beta} = (u_{-k,\alpha})^*_m U_{mn} (u_{k,\beta})^*_n$. Then,

$$\begin{aligned} B^*_{\alpha\theta} B_{\alpha\beta} &= [(u_{-k,\alpha})_p (u_{-k,\alpha})^*_m] U^*_{pr} (u_{k,\theta})_r U_{mn} (u_{k,\beta})^*_n \\ &= U^*_{pr} (u_{k,\theta})_r U_{pn} (u_{k,\beta})^*_n = (U^\dagger)_{rp} U_{pn} (u_{k,\theta})_r (u_{k,\beta})^*_n \\ &= \delta_{rn} (u_{k,\theta})_r (u_{k,\beta})^*_n = \delta_{\theta\beta}. \end{aligned} \tag{10.56}$$

Note that in the preceding we have summed over *all* α's, including both the occupied and unoccupied bands (we indeed used the full completeness relation). Because we make the assumption of no accidental degeneracies between members of the same Kramers' multiplet, the B matrix takes a block-diagonal form made out of N 2×2 matrices, and if the preceding is true for the full matrix, it will be true for the matrix of occupied states as well.

10.3.3 A New Pfaffian Z_2 Index

We now turn back to the TR polarization (eq. (10.49)):

$$\begin{aligned} P_T &= \frac{1}{2\pi} \left(\int_0^\pi dk A(k) - \int_{-\pi}^0 dk A(k) + 2i \log \left[\frac{\text{Pf}[w(\pi)]}{\text{Pf}[w(0)]} \right] \right) \\ &= \frac{1}{2\pi i} \left(\int_0^\pi dk \, \text{Tr}[B^\dagger \nabla_k B] - 2 \log \left[\frac{\text{Pf}[B(\pi)]}{\text{Pf}[B(0)]} \right] \right) \end{aligned} \tag{10.57}$$

The formula is now in a more $U(2N)$-symmetric form than the one we started with for P^{I}. We notice that $\text{Tr}[B^\dagger \nabla_k B]$ actually determines the $U(1)$ phase of the B_{mn}. In the representation chosen before, this can be proved by a brute-force calculation of the phase knowing that B_{mn} is a matrix block diagonal:

$$\begin{aligned} \text{Tr}[B^\dagger \nabla_k B] &= \text{Tr} \left[\begin{bmatrix} 0 & -e^{-i\chi_{-k,n}} \\ e^{-i\chi_{k,n}} & 0 \end{bmatrix} \begin{bmatrix} 0 & -\nabla_k e^{-i\chi_{-k,n}} \\ \nabla_k e^{-i\chi_{k,n}} & 0 \end{bmatrix} \right] \\ &= -i(\nabla_k \chi_k + \nabla_k \chi_{-k}) = \nabla_k \log(\text{Det}[B(k)]). \end{aligned} \tag{10.58}$$

We can actually prove this through a more general method that is representation independent. For a unitary matrix $B = e^{i\alpha} e^{i a_a x_a}$, where x_a is one of the Pauli generators for our block-diagonal case but can be, in general, an $SU(2N)$ traceless generator. By expanding the $SU(N)$ exponential and taking limits, we can prove the formula

$$\frac{\partial}{\partial a_b} e^{i a_a x_a} = \int_0^1 ds e^{i s a_a x_a} (i x_b) e^{i(1-s) a_c x_c}; \tag{10.59}$$

hence,

$$\text{Tr}[U^\dagger \partial_k U] = i \partial_k \alpha + \text{Tr}[e^{-i a_m x_m} \partial_k e^{i a_a x_a}]. \tag{10.60}$$

The second term is zero because

$$\begin{aligned}\mathrm{Tr}[e^{-ia_m x_m}\partial_k e^{ia_a x_a}] &= \frac{\partial a_b}{\partial k}\mathrm{Tr}[e^{-ia_m x_m}\int_0^1 ds e^{isa_a x_a}(ix_b)e^{i(1-s)a_c x_c}] \\ &= \frac{\partial a_b}{\partial k}\int_0^1 ds\ \mathrm{Tr}[e^{i(1-s)a_c x_c}e^{-ia_m x_m}e^{isa_a x_a}(ix_b)] = \frac{\partial a_b}{\partial k}\int_0^1 ds\ \mathrm{Tr}[(ix_b)].\end{aligned} \tag{10.61}$$

Hence, only the $U(1)$ part remains. We thus reach the following formula for the TR polarization:

$$P_T = \frac{1}{2\pi i}\left(\int_0^\pi dk\nabla_k \log[\mathrm{Det}[B(k)]] - 2\log\left[\frac{Pf[B(\pi)]}{Pf[B(0)]}\right]\right) \tag{10.62}$$

which is expressed only in terms of the winding of the phase of the matrix B between 0 and π. By using $\mathrm{Det}[B] = \mathrm{Pf}[B]^2$, we hence see that the TR polarization is

$$P_T = \frac{1}{\pi i}\log\left[\frac{\sqrt{\mathrm{Det}[B(\pi)]}}{\mathrm{Pf}[B(\pi)]}\frac{\mathrm{Pf}[B(0)]}{\sqrt{\mathrm{Det}[B(0)]}}\right] \tag{10.63}$$

The integer is defined modulo 2 because of the ambiguity of the logarithm. In other words, a gauge transformation will change the value of P_T only by an even integer. As such, even and odd values of P_T will go to even and odd values of P_T under a gauge transformation. This is intrinsically related to the fact that time-reversal symmetry exists in the system. Even and odd integers determine whether $\mathrm{Pf}(B(k))$ is on the same or opposite branch of $\sqrt{\mathrm{Det}[B(k)]}$ at $k = 0$ and π. The TR polarization is another nontrivial Z_2 index characterizing a TR insulator. We now analyze it further.

The TR polarization in equation (10.61) is defined for a 1-D system: it tells how the difference in the polarization of the elements of a Kramers' pair behave in a 1-D system. It depends only on the sewing matrices at the points 0 and 2π. However, the TR topological insulators are materials that also exist in two and three dimensions. In the case of the Hall conductance, in order to define the Chern number as an invariant in a pumping cycle, we looked at how the charge polarization varied in a pumping cycle from 0 to a period T when a periodic parameter was varied. In anticipation of the fact that the periodic parameter is another momentum, we can also denote this periodic parameter by k_y, the momentum in another direction. In a pumping cycle (between 0 and T—or between 0 and 2π if we are to talk about k_y), a TR Hamiltonian occurs at $t = 0$ and $t = T/2$ (or $k_y = 0, \pi$). This is clear because only these two points map into themselves under time reversal. When we consider the evolution of the Hamiltonian through the cycle, the change in TR polarization in half a cycle defines a Z_2 topological invariant.

The Z_2 invariant is related to the center of the occupied Wannier orbitals as a function of t. At $t = 0, T/2$, TR symmetry requires that the Wannier states come in TR pairs—the center of the occupied Wannier orbitals for each partner (the eigenvalues of the position operator) must be equal. However, in going from $t = 0$ to $t = T/2$, if the system is in a Z_2 nontrivial topological state, the Wannier states "switch partners," as in figure 10.2. While doing this, the TR polarization changes by 1 as it records the difference between the positions of the TR Wannier states. The partner switching has a nice explanation in terms of a system with edges: put the system on a cylinder with length in the y-direction of just a single unit cell. If we start with a state of all Wannier-occupied orbitals at $t = 0$, once we insert a flux of π (to go to $T/2$ or $k_y = \pi$), we see that one of the occupied Wannier orbitals closest to the edge has switched

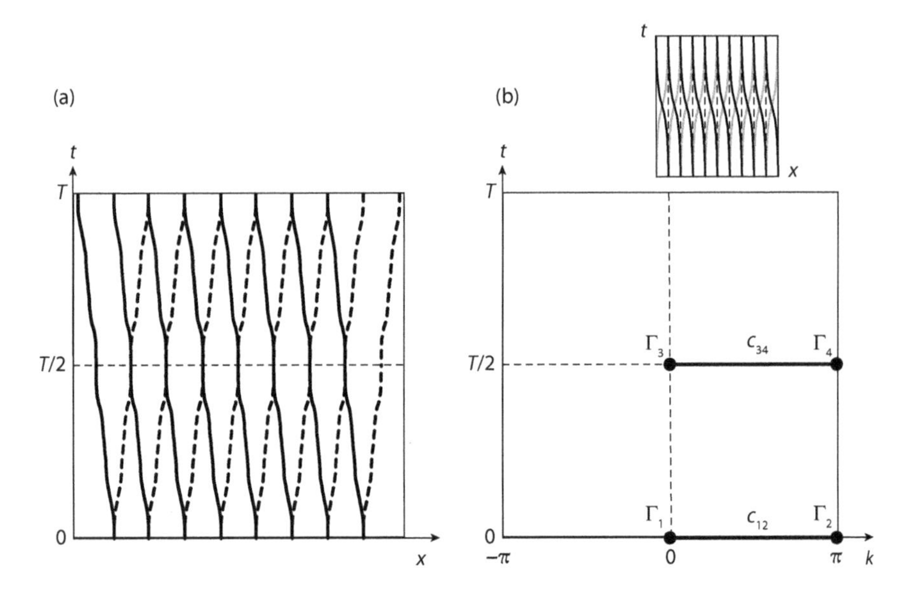

Figure 10.2. Position of the occupied Wannier orbitals' centers as a function of t. In going between two TR-invariant $H(k_x)$ Hamiltonians (the $k_y = 0$, π Hamiltonians), the TR polarization changes sign, a signal that the Wannier states switch partners (TR partners).

partners: its new, degenerate partner at $T/2$ is a previously *unoccupied* orbital. In the bulk, the partner switching does not result in much, except for the relabeling of the Wannier orbitals. On the edge, however, partner switching results in the occupied Wannier orbital at $t = 0$ switching partners and having as a partner at $t = T/2$, a formerly (before applying the flux π) unoccupied Wannier orbital. This means that the partner switching results in the appearance of an *unoccupied* Wannier orbital at each end (edge) of the sample. As such, the ground-state degeneracy will have changed: on one edge, the TR partner switching at $t = T/2$ results in an unoccupied Wannier orbital coming down and becoming degenerate with one of the the $t = 0$ occupied Wannier orbitals. Consequently, the state at $t = T/2$ is twofold degenerate on one edge, being a doublet (we have two states and occupation number 1—the occupied Wannier orbital at $t = 0$—which means we can occupy either of the two states at $t = T/2$). The total degeneracy, when we have two edges, is four, two from one edge and two from the other. We can see this mode clearly by again solving the Hamiltonian for a cylinder, where the length of the cylinder in the translationally invariant direction (along the edge) equals one lattice constant. In this case, the spectrum of states as a function of the momentum parallel to the surface is identical to the spectrum of states as a function of the flux added to the system. Figure 10.3 shows two possible boundary-state spectra as a function of the momentum along a path connecting the surface TR-invariant momenta Λ_a and Λ_b. The shaded region gives the bulk continuum states (only states from a single edge or boundary are shown; there are equivalent states on the *other* boundary). Time-reversal invariance requires the single-particle states at Λ_a and Λ_b to be twofold degenerate. However, there is still freedom in the way that these degenerate states connect up with each other. In figure 10.3a, the Kramers' partners switch pairs between Λ_a and Λ_b, whereas in figure 10.3b they do not. If the doublet at ϵ_{a1} is filled, then there are two options for what happens at the other TR-invariant point, ϵ_{b1}. The doublet Λ_b can be half-filled (Fig. 10.3a) or completely filled (Fig. 10.3b). If it is completely filled (same as the doublet ϵ_{a1}), then there is no change in the the ground state and there is no partner switching, because the ground state is singly degenerate at both TR-invariant points

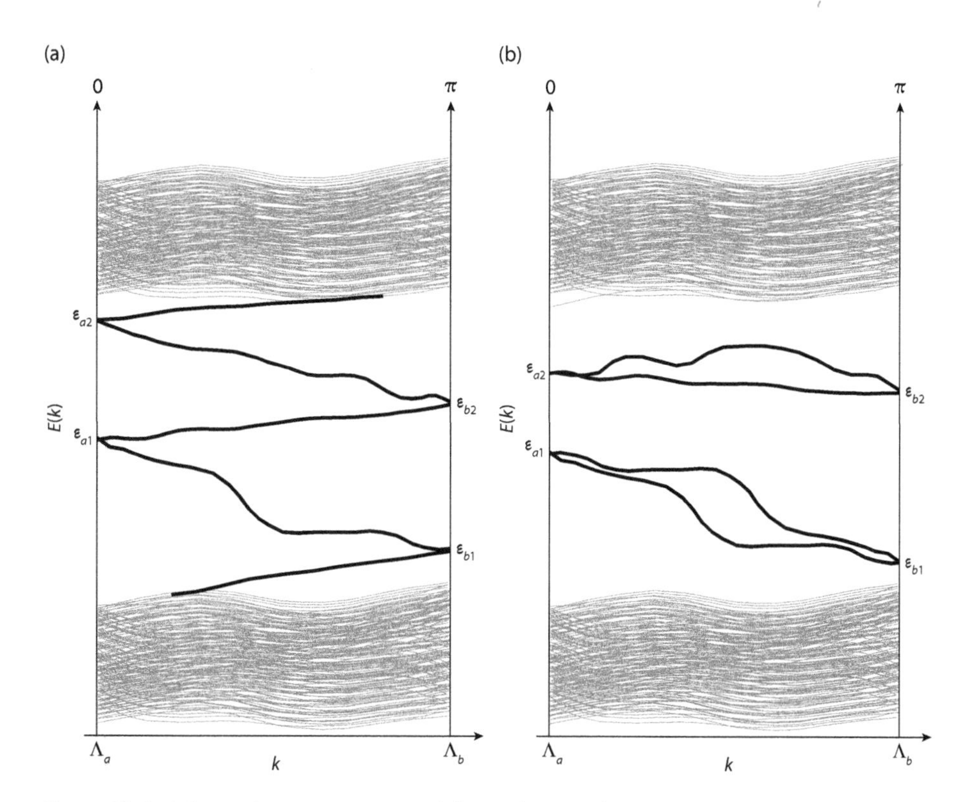

Figure 10.3. Schematic representation of the surface modes in the (a) partner-switching case and (b) trivial case. Explanations are given in the text.

Λ_a and Λ_b. If the doublet at Λ_b is half-filled, then it means that the TR invariant states have "switched partners," and the ground state at Λ_b is *doubly* degenerate. This is the hallmark of a TR-invariant topological insulator—the ground state is doubly degenerate either at threaded flux 0 or π but *not* at both. *This is also the interacting definition of a Z_2 TR-invariant topological insulator.* If the ground state is singly degenerate at threaded flux 0, then it is doubly degenerate at threaded flux π if the system is in a topologically nontrivial state. This corresponds to the other edge state coming down and becoming degenerate at Λ_b to form the doublet.

The ground state degeneracy aspect of topological insulators is very unusual. Typically, threading flux in a Chern TR-breaking insulator leads to no change in the ground-state degeneracy. We thread flux 2π in a Chern insulator and observe that a charge is pumped from one edge to the other. This nondegenerate ground state is a hallmark of noninteracting systems. Usually, ground-state degeneracy is obtained in interacting cases. The Laughlin $\nu = \frac{1}{3}$ state is 3-fold degenerate, and inserting flux 2π (on the torus) does not take us back into the same ground state, but into a degenerate one. Inserting three fluxes of 2π each takes us back into the ground state, and this is the version of the Laughlin gauge argument applied to the fractional quantum Hall effect. Topological insulators, though noninteracting, have—due to their discrete symmetry—this special characteristic that inserting *half* a unit flux takes us between ground states of different degeneracy. The degeneracy depends on the number of edges, but for a semi-infinite sample with one edge mode, it will be equal to 2.

There is in one-to-one connection between the TR polarization and the appearance of edge modes. If the periodic parameter t represents the k_y momentum, it means that, once a surface is placed perpendicular to the x-direction, we can project the two bulk TR-invariant momenta

$k_x = 0, \pi$ at $t = 0$ into one point denoted by $k_y = 0$ on the surface, while the other two bulk TR-invariant momenta $k_x = 0, \pi$ at $t = T/2$ go into one point on the surface $k_y = \pi$. If the difference between the TR polarization at $t = 0$ and the TR polarization at $t = T/2$ is unity, then there is a mismatch, which cannot be realized in a gapped system connected by an edge to the vacuum: an odd number of edge modes crossing the Fermi level between $t = 0$ and $t = T/2$ has to exist. Hence, the topological invariant is

$$P_T(T/2) - P_T(0) = P_T(k_y = \pi) - P_T(k_y = 0). \tag{10.64}$$

In two dimensions, this is equal to

$$\prod_{i=1}^{4} \frac{\sqrt{\mathrm{Det}[B(\Lambda_i)]}}{\mathrm{Pf}[B(\Lambda_i)]}, \tag{10.65}$$

where Λ_i are the TR-invariant momenta. This is the Z_2 topological invariant in two space dimensions.

These arguments indicate how to find the positions of the surface Fermi surfaces in the BZ. For example, suppose the surface state is perpendicular to a lattice vector **G**. Then, in the 2-D bulk BZ there are four TR-invariant points, which will, two by two, project onto two TR-invariant points for the edge BZ. Now, compute the TR polarization for each of those edge points—it will be the product of the $Pf[B]/\sqrt{\mathrm{Det}[B]}$ for the two bulk points that project onto the edge TR point. If the TR polarization changes sign in between the points, then there *must* be an edge mode crossing any Fermi level (situated in the gap) between the edge TR-invariant points. The edge mode crossing the bulk gap will give the TR-polarization change of 1.

10.4 Z_2 Index for 3-D Topological Insulators

Equation (10.65), the Z_2 index for 2-D TR-invariant topological insulators, strongly suggests that a similar index should exist in three dimensions—we could just replace the four TR momenta in two dimensions with the eight momenta in three dimensions. This line of reasoning suggests that a whole new class of 3-D topological insulators should exist. We now give the general arguments for this, and in chapters 11–14 we focus on specific models. In three dimensions we have eight time-reversal-invariant momenta. For each $k_z = 0$ or $k_z = \pi$, we can think of the 2-D Bloch Hamiltonians $h(k_x, k_y, 0)$ or $h(k_x, k_y, \pi)$ as being separate 2-D insulators. Each of these Hamiltonians has its own 2-D Z_2 index, as follows. We have two situations: first, the Z_2 index of the Hamiltonian $h(k_x, k_y, 0)$ equals that of the $h(k_x, k_y, \pi)$ Hamiltonian. This is the case of either the topologically trivial insulators or that of the weak topological insulators (which are simply stacks of 2-D topological insulators). Second, the Z_2 index of the Hamiltonian $h(k_x, k_y, 0)$ is different from that of the $h(k_x, k_y, \pi)$ Hamiltonian. This is the case of the *strong* topological insulator, the most interesting topological insulator.

In three dimensions, let a surface be perpendicular to a vector **G**, and let the eight TR-invariant points in the bulk BZ project onto four TR-invariant points on the surface BZ. We compute the TR polarization P_T of each of those points—each of the four TR points on the surface is a projection of a 1-D TR-invariant Hamiltonian in the direction perpendicular to the surface. This Hamiltonian has two TR-invariant points. For example, if the surface is perpendicular to the z-direction, the Hamiltonian $H(k_z, k_x = 0, k_y = 0)$ is a 1-D TR-invariant Hamiltonian whose time-reversal polarization is defined by $\frac{1}{\pi i} \log \left[\frac{\mathrm{Pf}(B(0,0,0))\mathrm{Pf}(B(\pi,0,0))}{\sqrt{\mathrm{Det}(B(0,0,0))\mathrm{Det}(B(\pi,0,0))}} \right]$ ($B(k_x, k_y, k_z)$ is the sewing matrix). This quantity is the TR polarization of the Λ point

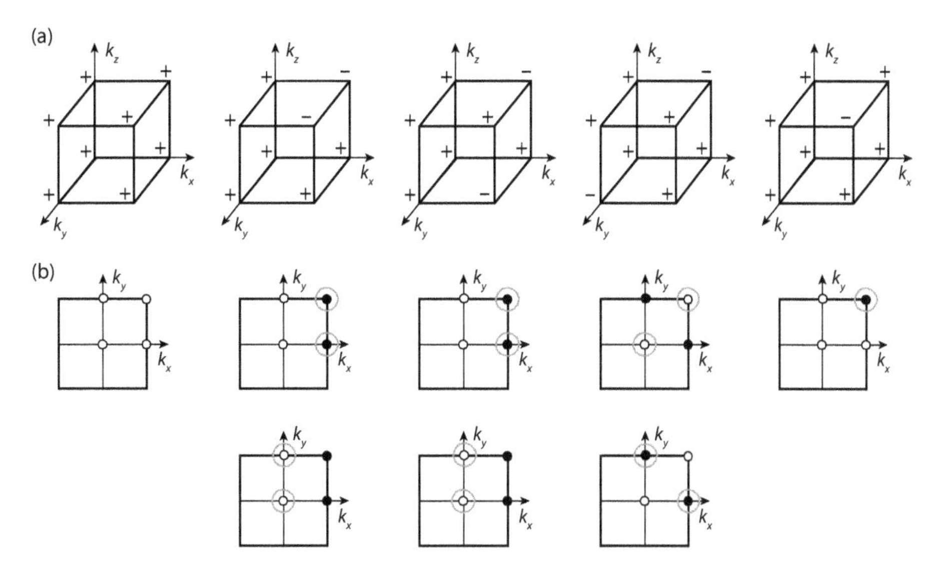

Figure 10.4. Schematic representation of the surface modes in the (a) partner-switching case and (b) trivial case. Detailed explanations on how to determine the locations of the Fermi surfaces are given in the text. Between an open and filled circle denoting a TR-invariant momentum, there always exists a gapless surface state, whereas between open and open or filled and filled circles there is none. Open and filled circles correspond to TR polarizations of 1 or −1, respectively, as presented in the text.

$k_x = k_y = 0$ of the surface perpendicular to z. Between any of the four TR-invariant points on the surface, if the TR polarization changes sign, there has to be a Fermi surface crossing there. This is enough to determine the position of the Fermi surfaces in 3-D bulk topological insulators, as in figure 10.4. This argument clearly shows us that for 3-D topological insulators, there is more than one different situation. If the product of the $\mathrm{Pf}(B(\mathbf{G}/2))/\sqrt{\mathrm{Det}(B(\mathbf{G}/2))}$ at all eight TR-invariant points in three dimensions is -1, then clearly we can have only one surface Fermi ring surrounding a TR-invariant point—this is the strong topological insulator. It is very robust to disorder and semirobust to interactions. A strong topological insulator has a surface state regardless of the surface on which the crystal is terminated.

However, if the product of all the TR-invariant polarizations is 1, then we can have several cases (see fig. 10.4). First is the case where all the four surface TR-invariant points have TR polarizations of the same sign (say, positive). Because there is no sign reversal between the polarizations, the surface will have no protected edge modes (there might be pairs of edge modes, because the trivial signature does not imply the nonexistence of an even number of surface states). However, it can happen that three out of the TR-invariant points have $+$ TR polarization coming from the product of two "bulk" TR point indices $\frac{\mathrm{Pf}(B)}{\sqrt{\mathrm{Det}(B)}}$ (which project onto the same surface TR point), which are $+, +$, whereas the fourth point has a TR polarization coming from the product of two bulk TR point indices $\frac{\mathrm{Pf}(B)}{\sqrt{\mathrm{Det}(B)}}$, which are $-, -$, such as in the case of the first drawings in figure 10.4. Here, if we are looking at a surface perpendicular to the z-direction, the three points $(k_x, k_y) = (0, 0), (0, \pi), (\pi, 0)$ have TR polarizations $+1$ coming from the products of bulk indices $+, +$. The fourth point (π, π) also has TR polarization $+1$ but coming from the product of bulk indices $-, -$. If this happens, then we can change the surface that ends the crystal to a perpendicular one from the previous one, and in this case, two TR invariant points have TR polarization 1, whereas the other two have TR polarization -1. For example, also in the case of the first drawings in figure 10.4, if we now look at the surface perpendicular to k_x, the points $(k_y, k_z) = (0, 0), (0, \pi)$ have TR polarization $+1$ coming from the product of two bulk indices $+, +$, whereas the points $(\pi, 0), (\pi, \pi)$ have TR polarization equal to -1 coming from the product of bulk indices $+ -$. It then means that on the yz-surface, we do have a surface state, but the surface state then necessarily includes an *even* number of TR-invariant points (fig. 10.4). The points that it includes are clear by the eigenvalue-change argument. This is a weak topological insulator. A weak topological insulator has surface states only on certain surfaces and not on others. It is equivalent to a stack of 2-D topological insulators.

Analyzing the possible Fermi surfaces of the surface states can clearly show the presence of a strong topological insulator. Imagine a surface perpendicular to a vector **G**, and then imagine the possible Femi surfaces on the surface BZ. The TR operator commutes with the position operator and so the bands on the surface must respect TR properties. As such, the Fermi surfaces of the surface states will surround TR invariant points. The remaining issue is the number of Fermi points the surface Fermi surface encloses. It can enclose zero TR-invariant points, in which case there does not need be any surface state and the insulator is trivial. It can enclose an even number of surface TR-invariant points; these are the weak topological insulators. The reason that these insulators are not as robust as the strong topological insulators is the following: suppose the Fermi surface of the surface states surrounds two TR-invariant surface momenta. Time-reversal invariance teaches us that as we lower the energy, we must hit a (generically) Dirac-type node at both TR-invariant points. This Dirac node is necessary due to time reversal, because at each TR-invariant momentum, there should be a Kramers' degeneracy. Then, assume a one-body term in the Hamiltonian, such as a charge-density wave term with wavevector connecting the two TR-invariant points surrounded by the Fermi surface of the surface state. This term could, for example, arise from a mean-field solution of a charge-density wave Hamiltonian. The result of a charge-density wave term at a wavevector equal to the difference between the two TR-invariant momenta folds the two into each other. We now have a double-Dirac point at the new, folded momentum and can open a gap without breaking time reversal because we would not be violating Kramers' theorem. We are led to the conclusion that, in a weak topological insulator, we can open a gap in the surface states without breaking time reversal. The last case, and the most robust one, is when the Fermi surface of the surface states encloses one (or three) TR-invariant point (S). In this case, we know that in lowering the surface Fermi level, we must hit a Dirac node located exactly at a TR-invariant point. The Dirac node cannot be gapped *by any means* because otherwise we would break time reversal. If the Dirac node gaps, then time reversal is broken. This is the most robust topological insulator. The folding scenario of gapping the weak topological insulator does not apply here—because we have only one Dirac cone, nothing can fold back into the Dirac cone, and hence it remains stable. If the Fermi surface of the surface state surrounds one TR-invariant point, our previous discussion of TR polarization switching shows the following: the product of the Pfaffian indices of the two TR-invariant bulk points that project onto the surface TR-invariant point surrounded by the Fermi surface must be different from the ones at the points not surrounded by the Fermi surface. This renders the product of the Pfaffian over all eight TR momenta indices,

$$Z_2 = \prod_{i=1}^{8} \frac{\sqrt{\text{Det}[B(\Lambda_i)]}}{\text{Pf}[B(\Lambda_i)]}, \tag{10.66}$$

to be negative. Λ_i are the eight TR-invariant momenta $(0, 0, 0)$, $(0, 0, \pi)$, $(0, \pi, 0)$, $(0, \pi, \pi)$, $(\pi, 0, 0)$, $(\pi, 0, \pi)$, $(\pi, \pi, 0)$, and (π, π, π).

Although the Pfaffian formulation of the Z_2 index brings out the nice physical interpretation of the nontrivial topological insulator, so far it is of academic interest. This is because its application to realistic situations requires choosing a smooth gauge. This is always possible in a TR-invariant topological insulator because the Chern number is zero, so in principle there should not be any problem with this prescription. In practice, however, choosing a smooth gauge is not always convenient. For example, in the HgTe model, the physics is most simply understood by taking advantage of the $U(1) \times U(1)$ S_z symmetry of the problem, in which each spin-$\uparrow$ or $\downarrow$ Hamiltonian has (opposite) nonzero Chern number. However, this gauge is problematic in terms of choosing smooth eigenstates: each spin $\uparrow$ and $\downarrow$ Hamiltonian has a nonzero Chern number—and, hence, smooth eigenstates are not available in this gauge. Choosing smooth eigenstates necessarily implies mixing the spins—and, hence, obscuring the nice picture of opposite Chern-number spins. As such, different expressions for the Z_2 invariants that might be easier to calculate are welcome.

10.5 Z_2 Number as an Obstruction

Earlier in this book, we learned that a nonzero Chern number is an obstruction to finding a smooth gauge for the Bloch wavefunctions over the whole BZ. If a smooth gauge could be defined, then Stokes' theorem would be valid and we would be able to equate the Chern number with the integral of the Berry potential over the full BZ boundary, which would be zero because the BZ does not have a boundary. This means that we can write the Chern number as an obstruction to Stokes' theorem:

$$C = \int_{BZ} d^2k\, F_{12} = \int_{BZ} d^2k\, F_{12} - \int_{\partial BZ} d\mathbf{k} \cdot \mathbf{A} \tag{10.67}$$

The second term, $\int_{\partial BZ} d\mathbf{k} \cdot \mathbf{A}$ vanishes, of course, because there is no such thing as a boundary of the full BZ, which is periodic. We now ask the same question for the topological insulator: can the invariant be written as an obstruction to defining a smooth gauge? We know that a smooth gauge can be defined because the Chern number is zero by virtue of having a TR-invariant system. However, we can define a "TR-smooth" gauge as the TR constraint:

$$\begin{aligned} \left|u_a^{\mathrm{I}}(-\mathbf{k}, -t)\right\rangle &= T \left|u_a^{\mathrm{II}}(\mathbf{k}, t)\right\rangle , \\ \left|u_a^{\mathrm{II}}(-\mathbf{k}, -t)\right\rangle &= -T \left|u_a^{\mathrm{I}}(\mathbf{k}, t)\right\rangle , \end{aligned} \tag{10.68}$$

or, in other words, $\chi_k = 0$ in the previous examples; in terms of the sewing matrix, $B = i\sigma_y$. This means that the gauges at $\pm(\mathbf{k}, t)$ are not independent and that $\mathrm{Det}[B(\mathbf{k}, t)] = 1 = \mathrm{Pf}[B(\Lambda_i)]$ for all k, t. This would mean that the Pfaffian invariant defined in this chapter is always equal to 1, which would mean that the insulator is always in a trivial phase. Thus, the nontrivial insulator Z_2 index must be an obstruction to defining the TR-smooth gauge. A similar procedure to that defined in the Chern insulator case must be performed: define wavefunctions that are smooth on patches, find the gauge transformation between them, and then patch them together to obtain the final result. Unlike in the case of the Chern insulator, in our case we must deal with the fact that the wavefunctions as defined on patches must satisfy time reversal. As such, we assume we have wavefunctions obeying equation (10.68), defined smoothly on two patches, A and B (fig. 10.5).

On patch A, we consider $\left|u_a^s(\mathbf{k}, t)\right\rangle_A$ smoothly defined everywhere in the upper-left and lower-right quadrants. On patch B, we have $\left|u_a^s(\mathbf{k}, t)\right\rangle_B$ smoothly defined everywhere in the upper-right and lower-left quadrants. Here, $s = \mathrm{I}, \mathrm{II}$ denote the Kramers' pairs, whereas $a = 1, \ldots, N$ denote the number of the occupied Kramers' pair. In the overlapping regions (or on

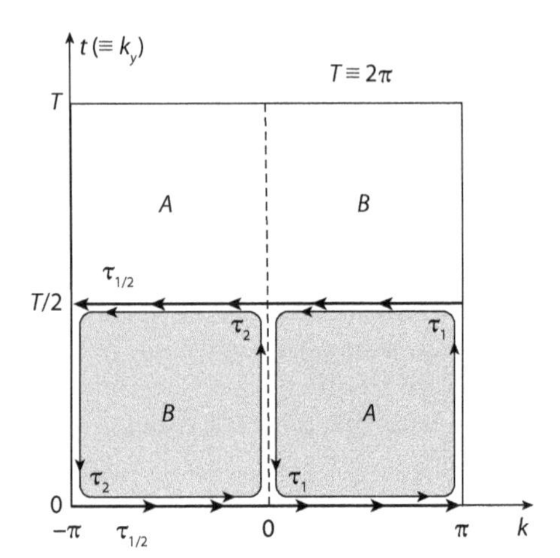

Figure 10.5. We define smooth wavefunctions satisfying equation (10.68), on two patches A and B, along with their TR counterparts. $\tau_{1/2}$ is the A, B half of the BZ represented by the shaded portion. $|u_a^s(\mathbf{k}, t)\rangle_A$ and, hence, the Berry potential A^A are well defined in region A but not in region B, and vice versa.

the boundary), these wavefunctions are related by a $U(2N)$ transition matrix:

$$|u_m(\mathbf{k}, t)\rangle_A = t_{mn}^{AB} \, |u_n(\mathbf{k}, t)\rangle_B \,, \tag{10.69}$$

where m, n are shorthand for s, a and run over all bands. Consider the change in the $U(1)$ phase of t^{AB} around the closed loop $\partial\tau_1$, which is the boundary of the τ_1 region A:

$$D = \frac{1}{2\pi i} \oint_{\partial\tau_1} d\mathbf{l} \cdot \mathrm{Tr}[t^{AB\dagger} \nabla t^{AB}] \tag{10.70}$$

This is an integer because it is equal to the winding number of the phase of $\mathrm{Det}[t^{AB}]$ around the loop $\partial\tau_1$. If D is nonzero and cannot be eliminated by a gauge transformation, then there is an obstruction to smoothly defining the wavefunctions on a single patch. We can show that D is defined only modulo 2 and, hence, the two classes of insulators, trivial and nontrivial, are characterized by $D = 0, 1 \pmod 2$. We can rewrite D as the boundary difference between the Berry potentials of the A and B gauges:

$$D = \frac{1}{2\pi i} \oint_{\partial\tau_1} d\mathbf{l} \cdot \mathrm{Tr}[t^{AB\dagger} \nabla t^{AB}] = \frac{1}{2\pi} \oint_{\partial\tau_1} d\mathbf{l}(\mathbf{A}^B(\mathbf{k}) - \mathbf{A}^A(\mathbf{k})), \tag{10.71}$$

where the Berry potentials $\mathbf{A}^A = \sum_n \langle u_n | \nabla | u_n \rangle_A$, $\mathbf{A}^B = \sum_n \langle u_n | \nabla | u_n \rangle_B$ are smoothly defined only in regions A and B, respectively. Because $|u_n\rangle_A$ is smoothly defined in the interior of region τ_1, we have that

$$\oint d\mathbf{l} \cdot \mathbf{A}^A = \int_{\tau_1} d\tau F^A. \tag{10.72}$$

Unfortunately, $\mathbf{A}^B$ is not smoothly defined in the interior of region τ_1, and so Stokes' theorem cannot be applied there. However, $\mathbf{A}^B$ is smoothly defined in the interior of region τ_2; hence,

$$\oint_{\partial\tau_1} d\mathbf{l} \cdot \mathbf{A}^B = -\oint_{\partial\tau_2} d\mathbf{l} \cdot \mathbf{A}^B + \oint_{\partial\tau_{1/2}} d\mathbf{l} \cdot \mathbf{A}^B = -\int_{\tau_2} d\tau F^B + \oint_{\partial\tau_{1/2}} d\mathbf{l} \cdot \mathbf{A}^B. \tag{10.73}$$

We hence obtain

$$D = -\int_{\tau_1} d\tau F^A - \int_{\tau_2} d\tau F^B + \oint_{\partial\tau_{1/2}} d\mathbf{l} \cdot \mathbf{A}^B. \tag{10.74}$$

The field strength is gauge invariant—$F^A = F^B = F$—and, hence,

$$D = -\int_{\tau_{1/2}} d\tau F + \oint_{\partial\tau_{1/2}} d\mathbf{l} \cdot \mathbf{A}^B. \tag{10.75}$$

We now show that, crucially, $\oint_{\partial\tau_{1/2}} d\mathbf{l} \cdot \mathbf{A}^B$ is gauge invariant only modulo 2. This is the basis for the Z_2 characterization of topological insulators as an obstruction. To do this, we would first like to prove that

$$\oint_{\partial\tau_{1/2}} d\mathbf{l} \cdot \mathbf{A}^B = \int_{\mathrm{I+\,II+\,III+\,IV+\,V+\,VI}} d\mathbf{l} \cdot \mathbf{A}^B, \tag{10.76}$$

where I+ II + III+ IV + V+ VI is the combination of the contours:

$$\begin{aligned}
&\mathrm{I}: k = 0 \to -\pi, \quad t = \frac{T}{2} \\
&\mathrm{II}: k = -\pi, \quad t = \frac{T}{2} \to 0 \\
&\mathrm{III}: k = -\pi \to 0, \quad t = 0 \\
&\mathrm{IV}: k = 0 \to \pi, \quad t = 0 \\
&\mathrm{V}: k = \pi, \quad t = 0 \to \frac{T}{2} \\
&\mathrm{VI}: k = \pi \to 0, \quad t = \frac{T}{2}
\end{aligned} \tag{10.77}$$

Because we are looking only at the boundary region, we have that A^B is well defined in this region. A^B is not well defined in region A, but it is well defined on the boundary of this region, which is actually the common boundary with region B. The key to proving gauge invariance modulo 2 is to realize that the contours I, VI or III, IV, or II, V are time reverses of each other, so that

$$\begin{aligned}
\oint_{\partial\tau_{1/2}} d\mathbf{l} \cdot \mathbf{A}^B &= \int_0^{-\pi} \left(\mathbf{A}^B\left(\mathbf{k}, \frac{T}{2}\right) + \mathbf{A}^B\left(-\mathbf{k}, \frac{T}{2}\right)\right) \\
&\quad + \int_{-\pi}^{0} (\mathbf{A}^B(\mathbf{k}, 0) + \mathbf{A}^B(-\mathbf{k}, 0)) + \int_0^{\frac{T}{2}} \left(\mathbf{A}^B\left(\pi, \frac{T}{2}\right) + \mathbf{A}^B\left(-\pi, \frac{T}{2}\right)\right),
\end{aligned} \tag{10.78}$$

where

$$\mathbf{A}^B(-\mathbf{k}, \frac{T}{2}) = -\sum_n \left\langle u_n^B\left(-\mathbf{k}, \frac{T}{2}\right)\right| \nabla_{\mathbf{k}} \left|u_n^B\left(-\mathbf{k}, \frac{T}{2}\right)\right\rangle = \mathbf{A}^B\left(\mathbf{k}, \frac{T}{2}\right) \tag{10.79}$$

due to the TR constraints in equation (10.68). Now, crucially, we can see that upon a gauge transformation, if $u^{\mathrm{I}}(\mathbf{k}, t) \to e^{i\theta_k} u^{\mathrm{I}}(\mathbf{k}, t)$ is the transformation of one of the doublets in a Kramers' pair, the transformation for the other doublet is *not* independent and is $u^{\mathrm{II}}(-\mathbf{k}, -t) \to e^{i\theta_{\mathbf{k}}} u^{\mathrm{II}}(-\mathbf{k}, -t)$. The full vector potential is a sum of the two Kramers' doublets I and II, so, under a gauge transformation, it transforms as

$$\oint_{\partial\tau_{1/2}} d\mathbf{l} \cdot \mathbf{A}^B \to \oint_{\partial\tau_{1/2}} d\mathbf{l} \cdot \mathbf{A}^B + 2\oint_{\partial\tau_{1/2}} d\mathbf{l} \cdot \nabla\theta_{\mathbf{k}}. \tag{10.80}$$

We can then remove the label B from $\mathbf{A}^B$ because we are counting things only mod 2, and we find

$$D = -\int_{\tau_{1/2}} d\tau F + \oint_{\partial\tau_{1/2}} d\mathbf{l} \cdot \mathbf{A} \qquad (\text{mod } 2) \tag{10.81}$$

If $D = 0$, it is obvious that there is really no obstruction to Stokes' theorem and the insulator is thus trivial (an even D can be deformed to zero by a gauge transformation). However, if $D = 1$, there is a clear obstruction to Stokes' theorem. This is the nontrivial topological insulator (strong). Equation (10.81) is also very suggestive physically: just as the Chern number in a Chern insulator is defined as an obstruction to Stokes' theorem over the *whole* BZ, the Z_2 invariant in a topological insulator is defined as an obstruction to Stokes' theorem over *half* the BZ. The halving of the BZ is due to TR invariance.

We also want to point out that we can now understand the criterion we proposed earlier for obtaining a topological insulator: start from the atomic limit and close an *odd* number of gaps in *half* the BZ. If we are far away from the boundary of the BZ and close the gap in an odd number of points in half the zone, we will get

$$\Delta(\int_{\tau_{1/2}} d\tau F) = 2n + 1 \tag{10.82}$$

because at every gap closing and reopening, the Chern curvature changes by 1. If we close and reopen an odd number of gaps in *half* the BZ starting from the atomic limit, we end up with a nontrivial topological insulator.

10.6 Equivalence between Topological Insulator Descriptions

We now prove that the two descriptions of the Z_2 invariants given in this chapter are equivalent. The first of the Z_2 invariants involved counting the number of zeros of the Pfaffian matrix,

$$m_{ij}(\mathbf{k}, t) = \langle u_i(\mathbf{k}, t) | \, T \, | u_j(\mathbf{k}, t) \rangle \,, \tag{10.83}$$

in half of the BZ. This matrix is antisymmetric at every point $\mathbf{k}$ in the BZ. The Pfaffian's vorticity modulo 2 in half the BZ determined whether the state is a topological insulator or not. The m_{ij} matrix has Pfaffian 1 at every TR-invariant point. These points belong to the "even" subspace. The matrix m_{ij} is *not* unitary. In contrast, the sewing matrix,

$$B_{ij}(\mathbf{k}, t) = \langle u_i(-\mathbf{k}, -t) | \, T \, | u_j(\mathbf{k}, t) \rangle \,, \tag{10.84}$$

is unitary everywhere in the BZ because it relates eigenstates at $-\mathbf{k}$ to the time reversal of the eigenstates at $\mathbf{k}$. It is *not* antisymmetric *everywhere* in the BZ. It is antisymmetric only at the special TR-invariant points. Its symmetry property is $B_{ij}(-\mathbf{k}) = -B_{ji}(\mathbf{k})$. The product of $\text{Pf}(B)/\sqrt{\text{Det}(B)}$ at all TR-invariant points gives us the Z_2 topological number. We can easily prove the identity $m(-\mathbf{k}, -t) = B(\mathbf{k}, t) m^*(\mathbf{k}, t) B(\mathbf{k}, t)^T$, which, by the identity $\text{Pf}(XAX^T) = \text{Det}[X]\text{Pf}[A]$ gives

$$\text{Det}[B(\mathbf{k}, t)] = \frac{\text{Pf}[m(\mathbf{k}, t)]}{\text{Pf}[m(-\mathbf{k}, -t)]^*}. \tag{10.85}$$

If we denote $p(k) = Pf[m(k, t^*)]$, where $t^* = 0, T/2$ are the two TR-invariant points ($t^* = -t$), we have

$$\log[\mathrm{Det}[B(k, t^*)]] = \log[p(k)] + \log[p(-k)], \tag{10.86}$$

which is true because $\mathrm{Pf}[m(-k, t^*)](\mathrm{Pf}[m(-k, t^*)])^* = 1$. At TR-invariant points Λ, it is obvious from their definition that the two matrices are equal: $B(\Lambda) = m(\Lambda)$. The TR polarization can then be expressed in terms of the matrix m:

$$\begin{aligned} P_T &= \frac{1}{2\pi i}\left[\int_0^{\pi} dk\nabla_k[\log(p(k)) + Log(p(-k))] - 2\log\left[\frac{Pf[B(\pi)]}{Pf[B(0)]}\right]\right] \\ &= \frac{1}{2\pi i}\left[\int_0^{\pi} d_k\nabla_k[\log(p(k))] - \int_{-\pi}^{0}\nabla_k\log(p(k))] - 2\int_0^{\pi} dk\nabla_k\log(p(k))\right] \\ &= -\frac{1}{2\pi i}\int_{-\pi}^{\pi} dk\nabla_k\log(p(k)) \end{aligned} \tag{10.87}$$

The preceding equation for the TR polarization is valid at both $t = 0$ and $t = T/2$. As usual, a modulo-2 ambiguity related to the ambiguity of the imaginary part of the logarithm is present. The Z_2 invariant, equal to the difference between the TR polarization at "times" 0 and $T/2$ (or momenta $k_y = 0, \pi$) can then be rewritten in a different way:

$$P_T\left(\frac{T}{2}\right) - P_T(0) = -\frac{1}{2\pi i}\oint_{\partial\tau_{1/2}} dk\nabla_k\log(Pf[m(k, t)]) \mod 2 \tag{10.88}$$

This is the phase winding of the m Pfaffian around the half $\partial\tau_{1/2}$ of the BZ, which counts the number of its zeros and their vorticity in half the BZ. This proves the connection between the two invariants.

10.7 Problems

1. *Zeros of the Pfaffian in the Kane-Mele and HgTe Models:* Compute explicitly the location of the zeros of the Pfaffian $Pf[\langle u_i(\mathbf{k})| T |u_j(\mathbf{k})\rangle]$ for both the Kane and Mele model and for the HgTe model of the QSH effect. Show that in the atomic limit, the Pfaffian does not have any zeros in the BZ. Show that, upon changing the parameters of the Hamiltonian, the Pfaffian acquires an odd number of zeros in the nontrivial QSH side *if all symmetries are broken.* If residual symmetries still remain (such as in the decoupled spin $\uparrow$–spin $\downarrow$ model of HgTe without any inversion-symmetry-breaking term that has a $U(1) \times U(1)$ symmetry), show that the Pfaffian acquires zeros along a set of lines rather than at isolated points. Show how, upon the introduction of terms breaking these symmetries, such as inversion-symmetry-breaking terms, the lines of zeros of the Pfaffian break up into isolated points.
2. *Projected Position Operator in a Lattice:* The position operator in a 1-D lattice is $X = \sum_j e^{i2\pi j/N_x} |j, \alpha\rangle \langle j, \alpha|$, where the j are the positions of the N_x sites and α is the on-site orbital (assuming a multi-orbital-per-site problem). (Since we are on the lattice, the position is the phase of X and contains an ambiguity of N_x, the periodicity of the lattice.) Find the expression for the projected position operator in the occupied bands and relate it to the density operator projected in the occupied bands at momentum $\Delta q = 2\pi/N_x$.
3. *Matrix Elements of the Position Operator:* Find the matrix elements of the operator PXP, where P is the projection operator in the occupied bands $P = \sum_{n\in\text{occupied bands},k} |n, k\rangle \langle n, k|$.

Show that the determinant of this operator is related to the exponential of the polarization $\int_0^{2\pi} A(k)\, dk$. This is the correct lattice version of the relation between polarization and Berry phase. Show that it is identical to the Wilson loop in the x-direction.

4. *Winding Number of the Phase of the Position Operator:* Compute the polarization (the phase of the PXP operator) in the x-direction as a function of k_y (for each k_y, the Hamiltonian as a function of k_x is a 1-D Hamiltonian whose polarization can be calculated) for the model $h(\mathbf{k}) = \sin k_x\sigma_x + \sin k_y\sigma_y + (M - \cos k_x - \cos k_y)\sigma_z$ and show that it winds by 2π as k_y goes from 0 to 2π if the model is in its nontrivial phase. The winding is the way the Wilson loop relates to the Chern number for 2-D systems.
5. *Maximally Localized Wannier Orbitals:* Find, explicitly, the expressions for the eigenstates of the projected position operator PXP. These eigenstates are called maximally localized Wannier orbitals. They have an analytic expression whose primary constituents are Wilson lines from 0 to any point k_x and the eigenvalues and eigenstates of the Wilson loop. Compare the analytic expression with numerical diagonalization. There will be $N_x \cdot N_{\text{occupied}}$ eigenstates of the projected position operator, where N_x is the number of sites in the x-direction and N_{occupied} is the number of occupied bands.
6. *Sewing Matrices for the HgTe Model:* Compute the sewing matrix $B(\mathbf{k})$ between states at $\mathbf{k}$ and $-\mathbf{k}$ explicitly for the HgTe model and show explicitly how the Berry potentials at the two points are related to one another through $a_i(-\mathbf{k}) = B(\mathbf{k})a_i^\star(\mathbf{k})B^\dagger(\mathbf{k}) - iB\partial_i B^\dagger$.
7. *Wannier Centers of a TR-Invariant System:* For the HgTe model, plot the Wannier centers (phases for each of the N_x eigenstates) of the projected position operator in the lower two bands as a function of k_y. Show that the Wannier centers exhibit time reversal as a function of k_y—they split in opposite pairs, which define the TR-invariant polarization defined in the text. Show that upon completion of a cycle k_y from 0 to 2π, the Wannier centers switch partners between different x-positions.
8. *TR Properties of the Wilson Loop in Two Dimensions:* Based on the preceding questions, compute the eigenvalues of the Wilson *loop* as a function of k_y. Show that are at least two such eigenvalues (because the TR topological insulator has at least two occupied bands) and that they exhibit a TR property, which tells us that we need consider only their spectrum from $k_y = 0$ to $k_y = \pi$. Finally, show that in the topologically nontrivial phase, one of these eigenvalues winds from 0 to π, whereas the other does not. This is another definition of a TR topological insulator. It bears resemblance to the definition of the Chern number through the Wilson loop winding number.
9. *Topological Invariant in Three Dimensions:* In 3-D translationally invariant periodic systems with Hamiltonian $H(k_x, k_y, k_z)$, show that the system is a 3-D nontrivial topological insulator if the $k_z = 0$ plane $H(k_x, k_y, 0)$ defines a trivial insulator, but the $k_z = \pi$ plane $H(k_x, k_y, k_z = \pi)$ is a nontrivial topological insulator, or vice versa.
10. *Position of Surface States and TR Pfaffian Eigenvalues:* Relate the positions of the Fermi surface of the surface state of 3-D topological insulators to the product of the Z_2 indices of the TR-invariant planes in the 3-D BZ of these insulators.

11

Crossings in Different Dimensions

In the previous chapter, we defined several ways of obtaining the Z_2 invariant of a TR-invariant topological insulator. There are two main prescriptions for determining whether a system is a trivial or nontrivial topological insulator. On the one hand, if we posses knowledge of the band structure over the whole BZ, we can use a smooth gauge choice and compute the invariants, as we did in the previous chapter. Knowledge of the bands over the whole BZ zone is rather hard to obtain. It requires detailed first-principle calculations of complex materials—most topological insulators contain spin-orbit coupling and, as such, the first-principle calculations get rather messy. On the other hand, a second route to understanding whether a system is a trivial or nontrivial insulator is to look at the phase transitions that occur when we take the system from the atomic limit with vanishing hopping and intracell hybridization to the experimental hopping parameters. As the hopping parameters are turned on from their zero values in the atomic limit, the system can undergo several transitions. First, the system could, in principle, not close any gap, and hence the topological state would remain trivial (the same as the initial atomic insulator). The system could also close a gap at the Fermi level and reopen it. Because the systems considered up to now are TR invariant, a gap closing at one point in the BZ is accompanied by another one at its TR point. Such gap closing and reopening in half the BZ is, of course, accompanied by a change in the topological character of the system. As such, it seems natural that we should study band crossings in TR-invariant systems as a way of inducing transitions between topologically trivial and nontrivial insulators. Part of this analysis has already been done in [48].

We have seen that the change of a topological quantum number occurs locally in $\mathbf{k}$ space. As we change an external parameter, the system may undergo a phase transition, which accompanies the closing of a bulk gap at a certain $\mathbf{k}$. When an external parameter is changed, a phase transition may occur in some cases, but in others it may not, due to level repulsion. In this chapter we investigate the question of what phase transitions can be induced by varying *only one* parameter in systems with different symmetries, such as time reversal or inversion. The question that we would like to ask is, How many parameters do we need to tune in a generic system in order to close the gap between two bands at some point in the BZ? Generically, gap closings should obtained by varying one parameter, which we call m. We exclude the gap closings achieved by tuning more than one parameter: in these cases the phase transition might be circumvented by perturbations, and extreme fine-tuning is necessary to induce a degeneracy. If we are required to fine-tune a large number of parameters to close a gap, then it is likely that the transition will not happen in a realistic material. We can think of the parameter m being varied, as induced by the tuning of the hopping parameters from the atomic limit to the limit of the lattice constant in the real material. The codimension of a degeneracy is defined by the number of tuned parameters necessary to achieve a degeneracy of levels (or a gap closing). We want to analyze different codimensions of systems with time reversal and/or inversion, see if we can induce phase transitions in these systems, and then try to understand how we go from the atomic limit to the nontrivial topological insulator. We start with the generic Hamiltonian for a spinfull system:

$$H(\mathbf{k}) = \begin{bmatrix} h_{\uparrow\uparrow}(\mathbf{k}) & h_{\uparrow\downarrow}(\mathbf{k}) \\ h_{\downarrow\uparrow}(\mathbf{k}) & h_{\downarrow\downarrow}(\mathbf{k}) \end{bmatrix}. \tag{11.1}$$

We consider the action of the time reversal on this system:

$$T = -i\sigma_y K \tag{11.2}$$

(note that sometimes in the literature the TR operator is defined as the negative of the preceding, which does not make a difference). Using the hermiticity property of the Hamiltonian, we have

$$TH(\mathbf{k})T^{-1} = \sigma_y H^T(\mathbf{k})\sigma_y = \begin{bmatrix} h^T_{\downarrow\downarrow}(\mathbf{k}) & -h^T_{\uparrow\downarrow}(\mathbf{k}) \\ -h^T_{\downarrow\uparrow}(\mathbf{k}) & h^T_{\uparrow\uparrow}(\mathbf{k}) \end{bmatrix}. \tag{11.3}$$

Hence, for TR-invariant Hamiltonians, which must satisfy $TH(\mathbf{k})T^{-1} = H(-\mathbf{k})$, we have

$$\begin{aligned} h_{\uparrow\uparrow}(\mathbf{k}) &= h^T_{\downarrow\downarrow}(-\mathbf{k}), \quad & h_{\downarrow\uparrow}(\mathbf{k}) &= -h^T_{\downarrow\uparrow}(-\mathbf{k}), \\ h_{\uparrow\downarrow}(\mathbf{k}) &= -h^T_{\uparrow\downarrow}(-\mathbf{k}), \quad & h_{\downarrow\downarrow}(\mathbf{k}) &= h^T_{\uparrow\uparrow}(-\mathbf{k}). \end{aligned} \tag{11.4}$$

The Hamiltonian matrix dimension is an arbitrary number—we now take it to be the number of bands (states) involved in the gap-closing-and-reopening transition. More specifically, we look for crossings that close and then reopen the bulk gap. Depending on the location of these crossings in the BZ and on the symmetries of the problem (such as inversion and time reversal), we find that crossings can be either between two bands (the generic case) or between four bands (when both time reversal and inversion are present or when the crossing happens at a TR-invariant point).

11.1 Inversion-Asymmetric Systems

The physics of a TR-invariant system is severely affected by whether the system has other symmetries, such as inversion. We first assume that the system does not have other symmetries besides time reversal. In this case, the gap in general cannot close at any TR-invariant point $\mathbf{k} = \mathbf{G}/2$. At $\mathbf{k} = \mathbf{G}/2$, both the conduction and the valence bands are doubly degenerate by time reversal, so the minimal Hamiltonian that would be able to describe a gap-closing transition at a TR-invariant point is 4×4, with doubly degenerate energies. Hence, need to use the Clifford algebra matrices—these are the only ones that give two sets of bands, each doubly degenerate:

$$H\left(\mathbf{k} = \frac{\mathbf{G}}{2}\right) = E_0 + \sum_{i=1}^{5} d_i \Gamma_i. \tag{11.5}$$

Note that the d_i are *constants*—they do not depend on k because we have fixed the k-point to be TR invariant. The Γ matrices have to be TR invariant and hence can be chosen as

$$\Gamma_1 = 1 \otimes \tau_x, \quad \Gamma_2 = \sigma_z \otimes \tau_y, \quad \Gamma_3 = 1 \otimes \tau_z, \quad \Gamma_4 = \sigma_y \otimes \tau_y, \quad \Gamma_5 = \sigma_x \otimes \tau_y, \tag{11.6}$$

where σ, τ operate in spin and orbital space explicitly. Note that, regardless of the representation chosen, there are only five such matrices that are TR invariant. The two doubly-degenerate bands will touch only when all d_i =0; hence, the codimension is 5. It is impossible to go through a phase transition by tuning only one parameter. Hence, generically, gap closings and reopenings in any dimensions cannot happen at TR-symmetric points in the absence of other symmetries (such as, for example, inversion). As such, crossings happen at generic points in the BZ, and their physics depends on the specific dimension of the space. We analyze this next.

11.1.1 Two Dimensions

We first consider the case of two dimensions. If $\mathbf{k} \neq \mathbf{G}/2$, then each band is generically nondegenerate (spin split) without inversion symmetry. At a gap closing, the effective Hamiltonian is a description of the valence and conduction band participating in the transition. Its most general form is that of a 2×2 matrix:

$$H(m, \mathbf{k}) = d_0(m, \mathbf{k}) + d_i(m, \mathbf{k})\sigma_i. \tag{11.7}$$

Degenerate eigenvalues are possible only if all three $d_i(m, \mathbf{k}) = 0$. We hence need to tune *all* three d_i's to zero in order to close the gap, so the codimension of a two-level system is 3. We are, however, on a lattice, and hence $\mathbf{k}$ scans the BZ. In two dimensions, we thus have two parameters ready to do their scanning job in the BZ: k_x and k_y. The only parameter we have to tune by hand is m. The gap can close when m is tuned to a critical value. The form of the Hamiltonian close to the gap closing is of massive Dirac form (up to a rotation). If we start from an insulator in the atomic limit, in the process of tuning the lattice constant (or the hoppings) from infinity (zero) to the actual values in the material, two things can happen. First, the gap does not close (or closes and reopens an even number of times in *half* the BZ), in which case we end up with a trivial insulator. Note that because the gap cannot close at a TR-invariant point by tuning only one parameter (as shown in the previous section, the codimension is 5 at such points), any gap closing at a generic $\mathbf{k}$-point is accompanied by a gap closing at $-\mathbf{k}$. Second, the nontrivial case occurs when gap closes and reopens an odd number of times in *half* the BZ, starting from the atomic limit. In that case, we reach a topological insulator, per the proof of the topological Z_2 index.

11.1.2 Three Dimensions

Just as in two dimensions, in three dimensions, for $\mathbf{k} \neq \mathbf{G}/2$, the bands are spin split and so the Hamiltonian for two bands crossing is 2×2. Close to a crossing point K_i, the Hamiltonian can be, in general, expanded in the Pauli matrices:

$$H(\mathbf{k}) = (k_i - K_i)A_{ij}\sigma_j, \tag{11.8}$$

where A_{ij} is the coupling matrix between momentum and the matrix formed by the conduction and valence bands. Now, try to gap this Hamiltonian: we cannot! If we add a "mass" term, we obtain

$$H(\mathbf{k}) = (k_i - K_i)A_{ij}\sigma_j + m\sigma_y. \tag{11.9}$$

We will move the gapless point only from $\mathbf{k} = \mathbf{K}$ to $\mathbf{k} = \mathbf{K} - m\hat{1}_y$. This impossibility to gap the system by the addition of a small mass comes from the fact that the codimension of 2×2 matrices is 3, but the space dimension is also 3. Therefore, in general, we do not need to tune any parameter to get a gapless system. Tuning three parameters, k_x, k_y, k_z, in the space m to make a transition will, in general, determine a curve. For a generic m, there will be a point $\mathbf{k}$ where the system is gapless. That point will move around with increasing value of m until it meets up with another point and annihilates thus becoming gapped again.

The reason why Dirac points in three dimensions are stable is rather deep and has to do with the fact that in three dimensions, we can actually define a Chern number for a Fermi surface. In two dimensions, only filled bands have integer Chern numbers because we needed to integrate over a 2-D manifold to obtain a Chern number. In three dimensions, Fermi surfaces define a 2-D manifold, so the integral of the Berry curvature over the whole Fermi surface gives a Chern number:

$$C_1 = \frac{1}{2\pi}\int_{FS} d\Omega^{ij}(\partial_i a_j(\mathbf{k}) - \partial_j a_i(\mathbf{k})), \tag{11.10}$$

where the integral is made over the Fermi surface and $a_i = -i\left\langle \mathbf{k}\right| \frac{\partial}{\partial k_i}\left|\mathbf{k}\right\rangle$ is the Berry potential. The Chern number vanishes if there is no crossing between the band whose Fermi surface we are looking at and any other bands. The Chern number comes from the monopole that stands at the band crossing. In the general case of the Hamiltonian $(k_i - K_i)A_{ij}\sigma_j$, the Chern number of the Fermi surface is $C_1 = \text{sign}(Det(A))$. Notice that in the current case, we are dealing with the Hamiltonian around an arbitrary degeneracy point $\mathbf{K}$. Therefore, the stability of the Dirac node in three dimensions *does not rely* on any other symmetries besides the charge $U(1)$ symmetry that protects Chern numbers (the absence of $c^\dagger c^\dagger$ terms). In a TR-invariant system, at $-\mathbf{K}$, we have another nodal point.

In three dimensions, the degeneracy point can be imagined to be the source of a monopole field. The Berry field strength of $\mathbf{B}(\mathbf{k}) = \nabla_{\mathbf{k}} \times \mathbf{A}(\mathbf{k})$ can be used to define a monopole density in $\mathbf{k}$ space $\rho = \nabla_{\mathbf{k}} \cdot \mathbf{B}(\mathbf{k})$. The monopole density can be nonzero only at degeneracy points—at any other points, the monopole density, by virtue of being the divergence of a curl, vanishes if $\mathbf{A}(\mathbf{k})$ can be well defined. The monopole density is a delta function of the momentum at the degeneracy point $\mathbf{K}$, $\rho(\mathbf{k}) = q\delta(\mathbf{k}-\mathbf{K})$, and has a monopole charge q, which cannot change under a local smooth change of the Hamiltonian parameters. This is another way of understanding the stability of the Dirac points in three spatial dimensions. No other symmetries besides $U(1)$ charge conservation are required. Two monopoles can annihilate *only* if they have opposite monopole charge. They can usually annihilate at any point in the BZ, and, therefore, 3-D Dirac crossings are not *globally* stable.

With TR invariance, there are more restrictions on the momentum space position where two monopoles can annihilate. With time reversal, we know that the Berry field strengths at points $\mathbf{K}$ and $-\mathbf{K}$ are negatives of each other: $\mathbf{B}(\mathbf{K}) = -\mathbf{B}(-\mathbf{K})$. As such, the monopole charges of the two monopoles at $\mathbf{K}$ and $-\mathbf{K}$ are identical (being the derivatives of the monopole Berry field strengths), and these monopoles cannot vanish by annihilating with each other at a TR-invariant point. What happens in three dimensions upon a band-closing-and-reopening transition is the following. Start from the atomic limit. To even be able to close the gap, we find that, at some arbitrary point $\mathbf{K}$, we must form a monopole-antimonopole charge. We cannot create a monopole charge out of nothing, and so to be able to close the gap at some arbitrary K, we must create a monopole-antimonopole pair. With TR symmetry, another monopole-antimonopole pair is created at $-\mathbf{K}$. Upon changing the parameters of the Hamiltonian, the monopoles or antimonopoles cannot disappear spontaneously unless they annihilate in pairs. The system is metallic, and there is no gap between the valence and conduction bands. Generically, this means that we cannot go from a trivial insulator to a topological one directly but need to go through a gapless phase. They cannot annihilate at TR-invariant points because monopoles of same charge meet there. As such, what happens is that the monopole created at K annihilates with the antimonopole created at $-\mathbf{K}$, and vice versa. The annihilations can happen at different points in the BZ, $\mathbf{K}'$.

We hence showed that going from a trivial insulator to a topological one in three-dimensions involves going through a gapless *phase* and not just a gapless point. The gapless phase shrinks to a point in the case of inversion-symmetric systems, as we now show.

11.2 Inversion-Symmetric Systems

If on top of TR symmetry we also have inversion symmetry, the gap closing and reopening cannot happen at any point *except* the TR-invariant points. The bands in an inversion-symmetric system are doubly degenerate. If we call the inversion operator P, then, when combined with TR symmetry, PT, we obtain a nonunitary operator that is a symmetry of the Bloch Hamiltonian: $[H(\mathbf{k}), PT] = 0$. Any Bloch wavefunction $|u(\mathbf{k})\rangle$ is accompanied by an orthogonal and degenerate one, $PT\,|u(\mathbf{k})\rangle$. If there is a gap-closing transition between conduction and valence bands, four bands must need at the transition point. The Hamiltonian is, hence, a 4×4 matrix that is two-fold degenerate in each of the two bands and so must be expandable in the Clifford algebra matrices:

$$H(\mathbf{k}) = E_0 + \sum_{i=1}^{5} d_i \Gamma_i \tag{11.11}$$

at *any* momentum k. Hence the codimension of our Hamiltonian is 5, there is strong level repulsion, and it would seem that there is no way to force a transition by changing only one parameter. This is not entirely true. We have inversion symmetry implemented by an operator P,

$$PH(\mathbf{k})P^{-1} = H(-\mathbf{k}), \tag{11.12}$$

which is a unitary matrix. Unitary operators are good because they support eigenvalues—unlike antiunitary operators such as time reversal. Applying inversion twice leads to the identity

$$P^2 = 1, \tag{11.13}$$

so the eigenvalues of P are ± 1. We now particularize to insulating systems with two occupied bands—this is the generic case because more bands can always be analyzed as the direct sum of two band subsystems. Inversion commutes with spin operators, so if we are to write inversion as a block matrix in spin $\uparrow, \downarrow$ basis, we would have

$$P = \begin{bmatrix} P_\uparrow & 0 \\ 0 & P_\downarrow \end{bmatrix}. \tag{11.14}$$

At inversion-symmetric points (which are the same as TR-invariant points), the Hamiltonian commutes with the inversion operator, and the bands have eigenvalues under the inversion operator. As such, in a basis where the operator is diagonal, we can write the inversion operator as

$$P_\uparrow = P_\downarrow = \begin{bmatrix} \eta_a & 0 \\ 0 & \eta_b \end{bmatrix}, \tag{11.15}$$

where $\eta_a = \pm 1$, $\eta_b = \pm 1$ are the parity eigenvalues of the wavefunctions at the specific inversion-symmetric point. One of the eigenvalues η_a, η_b corresponds to the conduction band and one corresponds to the valence band. The distinct combinations of η_a, η_b are $\eta_a = \eta_b$ or $\eta_a = -\eta_b$. We now analyze the two cases.

11.2.1 $\eta_a = \eta_b$

Because P acts on the orbital basis (and commutes with the spin basis), in this case, the two orbitals are of the same parity. For example, they could both be either s-orbitals or p-orbitals, *etc.*—remember that the parity operator depends on what orbitals you have in your problem.

In this case, the matrix P is unitarily equivalent to the 4×4 identity matrix (and anything unitarily equivalent to the identity matrix is the identity matrix $P = I$), and we have

$$H(\mathbf{k}) = PH(\mathbf{k})P^{-1} = H(-\mathbf{k}). \tag{11.16}$$

Hence, the Hamiltonian (which has both inversion and TR symmetry and, hence, doubly-degenerate bands) can be written as

$$H(\mathbf{k}) = E_0(\mathbf{k}) + \sum_{i=1}^{5} d_i(\mathbf{k})\Gamma_i, \tag{11.17}$$

where by the inversion operator chosen, $P = I$, we find the relations $d_i(\mathbf{k}) = d_i(-\mathbf{k})$, and $E_0(\mathbf{k}) = E_0(-\mathbf{k})$. Because, by time reversal and then inversion with $P = I$, we have $TH(\mathbf{k})T^{-1} = H(-\mathbf{k}) = H(\mathbf{k})$, then the matrices Γ_i have to be even under time reversal. As such, we can choose two of them to be the orbital x, z-Pauli matrices tensored with the identity in spin index, whereas the other three can be the orbital y-Pauli matrix tensored with the three Pauli matrices in spin space:

$$\Gamma_{1,2,3,4,5} = (\tau_x \mathrm{I},\ \tau_z \mathrm{I},\ \tau_y \sigma_x,\ \tau_y \sigma_y,\ \tau_y \sigma_z), \tag{11.18}$$

where σ is spin space and τ is orbital space. Hence, at any point in the BZ, including the TR-invariant points, the codimension is 5 and the gap cannot be closed by changing one parameter only. Even if we have three dimensions and, hence, three momenta to vary by themselves, the gap still cannot be closed without extreme fine-tuning because we would have to tune two parameters.

11.2.2 $\eta_a = -\eta_b$

In this case, P can be unitarily reduced to diagonal form of the z-Pauli matrix in orbital space:

$$P = P^{-1} = \begin{bmatrix} \tau_z & 0 \\ 0 & \tau_z \end{bmatrix}, \tag{11.19}$$

or $P = \tau_z I$. Note that the system described by this symmetry is composed of orbitals with opposite symmetry upon inversion (such as s and p). Let us find the transformation properties of the Hamiltonian:

$$H(\mathbf{k}) = E_0(\mathbf{k}) + \sum_{i=1}^{5} d_i(\mathbf{k})\Gamma_i. \tag{11.20}$$

We neglect the constant term, $E_0(\mathbf{k})$, because it does not influence the eigenstates nor the gap closing and reopening. We would now like to find the constraints that time reversal and the nontrivial inversion matrix impose on our Hamiltonian. We can write *any* Γ_i generically as a tensor product of matrices in orbital and spin space :

$$\Gamma_i = (x_{i1} I + x_{i2}\tau_x + x_{i3}\tau_y + x_{i4}\tau_z) \otimes (y_{i1} I + y_{i2}\sigma_x + y_{i3}\sigma_y + y_{i4}\sigma_z) \tag{11.21}$$

(in fact, we can write any of the matrices Γ_a, $\Gamma_{ab} = [\Gamma_a, \Gamma_b]/2i$ in such a manner), where x_i and y_i can take the values 1 or 0. If one of the x_i (y_i) is 1, the others x_i (y_i) are zero. The x_i, y_i have to be real because otherwise the Hamiltonian would not be Hermitian. Because we have both TR and inversion symmetry, the operator $PT = -i\tau_z\sigma_y K$ commutes with the Bloch Hamiltonian at each $\mathbf{k}$—$[PT, H(\mathbf{k})] = 0$. This severely reduces the possibilities for the matrices Γ_i: they have to commute with PT—$[PT, \Gamma_i] = 0$. There are exactly five Γ matrices that satisfy this constraint; they can be checked to be

$$\Gamma_1 = \tau_x \sigma_z, \quad \Gamma_2 = \tau_y I, \quad \Gamma_3 = \tau_x \sigma_x, \quad \Gamma_4 = \tau_x \sigma_y, \quad \Gamma_5 = \tau_z I = P. \tag{11.22}$$

It is easy to see that these matrices commute with PT. In $\Gamma_1, \Gamma_3, \Gamma_4$, we have the spin matrix, which changes sign upon time reversal; the spin matrix is tensored into the τ_x, τ_y orbital matrices, both of which anticommute with τ_z of the inversion matrix. Hence the tensor product commutes. The Γ_2 matrix contains only the τ_y in orbital space, but this matrix anticommutes with both P and T separately because it anticommutes with τ_z and it is imaginary and hence flipped in sign by K. Γ_5 is made of only τ_z, which commutes with both P and T separately. The five Γ matrices just obtained anticommute with each other. Notice that the Clifford algebra matrices are completely different in this case from the case where the inversion-symmetric matrix was a constant. Having found the Clifford algebra, we now obtain the symmetry of the Bloch functions $d_i(\mathbf{k})$. From $PH(\mathbf{k})P^{-1} = H(-\mathbf{k})$, which can be written $\Gamma_5 H(\mathbf{k})\Gamma_5 = H(-\mathbf{k})$, we obtain

$$d_a(\mathbf{k}) = -d_a(-\mathbf{k}), \; a = 1, 2, 3, 4; \qquad d_5(\mathbf{k}) = d_5(-\mathbf{k}). \tag{11.23}$$

We hence find that in this case of orbitals of opposite inversion-symmetry eigenvalues, d_5 is even in $\mathbf{k}$, whereas $d_{1,2,3,4}$ are odd in k. At generic $\mathbf{k} \neq \mathbf{G}/2$, this system has codimension 5, just like the previous case (there are five parameters to be tuned to zero to obtain a degeneracy). However, at a TR-invariant point, we have that

$$d_{1,2,3,4}\left(\frac{\mathbf{G}}{2}\right) = 0, \tag{11.24}$$

and the Hamiltonian is just

$$H\left(\frac{\mathbf{G}}{2}\right) = E_0\left(\frac{\mathbf{G}}{2}\right) + d_5\left(\frac{\mathbf{G}}{2}\right)\Gamma_5 = E_0\left(\frac{\mathbf{G}}{2}\right) + d_5\left(\frac{\mathbf{G}}{2}\right)P \tag{11.25}$$

The Hamiltonian at a TR-invariant point is, hence, just the inversion matrix. The bands have inversion eigenvalues ± 1 (Kramers' doublets have identical inversion eigenvalues by the fact that $[T, P] = 0$). The gap closes by tuning a single parameter, $d_5(\mathbf{G}/2)$, *only* in the case when the two orbitals have opposite inversion eigenvalues. We can now establish a relation between the topological nature of a system and its inversion eigenvalues. If we start from the atomic limit, all hoppings and spin-orbit coupling are tuned to zero, and the energy bands are flat, made out of the on-site energy of the two orbitals (and two spins). As such, the inversion eigenvalues at *all* TR-invariant points in the BZ must be the same—in the atomic limits, the Bloch eigenstates do not have momentum dependence and are identical at all points in the BZ. We now turn on the hoppings by varying one parameter. If we undergo a phase transition, it must be at a TR-invariant point—that is where the Hamiltonian codimension is 1; all other points experience strong level repulsion. When the Hamiltonian gap closes and reopens at a TR-invariant point, the Hamiltonian undergoes a transition from a trivial atomic limit insulator to a topological insulator. At the same time, the two inversion eigenvalues of the occupied Kramers' pair change from positive to negative (or vice versa). Because inversion eigenvalues of Kramers' pairs are identical, we can meaningfully talk about *half* the inversion eigenvalues. Upon the band closing and reopening *at one* of the TR-invariant points, the product of half of the inversion eigenvalues at those points becomes negative (from positive in the atomic limit).

Hence,

$$\prod_{i=1}^{8} \prod_{a \in \text{half of Kramers pairs}} \xi_a \left(\frac{\mathbf{G}}{2}\right) = -1 \tag{11.26}$$

is the condition for a nontrivial topological insulator in three dimensions.

11.3 Mercury Telluride Hamiltonian

We now use the simple example of the mercury telluride (HgTe) model for a 2-D TR-invariant topological insulator to understand how to apply the theory developed so far. We are looking at a Hamiltonian with s- and p-orbitals and spin $\uparrow, \downarrow$ on each site. Our Bloch basis is

$$\psi = (|s, \uparrow\rangle, |p, \uparrow\rangle, |s, \downarrow\rangle, |p, \downarrow\rangle)^\dagger, \tag{11.27}$$

where s, p denote the different orbitals. The TR matrix is

$$T = -i\sigma_y \times IK = \begin{bmatrix} 0 & 0 & -1 & 0 \\ 0 & 0 & 0 & -1 \\ 1 & 0 & 0 & 0 \\ 0 & 1 & 0 & 0 \end{bmatrix} K, \tag{11.28}$$

where K is complex conjugation. Because in our model we have s- and p-orbitals on site, the inversion matrix will be different than in the case of graphene, which has two inequivalent sublattices; in that case inversion sends one of the atoms in the unit cell into another and is given by the t_x matrix, as we explained in previous chapters. In our case, the s-orbitals are even under inversion, whereas the p are odd, so the inversion matrix contains the Pauli t_z (in effect, in the real material, the s-orbitals form an antibonding orbital, whereas the p-orbitals form a bonding orbital, renders their eigenvalues exactly the opposite of the ones here, but that is a material detail—the important property is that the two orbitals have opposite inversion eigenvalues). Inversion also commutes with spin, so it must have the same form for both spin $\uparrow, \downarrow$. It is

$$P = I \times t_z \begin{bmatrix} 1 & 0 & 0 & 0 \\ 0 & -1 & 0 & 0 \\ 0 & 0 & 1 & 0 \\ 0 & 0 & 0 & -1 \end{bmatrix}. \tag{11.29}$$

From now on, spin matrices will be σ_i, whereas the orbital matrices will be t_i. As we have learned, the difference between P and T is that $P^2 = 1$, but $T^2 = -1$. To connect with the matrix basis used in the original paper, we can define the following Clifford-algebra matrices and their transformation properties under time reversal and inversion:

$$\Gamma_1 = \sigma_z \times t_x, \quad T\Gamma_1 T^{-1} = -\Gamma_1, \quad P\Gamma_1 P^{-1} = -\Gamma_1, \tag{11.30}$$

$$\Gamma_2 = I \times t_y, \quad T\Gamma_2 T^{-1} = -\Gamma_2, \quad P\Gamma_2 P^{-1} = -\Gamma_2, \tag{11.31}$$

$$\Gamma_3 = I \times t_z = P, \quad T\Gamma_3 T^{-1} = \Gamma_3, \quad P\Gamma_3 P^{-1} = \Gamma_3, \tag{11.32}$$

$$\Gamma_4 = \sigma_x \times t_x, \quad T\Gamma_4 T^{-1} = -\Gamma_4, \quad P\Gamma_4 P^{-1} = -\Gamma_4, \tag{11.33}$$

$$\Gamma_5 = \sigma_y \times t_x, \quad T\Gamma_5 T^{-1} = -\Gamma_5, \quad P\Gamma_5 P^{-1} = -\Gamma_5. \tag{11.34}$$

These matrices satisfy the Clifford property $\{\Gamma_a, \Gamma_b\} = 2\delta_{ab}$, which implies that any Hamiltonian built from a combination of them,

$$H(\mathbf{k}) = \sum_{a=1}^{5} d_a(\mathbf{k})\Gamma_a, \tag{11.35}$$

where $d_a(\mathbf{k})$ are Bloch-periodic functions of the 2-D momentum, will have doubly-degenerate bands. We now ask what we require for a model built out of these matrices to respect time reversal $TH(\mathbf{k})T^{-1} = H(-\mathbf{k})$:

$$d_i(\mathbf{k}) = -d_i(-\mathbf{k}), \quad i = 1, 2, 4, 5; \qquad d_3(\mathbf{k}) = d_3(-\mathbf{k}). \tag{11.36}$$

Notice that the TR constraint, as well as the (artificial, soon to be relaxed) constraint that the Hamiltonian be spanned by the five Clifford matrices, implies that the Hamiltonian is also inversion symmetric: $PH(\mathbf{k})P^{-1} = H(-\mathbf{k})$. We can ask if, *while keeping the spectrum doubly degenerate at each k*, can we add any other terms to the Hamiltonian that break P but keep TR? The answer, as before, is obviously no: to keep the spectrum doubly degenerate, we must use Clifford matrices, and time reversal on those implies P. This is just an explicit example of the theorem that a system with both TR and inversion symmetry must have doubly-degenerate bands.

Notice how the Hamiltonian in equation (11.35), when it respects TR invariance, turns into exactly the inversion matrix at TR-invariant points:

$$H\left(\frac{\mathbf{G}}{2}\right) = d_3\left(\frac{\mathbf{G}}{2}\right)\Gamma_3 = d_3\left(\frac{\mathbf{G}}{2}\right)P. \tag{11.37}$$

This means that the Bloch wavevectors at the TR-invariant points can be classified by inversion (P) quantum numbers or eigenvalues (because we are working with orbitals of opposite inversion quantum numbers, it means that, as we've just learned, the transition at the TR-invariant point has codimension 1 and thus can happen by tuning one parameter). The change in the inversion quantum number, sign($d_3(\mathbf{G}/2)$), of the occupied and empty bands can occur only through a phase transition, which, as we've learned, changes the topological class of the insulator. Let's again see this through an even more explicit example. The inversion-symmetric Hamiltonian of HgTe is

$$H(\mathbf{k}) = \sin(k_x)\Gamma_1 + \sin(k_y)\Gamma_2 + (2 + M - \cos(k_x) - \cos(k_y))\Gamma_3 = \sum_{i=1}^{3} d_i(\mathbf{k})\Gamma_i. \tag{11.38}$$

This is the lattice generalization of the continuum HgTe Hamiltonian developed earlier around the (0, 0) point in the BZ. This simple Hamiltonian is even more than inversion symmetric: its symmetry class is $U(1) \times \bar{U}(1)$, because it is just two Chern insulators (physically equivalent to Haldane's model), one with spin $\uparrow$ and one with spin $\downarrow$, which have opposite Hall conductances. As we have learned, each separate Chern insulator has a trivial phase for $M > 0$ and $M < -4$, with phase transitions separating phases of different Chern insulators with nonzero Hall conductance at $M = 0, -2, -4$. Close to the TR-invariant points, the Hamiltonian is of Dirac type in k_x, k_y and has a mass matrix given by the term Γ_3:

$$H\left(\frac{\mathbf{G}}{2}\right) = d_3\left(\frac{\mathbf{G}}{2}\right)\Gamma_3 = d_3\left(\frac{\mathbf{G}}{2}\right)P \tag{11.39}$$

An eigenstate of $H(\mathbf{G}/2)$ will be an eigenstate of P with value

$$H\left(\frac{\mathbf{G}}{2}\right)\psi = E\psi \rightarrow d_3\left(\frac{\mathbf{G}}{2}\right)P\psi = E\psi, \rightarrow P\psi = \frac{E}{d_3(\mathbf{G}/2)}\psi. \tag{11.40}$$

But we have that $E_{\pm} = \pm\sqrt{d_3(\mathbf{G}/2)^2}$; hence,

$$P\psi = \pm\text{sign}\left(d_3\left(\frac{\mathbf{G}}{2}\right)\right)\psi. \tag{11.41}$$

For the occupied band, the eigenvalue of P is $-d_3(\mathbf{G}/2)$. If the mass of the Dirac fermion is identical at all the TR-invariant points, it means that the Hall conductance of both the spin $\uparrow$ and of the spin $\downarrow$ fermions is zero (the ferromagnetic situations explained in chap. 8), and the system has no spin-Hall conductance either. However, if the bands are such that the eigenvalue of P has one sign at one point and has the other sign at all the other TR-invariant points, it *must* be that a Dirac fermion changed sign at some TR-invariant point (as proved before, this can happen only at a TR-invariant point because otherwise the codimension would be too large).

Before we analyze all the other symmetries of the matrices that can form the Hamiltonian we remark that inversion can easily be broken by introducing one of the $SU(4)$ matrices:

$$\Gamma_{ab} = -\frac{1}{2i}[\Gamma_a, \Gamma_b], \tag{11.42}$$

with the following transformation properties under inversion and time reversal: $\Gamma_{12} = -\sigma_z \times t_z$, T even, P odd; $\Gamma_{13} = \sigma_z \times t_y$, T even, P odd; $\Gamma_{14} = -\sigma_y \times 1$, T odd, P even; $\Gamma_{15} = -\sigma_x \times 1$, T odd, P even; $\Gamma_{23} = -1 \times t_x$, T even, P odd; $\Gamma_{24} = \sigma_x \times t_z$, T odd, P even; $\Gamma_{25} = -\sigma_y \times t_z$, T odd, P even; $\Gamma_{34} = -\sigma_x \times t_y$, T even, P odd; $\Gamma_{35} = \sigma_y \times t_y$ T even, P odd; $\Gamma_{45} = \sigma_z \times 1$, T odd, P even. We are now in position to introduce terms that will break the inversion symmetry but maintain time reversal. The Hamiltonian

$$H(\mathbf{k}) = \sin(k_x)\Gamma_1 + \sin(k_y)\Gamma_2 + (2 + M - \cos(k_x) - \cos(k_y))\Gamma_3 + \Delta\Gamma_{35} \tag{11.43}$$

will break P but maintain time reversal, hence being an apt description of HgTe quantum wells, which have slightly broken inversion symmetry. It will have spin-split bands. Rotation symmetry can be broken by adding different coefficients to the $\sin(k_x)$, $\sin(k_y)$ terms. Δ is the amount of P-breaking in the Hamiltonian that is due to the tetrahedral lattice structure with different atoms (Hg, Te) making up the crystal. The time reversal can be broken by many terms, among which are the Zeeman terms, which in HgTe have the form

$$\begin{aligned} g_x &= \Gamma_{24} - \Gamma_{15}, & Tg_xT^{-1} &= -g_x, & Pg_xP^{-1} &= g_x, \\ g_y &= \Gamma_{25} - \Gamma_{14}, & Tg_yT^{-1} &= -g_y, & Pg_yP^{-1} &= g_y, \\ g_z &= \Gamma_{45}, & Tg_zT^{-1} &= -g_z, & Pg_zP^{-1} &= g_z. \end{aligned} \tag{11.44}$$

11.4 Problems

1. *Monopole Charge in Three Dimensions:* From the explicit form of the eigenstates of the system, obtain the expression for the monopole charge at the gapless point of the Weyl Hamiltonian $k_i\sigma_i, i = 1, 2, 3$, and show that it is indeed a delta function localized at $k_i = 0$.

2. *Explicit Band Crossing in Three Dimensions with Time Reversal and Inversion:* Choose a simple 3-D Hamiltonian $\sin(k_x)\tau_x \otimes \sigma_x + \sin(k_y)\tau_x \otimes \sigma_y + \sin(k_z)\tau_x \otimes \sigma_z + (M - \cos(k_x) - \cos(k_y) - \cos(k_z))\tau_z \otimes I$ and show that it is a topological insulator upon a gap-closing-and-reopening transition at one of the TR-invariant points. (Which one?)

3. *Explicit Band Crossing in Three Dimensions without Inversion:* Now break inversion symmetry by adding several terms, $\Delta_1\tau_x \otimes I + \Delta_2\tau_z \otimes I$ (add other terms as well), to the Hamiltonian in the previous problem. For $\Delta << 1$ small, how does the phase transition between the phases $M \to \infty$ and $M = 2$ (a nontrivial phase) occur? Does the system go through a metallic phase? Where in the BZ is the gap-closing point(s) (you can diagonalize the Hamiltonian numerically if an analytic solution is not possible).

4. *Monopole-Antimonopole Evolution:* Show explicitly the evolution of the monopole-antimonopole pairs in the gap-closing-reopening transition of a 3-D Hamiltonian with the small inversion symmetry breaking in the previous problem (when inversion symmetry is present, we can close the gap and reopen it at one point, thereby not observing the nontrivial effect of the monopole-antimonopole evolution). Where do the monopoles form, how many of them are there (generically), and how do they recombine to create the topological insulating state?

5. *Weyl Fermions and Fermi Arcs:* Now solve the same Hamiltonian as in the past three problems but with *open* boundary conditions in the z-direction. How does the surface electronic structure evolve from no surface state in the case $M \to \infty$ to the Dirac surface state for $M = 2$? Are there surface states in the metallic phase? What happens to them as the parameters are tuned toward the topological insulator? Use translational invariance in the x- and y-directions and plot the dispersion of the spectrum in the 2-D BZ of the surface. (The answer is yes: in the metallic case, there are *surface Fermi arcs* connecting the bulk monopole-antimonopole pair. There are at least two such surface Fermi arcs, one for each monopole-antimonopole pair, and they exhibit TR invariance. Upon the monopole-antimonopole pair that originated through a gap closing at some k-point annihilating with the antimonopole-monopole pair from the $-k$-point, the TR Fermi arcs connect and form the surface state of the topological insulator. These Fermi arcs are among the hottest research topics.

12

Time-Reversal Topological Insulators with Inversion Symmetry

We have given strong arguments that the topological classification of insulators with both TR and inversion symmetry should depend on the inversion eigenvalues of the system. It is now time to prove this. (We will prove this through another method, that of dimensional reduction, in the next chapter.) In this chapter we focus on insulators with both TR symmetry and inversion symmetry:

$$PH(\mathbf{k})P^{-1} = H(-\mathbf{k}), \tag{12.1}$$

where the inversion matrix is unitary and squares to the identity $P^{\dagger}P = 1,\ PP = 1$. At inversion-symmetric momenta (which are also TR-invariant momenta if we have TR invariance), $\mathbf{k} = \mathbf{G}/2$, the inversion operator commutes with the Hamiltonian $[P,\ H(G/2)] = 0$, and hence the bands at $\mathbf{G}/2$ can be classified by an inversion eigenvalue:

$$P\left|u_{i,\mathbf{k}=\mathbf{G}/2}\right\rangle = \zeta_i(\mathbf{G}/2)\left|u_{i,\mathbf{k}=\mathbf{G}/2}\right\rangle. \tag{12.2}$$

In a 1-D system there are two eigenvalues for each band i, $\zeta_i(0)$ and $\zeta_i(\pi)$. Assume $\left|u_i(\mathbf{k})\right\rangle$ is an eigenstate of the Hamiltonian at energy $E_i(\mathbf{k})$, then $P\left|u_i(\mathbf{k})\right\rangle$ is necessarily an eigenstate at $-\mathbf{k}$ of the same energy:

$$H(-\mathbf{k})P\left|u_i(\mathbf{k})\right\rangle = PH(\mathbf{k})\left|u_i(\mathbf{k})\right\rangle = E_i(\mathbf{k})P\left|u_i(\mathbf{k})\right\rangle. \tag{12.3}$$

We define the sewing matrix B_{ij} as the "sewing" matrix connecting the bands at k with the ones at $-\mathbf{k}$ through the inversion operator:

$$\left|u_i(-\mathbf{k})\right\rangle = B^*_{ij}(\mathbf{k})P\left|u_j(\mathbf{k})\right\rangle, \tag{12.4}$$

where $i,\ j$ run over the occupied bands $1, \ldots, N$. If the bands are singly degenerate, then obviously the matrix B_{ij} has only diagonal terms, which are phases. Let $(u_{i,\mathbf{k}})_\alpha$ be the αth component of the ith eigenstate of $H(\mathbf{k})$. The matrix P is then a unitary matrix $P = P_{\alpha\beta}$. We rewrite equation (12.4) component by component as $(u_{i,-\mathbf{k}})_\alpha = B^*_{ij}(\mathbf{k})P_{\alpha\beta}(u_{j,k})_\beta$ and find

$$\left\langle u_{i,-\mathbf{k}}\right| P\left|u_{j,\mathbf{k}}\right\rangle = (u_{i,-\mathbf{k}})^*_\alpha P_{\alpha\beta}(u_{i,\mathbf{k}})_\beta = B_{il}(\mathbf{k})P^*_{\alpha\theta}(u_{l,\mathbf{k}})^*_\theta P_{\alpha\beta}(u_{j,\mathbf{k}})_\beta = B_{ij}, \tag{12.5}$$

where we have successively used the identities $P^*_{\alpha\theta}P_{\alpha\beta} = (P^{\dagger}P)_{\theta\beta} = \delta_{\theta,\beta}$ and $(u_{l,\mathbf{k}})^*_\theta(u_{j,\mathbf{k}})_\theta = \delta_{lj}$.

The matrix B_{ij} relates eigenstates to eigenstates, and hence it is unitary. This can be proved mathematically by an extension of the argument used for TR sewing matrices. That B_{ij} has to be unitary can easily be seen by extending the matrix to belong to *all* bands, occupied *and* unoccupied. However, since we have a full gap in the system, none of the occupied bands at $\mathbf{k}$ can transform to unoccupied bands at $-\mathbf{k}$, and vice versa (otherwise we would not have an insulator). This means that the full sewing matrix is diagonal in the block of occupied and unoccupied bands: $B^{\text{full}} = \text{diag}(B^{\text{occupied}}, B^{\text{unoccupied}})$. Hence, if the full matrix B is unitary, so are the $B^{(\text{un})\text{occupied}}$. To prove that the full B is unitary is easy by using completeness of the eigenstates $\sum_{i\in\text{all bands}} |u_i(\mathbf{k})\rangle \langle u_i(\mathbf{k})| = I$. We have (double index means summation)

$$(B^\dagger B)_{ij} = (B^\dagger)_{im} B_{mj} = \langle u_{i,\mathbf{k}}| P^\dagger |u_{m,-\mathbf{k}}\rangle \langle u_{m,-\mathbf{k}}| P |u_{j,\mathbf{k}}\rangle$$

$$= \langle u_{i,\mathbf{k}}| P^\dagger P |u_{j,\mathbf{k}}\rangle = \langle u_{i,\mathbf{k}} |u_{j,\mathbf{k}}\rangle = \delta_{ij}. \tag{12.6}$$

We now revert to the notation where B_{ij} has i, j taking values over the occupied bands. We note that at inversion-invariant points, the sewing matrix is diagonal and takes the form of the inversion eigenvalues of the bands:

$$B_{ij}\left(\frac{\mathbf{G}}{2}\right) = \langle u_{i,\mathbf{G}/2}| P |u_{j,\mathbf{G}/2}\rangle = \zeta_j \left(\frac{\mathbf{G}}{2}\right) \delta_{ij}, \tag{12.7}$$

and hence the determinant of the inversion sewing matrix at $\mathbf{G}/2$ is the product of the eigenvalues over all the occupied bands at that inversion-invariant point:

$$\text{Det}\left(B\left(\frac{\mathbf{G}}{2}\right)\right) = \prod_{j=1}^{N_1} \zeta_j \left(\frac{\mathbf{G}}{2}\right). \tag{12.8}$$

12.1 Both Inversion and Time-Reversal Invariance

We have briefly analyzed the bands with inversion symmetry. We now add TR symmetry to the system to find that the expression of a Z_2 invariant for the topological insulator becomes much simpler. With inversion *and* time reversal, the bands are doubly degenerate, and the operator PT commutes with the Bloch Hamiltonian at each $\mathbf{k}$. We can define another unitary matrix that related bands *at the same k*:

$$|u_m(\mathbf{k})\rangle = v^*_{mn}(\mathbf{k}) PT |u_n(\mathbf{k})\rangle, \tag{12.9}$$

or—in the notation of [9]—$v_{mn} = \langle u_m(k)| PT |u_n(k)\rangle$. Due to the presence of the antiunitary $T = UK$ operator with U a unitary antisymmetric matrix and because $P^\dagger = P$, the preceding matrix has vanishing diagonal entries:

$$\langle u_n(\mathbf{k})| PT |u_n(\mathbf{k})\rangle = (u_{n\mathbf{k}})^*_\alpha P_{\alpha\beta} U_{\beta\gamma} (u_{n\mathbf{k}})^*_\gamma = (u_{n\mathbf{k}})^*_\gamma P_{\alpha\beta} U_{\beta\gamma} (u_{n\mathbf{k}})^*_\alpha$$

$$= (u_{n\mathbf{k}})^*_\gamma U_{\alpha\beta} P^*_{\beta\gamma} (u_{n\mathbf{k}})^*_\alpha = -(u_{n\mathbf{k}})^*_\gamma P^*_{\beta\gamma} U_{\beta\alpha} (u_{n\mathbf{k}})^*_\alpha$$

$$= -(u_{n\mathbf{k}})^*_\gamma P_{\gamma\beta} U_{\beta\alpha} (u_{n\mathbf{k}})^*_\alpha = -\langle u_n(\mathbf{k})| PT |u_n(\mathbf{k})\rangle = 0, \tag{12.10}$$

where we used the fact that U is an antisymmetric matrix. The matrix v is, of course, unitary as it relates eigenstates to eigenstates. The proof is simple, follows the one for the inversion sewing matrix, and will not be presented here. We now want to find the Berry-phase vector potential (double index implies summation):

$$\begin{aligned}\mathbf{A}(\mathbf{k}) &= -i(u_{n\mathbf{k}})^*_a \nabla (u_{n\mathbf{k}})_a = -i v_{mn}(\mathbf{k}) P^*_{\alpha\beta} U^*_{\beta\gamma}(u_{n\mathbf{k}}) \nabla (v^*_{mq}(\mathbf{k}) P_{\alpha\delta} U_{\delta\chi}(u_{q\mathbf{k}})^*_\chi) \\ &= -i(v_{mn}(\mathbf{k}) \nabla v^*_{mq}(\mathbf{k})) P^*_{\alpha\beta} P_{\alpha\delta} U^*_{\beta\gamma} U_{\delta\chi}(u_{n\mathbf{k}})_\gamma (u_{q\mathbf{k}})^*_\chi \\ &\quad -i v_{mn}(\mathbf{k}) v^*_{mq}(\mathbf{k}) P^*_{\alpha\beta} U^*_{\beta\gamma} P_{\alpha\delta} U_{\delta\chi}(u_{n\mathbf{k}})_\gamma \nabla (u_{q\mathbf{k}})^*_\chi \\ &= -i v_{mn} \nabla v^*_{mn} - i(v^\dagger)_{qm} v_{mn} (u_{n\mathbf{k}})_\gamma \nabla (u_{q\mathbf{k}})^*_\gamma = \\ &= i \mathrm{Tr}[v^\dagger \nabla v] - \mathbf{A}(\mathbf{k}), \end{aligned} \tag{12.11}$$

and hence the Abelian Berry potential takes a nice form in terms of the sewing matrix v between states at the same k:

$$\mathbf{A}(\mathbf{k}) = \frac{i}{2} \mathrm{Tr}[v^\dagger \nabla v] = \frac{i}{2} \nabla \log[\mathrm{Det}[v(\mathbf{k})]] = i \nabla \log[Pf[v(\mathbf{k})]] \tag{12.12}$$

Because the matrix v is antisymmetric, we can take its Pfaffian. Under a gauge transformation of only one of the occupied eigenstates, we have

$$|u_{n,\mathbf{k}}\rangle \to e^{i\theta_\mathbf{k}} |u_{n,\mathbf{k}}\rangle \quad \text{for } n = 1, \qquad |u_{n,\mathbf{k}}\rangle \to |u_{n,\mathbf{k}}\rangle \quad \text{for } n \neq 1, \tag{12.13}$$

and the Pfaffian of v transforms into

$$Pf[v(\mathbf{k})] \to Pf[v(\mathbf{k})] e^{-i\theta_\mathbf{k}}. \tag{12.14}$$

Hence, we can make the Pfaffian of $v(\mathbf{k})$ unity everywhere (we can cancel its phase everywhere). Making the Pfaffian unity is a matter of gauge choice. Had we picked a gauge where the Pfaffian of the TR sewing matrix w (for the remainder of this chapter, we call the TR sewing matrix w to avoid confusion with the inversion sewing matrix defined at the beginning of this chapter, which was also denoted by B) is unity, the square root of the determinant of w would switch branches in the BZ in the case of a nontrivial TR topological insulator.

We aim to link the topological character of a TR and inversion-symmetric topological insulator with the inversion eigenvalues that its band exhibit at the TR (and inversion) symmetric points $\mathbf{G}_i/2$. We then need to express the Z_2 invariant characterizing a TR topological insulator ($\prod_i Pf(w(\mathbf{G}_i/2))/\sqrt{\mathrm{Det}(w(\mathbf{G}_i/2))}$) with the inversion eigenvalues of the bands. We proceed to do that. In the gauge where $Pf[v(\mathbf{k})] = 1$, $\mathrm{Det}[w(\mathbf{k})] = 1$ due to the identity (we can take the Pfaffian of the left- and right-hand sides)

$$v(-\mathbf{k}) = w(\mathbf{k}) v^*(\mathbf{k}) w^T(\mathbf{k}). \tag{12.15}$$

To prove this identity, we note that it is equivalent to

$$w^\dagger(\mathbf{k}) v(-\mathbf{k}) = v^*(\mathbf{k}) w^T(\mathbf{k}), \tag{12.16}$$

or, for the matrix elements of the matrices, $(w^*(\mathbf{k}))_{nm}(v(-\mathbf{k}))_{nr} = (v^*(\mathbf{k}))_{mn}(w(\mathbf{k}))_{rn}$. The strategy to prove the preceding is the following: we are going to prove it as if we were summing over all the bands. We let the double-index summation n in the preceding equation run over all the bands, not only the occupied ones; we will prove it as if the matrices v, w had indices running over all bands, occupied and unoccupied. We then realize that both matrices w, v are block diagonal in the occupied and unoccupied bands, and, as such, the preceding relation will also be true for the truncated matrices v and w of the occupied bands. We call the left-hand side of equation (12.16) LHS and the right-hand side, RHS. We use the identity $(u_{n,-\mathbf{k}})_\alpha (u^*_{n,-\mathbf{k}})_\theta = \delta_{\alpha\theta}$ to obtain

$$\text{LHS} = U^*_{\theta\beta}(u_{m\mathbf{k}})_\beta P_{\theta\gamma} U_{\gamma\delta}(u_{\gamma,-\mathbf{k}})^*_\delta, \tag{12.17}$$

$$\text{RHS} = (u_{m,\mathbf{k}})_\theta P^*_{\theta\gamma} U^*_{\gamma\delta}(u_{\gamma,-\mathbf{k}})^*_\alpha U_{\alpha\delta}. \tag{12.18}$$

We then use $P_{\theta\gamma} U_{\gamma\delta} = U_{\theta\gamma} P^*_{\gamma\delta}$ to obtain

$$\text{LHS} = (u_{m,\mathbf{k}})_\beta (u_{r,-\mathbf{k}})^*_\delta U^*_{\theta\beta} U_{\theta\gamma} P^*_{\gamma\delta} = (u_{m,\mathbf{k}})_\beta (u_{r,-\mathbf{k}})^*_\delta P^*_{\beta\delta}, \tag{12.19}$$

$$\text{RHS} = (u_{m,\mathbf{k}})_\theta (u_{r,-\mathbf{k}})^*_\alpha P^*_{\theta\gamma} U_{\alpha\delta}(U^\dagger)_{\delta\gamma} = (u_{m,\mathbf{k}})_\theta (u_{r,-\mathbf{k}})^*_\alpha P^*_{\theta\alpha} = \text{LHS}, \tag{12.20}$$

which proves that the RHS equals the LHS and, hence, completes the proof of $\text{Det}[w(k)] = 1$. We rewrite the matrix w at a TR and inversion-symmetric point as

$$\begin{aligned} w_{mn}\left(\frac{\mathbf{G}_i}{2}\right) &= \left\langle u_m\left(\frac{\mathbf{G}_i}{2}\right)\right| PPT \left| u_n\left(\frac{\mathbf{G}_i}{2}\right)\right\rangle \\ &= \zeta_m\left(\frac{\mathbf{G}_i}{2}\right)\left\langle u_m\left(\frac{\mathbf{G}_i}{2}\right)\right| PT \left| u_n\left(\frac{\mathbf{G}_i}{2}\right)\right\rangle = \zeta_m\left(\frac{\mathbf{G}_i}{2}\right) v_{mn}, \end{aligned} \tag{12.21}$$

where $\zeta_m(\Gamma_i)$ is the inversion eigenvalue of band m at the TR-invariant point $G_i/2$. Hence

$$\text{Pf}[w]^2 = \text{Det}[w] = \left(\prod_{n=1}^{2N} \zeta_n\right) \text{Det}[v] = \left(\prod_{n=1}^{2N} \zeta_n\right) \tag{12.22}$$

However, TR-invariant bands belonging to a Kramers' pair have the same inversion eigenvalue under inversion because $[P, T] = 0$; hence, we have

$$\text{Pf}[w]^2 = \left(\prod_{n=1}^{N} \zeta_n\right)^2 \tag{12.23}$$

and, thus, the natural choice

$$\text{Pf}[w] = \left(\prod_{n=1}^{N} \zeta_n\right). \tag{12.24}$$

Hence, the Z_2 topological invariant $\prod_i Pf(w(\mathbf{G}_i/2))/\sqrt{\text{Det}(w(\mathbf{G}_i/2))}$ is a product over all TR-symmetric points $(\mathbf{G}_i/2)$ of the inversion eigenvalues of *half* of the Kramers' doublets. The astute reader will notice that there is a sign ambiguity in equation (12.24). We could have taken $\text{Pf}[w] = -\left(\prod_{n=1}^{N} \zeta_n\right)$. In this case, we would have $\text{Pf}[w] = -1$ in the atomic limit, where the eigenvalues are direct correspondents of the atomic orbitals and the product of inversion eigenvalues of the Kramers' doublets has to be 1. This would say that a topological insulator $(\text{Pf}[w] = -1)$ is actually the trivial insulator, obviously not correct. Hence, the sign chosen in equation (12.24) is the correct one. We have hence proved our assertion that the product over *half* of the inversion eigenvalues in an inversion and TR-invariant insulator gives the Z_2 topological index of the insulator.

12.2 Role of Spin-Orbit Coupling

Our preceding discussion might seem a little paradoxical: nowhere in this discussion did the role of spin-orbit coupling seem to become important. The values of the inversion eigenvalues ζ_i do not seem to have much to do with spin-orbit interaction. Bands will have inversion eigenvalues if inversion symmetry is present, irrespective of whether there is spin-orbit coupling or not. The question then arises: do we even need spin-orbit coupling to have a topological insulator? For TR topological insulators with inversion, the answer, proven in [9], is yes.

Spin-orbit coupling ensures that an energy gap exists and is finite everywhere in the BZ. We will now prove in more detail the following statement found in [9]: in a parity and TR-invariant system, if the spin-orbit interaction is absent, the negative product of all ζ_i implies that the energy gap must vanish somewhere in the BZ. We begin the proof by noting that in the absence of spin-orbit coupling, we can consider spin $\uparrow$ and spin $\downarrow$ fermions separately. We will then prove that spin $\uparrow$s electrons have a vanishing gap somewhere in the BZ. Spin $\downarrow$ electrons have vanishing gaps at same point in the BZ, but as there is no spin-orbit coupling, they cannot mix and open a gap. Zeeman terms are forbidden by time reversal. We hence look at $\uparrow$ spins—for these to be nontrivial, the value of the product

$$\prod_{i=1}^{4} \zeta_i = -1 \tag{12.25}$$

implies the violation of Stokes' theorem. We want to prove that

$$\oint_{\partial\tau_{1/2}} \mathbf{A}(\mathbf{k})d\mathbf{k} - \int F(\mathbf{k})d^2k = \oint_{\partial\tau_{1/2}} \mathbf{A}(\mathbf{k})d\mathbf{k} = \pi \prod_{i=1}^{4} \zeta_i = -\pi, \tag{12.26}$$

where we used $F(\mathbf{k}) = 0$ for both TR and I and where the contour is *half* of the BZ—cut in such way that the half never contains both points $\mathbf{k}$, $-\mathbf{k}$. One example of such a contour is $k_y \in [0, \pi]$, $k_x \in [-\pi, \pi]$. If equation (12.26) is true, then there is an obstruction to defining Stokes' law for the half BZ; hence, one of our starting assumptions was wrong. Because TR and P were not assumptions and because *if there is a gap*, $F(\mathbf{k})$ is well defined and vanishes in the presence of TR and P, it *must* mean that the only other assumption is wrong—that the system can have a gap. The system is gapless, and the π Berry curvature in half the BZ in equation (12.26) is created by the gapless bulk Dirac fermion.

We now want to prove equation (12.26). To do this, we need to obtain a connection between $A(-\mathbf{k})$ and $A(\mathbf{k})$ (A is the sum over the diagonal part of the non-Abelian Berry potential—double index means summation). Such a connection is easy to prove in the sewing matrix formalism with sewing matrix $B_{ij}(\mathbf{k}) = \langle u_i(\mathbf{k})|\, P\, |u_j(\mathbf{k})\rangle$. In this chapter, we present only the result:

$$\mathrm{Tr}[B^\dagger(\mathbf{k})\nabla B(\mathbf{k})] = i(\mathbf{A}(-\mathbf{k}) + \mathbf{A}(\mathbf{k})). \tag{12.27}$$

For unitary matrices, we can relate the preceding trace to a determinant $\mathrm{Tr}[B^\dagger(\mathbf{k})\nabla_k B(\mathbf{k})] = \nabla_k \log[\mathrm{Det}[B(\mathbf{k})]]$. The easiest way to prove this is go to the representation in which B_{ij} is a matrix of phases $B_{ii} = e^{i\phi_i}$ (when each of the bands are nondegenerate): $\mathrm{Det}[B_{ij}] = e^{i\sum_{i=1}^N \phi_i}$, $\nabla_k \log[\mathrm{Det}[B]] = i\sum_{i=1}^N \nabla_k\phi_i = \sum_i e^{-i\phi_i}\nabla_k e^{i\phi_i} = \mathrm{Tr}[B^\dagger\nabla_k B]$. But, more importantly and without picking this representation, the preceding relation is easy to prove because $\mathrm{Tr}[B^\dagger(\mathbf{k})\nabla_k B(\mathbf{k})]$ is the gradient of the $U(1)$ phase of the matrix $B(\mathbf{k})$ (which we remember is a $U(N)$ matrix), whereas Det$[B(\mathbf{k})]$ is exactly the phase of the matrix because a $U(N)$ matrix is decomposable into a phase and an $SU(N)$ is part of unit determinant. The integral of equation (12.27) between two inversion-symmetric points yields

$$\int_0^{\pi} d\mathbf{k} \cdot \nabla \log[\mathrm{Det}[B(\mathbf{k})]] = i \int_0^{\pi} d\mathbf{k} \cdot (\mathbf{A}(-\mathbf{k}) + \mathbf{A}(\mathbf{k})). \tag{12.28}$$

The LHS is

$$\mathrm{LHS} = \log \left[\frac{\mathrm{Det}[B(\pi)]}{\mathrm{Det}[B(0)]} \right] = \log \left[\prod_{i=1}^{N} \zeta_i(\pi)\zeta_i(0) \right], \tag{12.29}$$

where we used the fact that the inversion eigenvalues can be only ± 1 (because $P^2 = 1$). The RHS is

$$i \int_0^{\pi} dk(A(-k) + A(k)) = i \left(\int_0^{\pi} dk A(k) + \int_{-\pi}^{0} dk A(k) \right) = i \int_{-\pi}^{\pi} dk A(k) = i 2\pi P_{\rho}, \tag{12.30}$$

where P_ρ is the charge polarization. Thus, we obtained the relation for the charge polarization:

$$P_\rho = \frac{1}{2\pi i} \log \left[\prod_{i=1}^{N} \zeta_i(\pi)\zeta_i(0) \right], \tag{12.31}$$

where the integer ambiguity of the logarithm is identical to the integer ambiguity of the polarization. We now let our Hamiltonian depend on a control parameter, $t = k_y$, and look at the variation of the charge polarization over tuning this parameter from 0 to half-cycle π. When we have an *inverted* band structure, $\prod_{i=1}^{N} \zeta_i(\pi)\zeta_i(0) = -1$, the integral over

$$\oint_{\partial\tau_{1/2}} \mathbf{A}(k) \cdot d\mathbf{k} = P_\rho \left(\frac{T}{2} \right) - P_\rho(0) = \pi \prod_{i=1}^{4} \delta_i \tag{12.32}$$

because $T/2$ means $k_y = \pi$ and $t = 0$ means $k_y = 0$. As such, Stokes' theorem equation (12.26) is violated, and the assumption of a gap was wrong.

12.3 Problems

1. *Mirror Symmetry and the 3-D Topological Insulator:* Show that mirror symmetry (only in spatial coordinates, the spin is left invariant) prevents the existence of a 3-D topological insulator. (*Hint:* First show that it prevents the existence of a 2-D topological insulator and then generalize to three dimensions. We can adiabatically transform the 2-D topological insulator into two uncoupled spin species—up and down—with opposite Chern numbers. What does mirror symmetry imply about the Chern number?)

2. *Theorem Relating Inversion Eigenvalues and the Gaplessness of a 3-D System:* Show that an inversion-symmetric system (with or without time reversal) that has the product of inversion eigenvalues at the eight inversion-invariant points in the BZ equal to -1 has to be gapless.

13

Quantum Hall Effect and Chern Insulators in Higher Dimensions

In previous chapters we have analyzed the behavior of topological insulators from the perspective of topological band theory. We have seen that there exist new types of insulators, in both two and three dimensions, that are TR invariant and support gapless boundary states. The topological insulators studied so far are part of a large series of insulators, in many dimensions, stabilized by various symmetries such as time reversal, charge conjugation, inversion, etc. A periodic table exists and classifies the topological insulators with charge conjugation, TR, and chiral symmetries in all space-time dimensions but repeats itself every eight dimensions. This table is outside the purpose of this book, but in the current chapter we will understand a small part of this table explicitly. We will see how 3-D TR-invariant topological insulators can be obtained from 4-D Chern insulators through dimensional reduction. The procedure will be useful for several things: we will obtain a field theory of the topological insulator by dimensionally reducing the Chern-Simons field theory of the 4-D Chern insulator. We will then obtain a simple proof of the relation between the Z_2 index and the inversion symmetry eigenvalues.

13.1 Chern Insulator in Four Dimensions

In a previous chapter, we argued that the physics of a Chern insulator in two dimensions can be understood through the 2-D Chern-Simons action of the background field A:

$$S_{\text{eff}} = \frac{\sigma_H}{2} \int d^2x \int dt \epsilon^{\mu\nu\tau} A_\mu \partial_\nu A_\tau. \tag{13.1}$$

The equations of motion for this action give both the Streda formula and the Hall current equation. However, this formula could have been guessed without knowing any of the physics of the Hall effect. In one time and two space dimensions, there are only two isotropic tensors using the space-time indices: $\delta_{\alpha\beta}$ and $\epsilon_{\mu\nu\tau}$. We want to build an electromagnetic Lagrangian made out of the background external field A_μ. The Lagrangian has to be a scalar and also gauge invariant in a system without boundaries (with boundaries, there could exist edge modes whose action under gauge invariance should also be considered). The leading-order gauge-invariant terms we can build are quadratic in A_μ. For example, the first-order gauge-invariant term in A_μ vanishes upon contraction with $\delta_{\mu\nu} F_{\mu\nu} = 0$. The second-order gauge-invariant terms in A_μ are of two kinds: First is the electromagnetic term $F_{\mu\nu}\delta_{\mu\rho}\delta_{\nu\theta}F_{\rho\theta} = F_{\mu\nu}F_{\mu\nu}$. The remaining second-order gauge-invariant term is the Chern-Simons term, where we contract the indices of $A_\mu \partial_\nu A_\rho$ with the $\epsilon^{\mu\nu\rho}$ symbol. In high-energy physics, the Chern-Simons Lagrangian is sometimes written in "wedge" form: $A \wedge dA$.

As defined by low-energy effective Lagrangians, indexology (or the art of writing a Lagrangian by just knowing how many dimensions it has and how to contract tensors) shows us that Chern classes—or insulators described by Chern numbers—can exist only in even space-time dimensions. The Chern numbers are the coefficients of the Chern-Simons Lagrangians of actions:

$$S = \frac{C}{d!} \int d^d k\, dt \epsilon^{\mu_1\mu_2\mu_3\mu_4\mu_5\cdots\mu_d\mu_{d+1}} A_{\mu_1} \partial_{\mu_2} A_{\mu_3} \partial_{\mu_4} A_{\mu_5} \cdots \partial_{\mu_d} A_{\mu_{d+1}}, \tag{13.2}$$

where the constant C is the $(d/2)$nd Chern number. The preceding Lagrangian makes sense only in even space dimensions. How do we obtain this effective Lagrangian? In condensed matter, we are interested in the physics of electrons subject to external fields. The starting Lagrangian must contain fermionic degrees of freedom. If the system is an insulator and the fermions are gapped, we can integrate them out to obtain an effective local action for the electromagnetic field. Of course, many insulators are trivial and lead to $C = 0$. However, as we have seen through out this book, there exist a series of nontrivial topological insulators that lead to $C \neq 0$. The Chern-Simons Lagrangians such as the one in equation (13.2) come about by integrating out Dirac fermions coupled to an electromagnetic field in even d-dimensions. The Lagrangian in equation (13.2) is, in fact, the parity anomaly of Dirac fermions, and the response function to compute when integrating out fermions (which as we have seen, is linear-response current-current correlations in two space dimensions) becomes nonlinear response in $d > 2$. The Dirac fermions need to be massive so that the system has a gap. As such, in any even dimension d, the Hamiltonian for the Dirac fermions will contain *all* the $d + 1$ Dirac (Clifford) matrices (d matrices from the coordinates, and the $(d + 1)$st matrix coupling to the mass. Generally speaking, the response function, which gives the coefficient C in front of the action, involves taking traces over the maximum product of Dirac (Clifford) matrices available in the considered dimension. The trace over the product of all Dirac (Clifford) matrices gives $\epsilon^{\mu_1\mu_2\mu_3\mu_4\ldots\mu_d\mu_{d+1}}$. The only matrices that give these kind of traces are the Clifford-algebra matrices, which always come in odd numbers. In two space dimensions the Clifford matrices are the same as the Pauli matrices, with the usual identity:

$$\mathrm{Tr}[\sigma_\mu \sigma_\nu \sigma_\theta] = 2i\epsilon^{\mu\nu\theta}, \tag{13.3}$$

whereas in four dimensions we have the 5-D Clifford algebras:

$$\mathrm{Tr}[\Gamma_{\mu_1}\Gamma_{\mu_2}\Gamma_{\mu_3}\Gamma_{\mu_4}\Gamma_{\mu_5}] = 4i\epsilon^{\mu_1\mu_2\mu_3\mu_4\mu_5} \tag{13.4}$$

For a translationally invariant insulator, with periodic boundary conditions, the coefficient C (the Chern number) has an expression in terms of the Bloch wavefunctions of the insulating band structure. In two dimensions, we found the form of the Chern number in terms of the current-current correlation function and then expressed it in terms of the projectors onto the occupied bands. We now present the calculation of the coefficient of the Chern-Simons action for higher-dimensional Chern insulators. We particularize to four space dimensions and consider the Hamiltonian of a $(4 + 1)$-D insulator coupled to a $U(1)$ gauge field giving the Peierls phases A_{mn}, $A_{0,m}$:

$$H = \sum_{m,n} \left(c^\dagger_{m\alpha} h^{\alpha\beta}_{mn} e^{iA_{mn}} c_{n\beta} + \text{h.c.} \right) + \sum_m A_{0m} c^\dagger_{m\alpha} c_{m\alpha}, \tag{13.5}$$

where m, n are sites on the 4-D lattice. In the standard approach, the effective action of gauge field A_μ is obtained by integrating out the fermions in the path

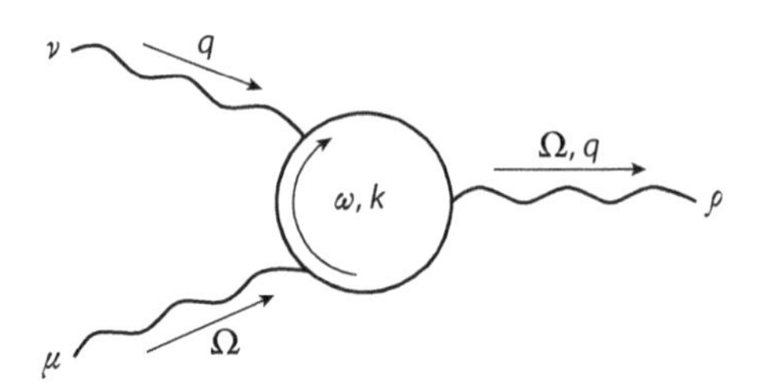

Figure 13.1. The nonlinear-response Feynman bubble diagram that gives the topological term eq. (13.7) of the gauge-field action. The loop is a fermion loop, and solid lines are fermion propagators. The wavy lines are external legs corresponding to the gauge field. Due to the fact that we have nonzero external momentum entering the loop, the response will involve both electric and magnetic fields, unlike in two space dimensions, where it involves only electric fields.

integral:

$$\begin{aligned} e^{iS_{\text{eff}}[A]} &= \int D[c]D[c^\dagger]e^{i\int dt\left[\sum_m c^\dagger_{ma}(i\partial_t)c_{ma} - H[A]\right]} \\ &= \text{Det}\left[(i\partial_t - A_{0m})\,\delta^{\alpha\beta}_{mn} - h^{\alpha\beta}_{mn}e^{iA_{mn}}\right] \end{aligned} \tag{13.6}$$

As elaborated earlier in the book, for (2 + 1)-D insulators, the effective action S_{eff} contains a Chern-Simons term $(C_1/4\pi)A_\mu\epsilon^{\mu\nu\tau}\partial_\nu A_\tau$, which shows up in the second-order expansion of the preceding functional determinant (the second-order expansion is equivalent to the current-current correlation function, a two-point function, or linear response). The first Chern number, C_1, is the integral over the full BZ of the Berry curvature of the occupied bands. For the (4 + 1)-D system, integrating out the fermions generates a similar topological term in the effective action, but only when the determinant is expanded to higher order in A, a sign that nonlinear response needs to be taken into account). This is the second Chern-Simons action:

$$S_{\text{eff}} = \frac{C_2}{24\pi^2}\int d^4x\, dt\epsilon^{\mu\nu\rho\sigma\tau}A_\mu\partial_\nu A_\rho\partial_\sigma A_\tau, \tag{13.7}$$

where $\mu, \nu, \rho, \sigma, \tau = 0, 1, 2, 3, 4$. The Chern-Simons action determines the Hall response of the fermionic system to an applied field through the equation of motion: $j_\mu(\mathbf{x}) = \delta S_{\text{eff}}[A]/\delta A_\mu(\mathbf{x})$. Due to the nonlinear nature of the Lagrangian, the response to a field will also be nonlinear. The coefficient C_2 is obtained in terms of integrals over the fermionic Green's functions as the nonlinear one-loop Feynman diagram in figure 13.1. It contains two more derivatives than its 2-D counterpart and is

$$C_2 = -\frac{\pi^2}{15}\epsilon^{\mu\nu\rho\sigma\tau}\int\frac{d^4k d\omega}{(2\pi)^5}\text{Tr}\left[G\frac{\partial G^{-1}}{\partial q^\mu}G\frac{\partial G^{-1}}{\partial q^\nu}G\frac{\partial G^{-1}}{\partial q^\rho}G\frac{\partial G^{-1}}{\partial q^\sigma}G\frac{\partial G^{-1}}{\partial q^\tau}\right], \tag{13.8}$$

where $q^\mu = (\omega, k_1, k_2, k_3, k_4)$ is the shorthand notation of the frequency-momentum vector and $G(q^\mu) = [\omega + i\delta - h(k)]^{-1}$ is the single-particle Green's function. One of the derivatives in the preceding is the derivative of the inverse Green's function with respect to ω, and hence we are left with the momentum and frequency integral of a trace over the four spatial derivatives of the momentum.

13.2 Proof That the Second Chern Number Is Topological

The form of the Chern number in terms of the Green's functions in equation (13.8) makes it easy to see that it is topological and does not depend on small deformations of the Hamiltonian as long as the Green's function has no zeros (as long as an inverse of the Green's function exists). A similar conclusion is reached by finding the expression of the second Chern number in terms of band projectors. We assume an infinitesimal deformation (which does not close the gap) of the Hamiltonian $H(k)$. This transformation then is reflected in an infinitesimal deformation of the Green's function $G(\mathbf{k}, \omega)$. Under a variation $\delta G(\mathbf{k}, \omega)$ of $G(\mathbf{k}, \omega)$ we have

$$\begin{aligned}\delta\left(G\partial_\mu G^{-1}\right) &= \delta G\partial_\mu G^{-1} - G\partial_\mu\left(G^{-1}\delta G G^{-1}\right) \\ &= -G\left(\partial_\mu G^{-1}\right)\delta G G^{-1} - \partial_\mu\left(\delta G\right)G^{-1}.\end{aligned} \tag{13.9}$$

Plugging the preceding variation into the formula for the second Chern number and performing some tedious but straightforward algebra leads us to the result that the second Chern number variation is related to a total derivative [13]:

$$\begin{aligned}\delta C_2 = \frac{\pi^2}{15}\epsilon^{\mu\nu\rho\sigma\tau}\int\frac{d^4k d\omega}{(2\pi)^5} \\ \times\partial_\mu \mathrm{Tr}\left[\left(G^{-1}\delta G\right)\left(\partial_\nu G^{-1}G\right)\left(\partial_\rho G^{-1}G\right)\left(\partial_\sigma G^{-1}G\right)\left(\partial_\tau G^{-1}G\right)\right] = 0.\end{aligned} \tag{13.10}$$

Thus, under periodic boundary conditions, the second Chern number is invariant to small deformations of the Hamiltonian. Again, we remind the reader that the derivation above is valid only when the Green's function has no zeros.

13.3 Evaluation of the Second Chern Number: From a Green's Function Expression to the Non-Abelian Berry Curvature

Just as the first Chern number was related to the Abelian Berry curvature, the coefficient C_2 defined in equation (13.8) is related to the non-Abelian Berry's phase gauge field in momentum space. We now want to prove that the nonlinear response function is identical to the Berry curvature form $\mathrm{Tr}[f \wedge f]$ integrated over the full 4-D BZ. It is the second Chern number of the non-Abelian Berry's phase gauge field in the BZ:

$$C_2 = \frac{1}{32\pi^2}\int d^4k\epsilon^{ijk\ell}\mathrm{Tr}\left[f_{ij}f_{k\ell}\right] \tag{13.11}$$

$$f_{ij}^{\alpha\beta} = \partial_i a_j^{\alpha\beta} - \partial_j a_i^{\alpha\beta} + i\left[a_i, a_j\right]^{\alpha\beta}, \quad a_i^{\alpha\beta}(\mathbf{k}) = -i\left\langle\alpha, \mathbf{k}\right|\frac{\partial}{\partial k_i}\left|\beta, \mathbf{k}\right\rangle,$$

where $i, j, k, \ell = 1, 2, 3, 4.$ are now spatial indices. The index α refers to the occupied bands. For a multiband model with N occupied bands, $a_i^{\alpha\beta}$ is a $U(N)$ non-Abelian gauge field, and $f_{ij}^{\alpha\beta}$ is its non-Abelian field strength. Its trace gives the Abelian field strength used in the calculation of the first Chern number, i.e., $C_1 = \frac{1}{2\pi}d^2k\sum_\alpha f^{\alpha\alpha}(k)$.

In order to prove equation (13.12), similar to the case of the first Chern number, we first go to the flat-band Hamiltonian limit:

$$h_F(\mathbf{k}) = \epsilon_G P_G(\mathbf{k}) + \epsilon_E P_E(\mathbf{k}), \tag{13.12}$$

where $P_G(\mathbf{k})$ and $P_E(\mathbf{k})$ are the projection operators to the occupied and unoccupied bands, respectively. We prove equation (13.12) for the flat-band Hamiltonian and argue that because the result does not depend on the eigenvalues ϵ_G, ϵ_P but only on the eigenstates, it will remain unchanged for the original Hamiltonian (the flat-band Hamiltonian and the original

Hamiltonian have identical eigenstates). As found in the case of the first Chern number, the Green's function and derivatives of the inverse of the Green's function with respect to frequency and momentum have a simple expression in terms of the flat-band Hamiltonian see equation (3.62). Hence, equation (13.8) can be expressed in terms of projection operators only:

$$C_2 = -\frac{\pi^2}{3}\epsilon^{ijk\ell}\int\frac{d^4kd\omega}{(2\pi)^5}\times\sum_{n,m,s,t=1,2}\frac{\mathrm{Tr}\left[P_n\frac{\partial P_G}{\partial k_i}P_m\frac{\partial P_G}{\partial k_j}P_s\frac{\partial P_G}{\partial k_k}P_t\frac{\partial P_G}{\partial k_\ell}\right](\epsilon_E-\epsilon_G)^4}{(\omega+i\delta-\epsilon_n)^2(\omega+i\delta-\epsilon_m)(\omega+i\delta-\epsilon_s)(\omega+i\delta-\epsilon_t)} \tag{13.13}$$

where the summation is over all the poles $\epsilon_{1,2}=\epsilon_{G,E}$ and $P_{1,2}(\mathbf{k})=P_{G,E}(\mathbf{k})$, respectively, the projectors into the occupied and empty bands. From the identity $P_E+P_G\equiv\mathbb{I}$ (with $\mathbb{I}$ the identity matrix) and $P_E^2=P_E,\ P_G^2=P_G$, we earlier found nice identities involving projectors, as a result of which we found that

$$P_G\partial_i P_G P_G = P_E\partial_i P_G P_E = 0. \tag{13.14}$$

This implies that the trace in equation (13.13) can be nonzero only when $n\neq m,\ m\neq s,\ s\neq t,\ and\ t\neq n$. Hence, only 2 of the 16 terms summed over in equation (13.13) survive the tracing operation: $\mathrm{Tr}\left[P_G\partial_i P_G P_E\partial_j P_G P_G\partial_k P_G P_E\partial_\ell P_G\right]$ and $\mathrm{Tr}\left[P_E\partial_i P_G P_G\partial_j P_G P_E\partial_k P_G P_G\partial_\ell P_G\right]$. Carrying out the integral over ω and using the projector identities proved in chapter 3, we find

$$C_2=\frac{1}{8\pi^2}\int d^4k\epsilon^{ijk\ell}\mathrm{Tr}\left[P_E\frac{\partial P_G}{\partial k_i}\frac{\partial P_G}{\partial k_j}P_E\frac{\partial P_G}{\partial k_k}\frac{\partial P_G}{\partial k_\ell}\right]. \tag{13.15}$$

This is the formula we use to numerically implement the calculation of C_2—the projectors in the upper and lower bands are gauge invariant, a major simplification because no gauge choice is necessary.

Equation (13.15) is just the second Chern number. The Berry phase gauge field is defined as usual by $a_i^{\alpha\beta}(\mathbf{k})=-i\left\langle\alpha,\mathbf{k}\right|\partial_{k_i}\left|\beta,\mathbf{k}\right\rangle$, in which $\alpha,\beta=1,2,\ldots,N_{\mathrm{occupied}}$ are the occupied bands. The $U(N_{\mathrm{occupied}})$ gauge field strength $\partial_i a_j^{\alpha\beta}-\partial_j a_i^{\alpha\beta}+i\left[a_i,a_j\right]^{\alpha\beta}$ is given by

$$\begin{aligned} f_{ij}^{\alpha\beta} &= -i\frac{\partial\left\langle\alpha,\mathbf{k}\right|}{\partial k_i}\frac{\partial\left|\beta,\mathbf{k}\right\rangle}{\partial k_j}+i\frac{\partial\left\langle\alpha,\mathbf{k}\right|}{\partial k_i}\sum_{\gamma=1}^{N_{\mathrm{occupied}}}\left|\gamma,\mathbf{k}\right\rangle\left\langle\gamma,\mathbf{k}\right|\frac{\partial\left|\beta,\mathbf{k}\right\rangle}{\partial k_j}-(i\leftrightarrow j)\\ &= -i\left(\frac{\partial\left\langle\alpha,\mathbf{k}\right|}{\partial k_i}P_E(\mathbf{k})\frac{\partial\left|\beta,\mathbf{k}\right\rangle}{\partial k_j}-(i\leftrightarrow j)\right), \end{aligned}$$

where $(i\leftrightarrow j)$ denotes the term with $i,\ j$ exchanged. Because $\left\langle\alpha,\mathbf{k}\right|P_E(\mathbf{k})=0$ for α, the index of one of the occupied bands, we have

$$\sum_{\alpha,\beta=1}^{N_{\mathrm{occupied}}}\left|\alpha,\mathbf{k}\right\rangle f_{ij}^{\alpha\beta}\left\langle\beta,\mathbf{k}\right| = -i\frac{\partial P_G(\mathbf{k})}{\partial k_i}P_E\frac{\partial P_G(\mathbf{k})}{\partial k_j}-(i\leftrightarrow j).$$

Thus, equation (13.15) can be written in terms of the Berry phase field strength as $F\wedge F$ [13]:

$$C_2=\frac{1}{32\pi^2}\int d^4k\epsilon^{ijk\ell}\mathrm{Tr}\left[f_{ij}f_{k\ell}\right]. \tag{13.16}$$

We proved the equivalence of the nonlinear correlation function eq. (13.8), and the second Chern number eq. (13.11) for flat-band models of the form of equation (13.12). We find that the second Chern number depends only on the eigenstates of the flat-band Hamiltonian and *not* on the energies. This means that the Chern number is the same for the two Hamiltonians. It means that even a non-flat-band Hamiltonian will have the same second Chern number. As mentioned at the beginning of this chapter, by indexology it is clear that the current procedure in this section can be easily generalized to any odd space-time dimensions. The coefficient of the nth Chern-Simons term in the effective action of an external gauge field is obtained from integrating out the fermions. In $2n+1$ space-time dimensions, an nth Chern number C_n is defined as the integral over n wedge products $f \wedge f \cdots \wedge f$ of non-Abelian Berry curvature f integrated over the BZ of the band insulator. This number appears as a topological response coefficient to an external gauge field. Notice that even though in $2+1$ space-time dimensions we defined the Chern number as the integral of the Abelian Berry curvature over the BZ, this can be interpreted as a particularization of the general formula given before because the Abelian Berry curvature is the trace of the non-Abelian one.

13.4 Physical Consequences of the Transport Law of the 4-D Chern Insulator

We saw that the physical consequence of the $2+1$ space-time Chern-Simons action of chapter 3 was the existence of a quantum Hall effect—a transversal current response to an applied magnetic field. We now want to find the physical consequences of the $4+1$ space-time Chern-Simons action. In $4+1$ space time dimensions, by applying the Lagrangian equation of motion we obtain the following relationship between the current and the applied fields:

$$j^{\mu} = \frac{C_2}{8\pi^2} \epsilon^{\mu\nu\rho\sigma\tau} \partial_\nu A_\rho \partial_\sigma A_\tau. \tag{13.17}$$

This is the nonlinear response to the external field A_μ. It involves both electric and magnetic fields. In the next chapter, we will analyze several special cases of such an equation of motion. This is the equation of motion of the quantum Hall effect in four spatial dimensions. Notice that, unlike in the 2-D quantum Hall effect, which needed TR breaking to exist, the 4-D quantum Hall effect can exist without any breaking of TR symmetry. The reason is that the equation now relates the current operator, which is odd in time (as a velocity) with the product of an electric field (even in time) and a magnetic field (odd in time). The proportionality constant (the 4-D Hall conductance) is then even in time and does not need TR breaking to be nonzero. This is reflected in the Chern-Simons Lagrangian, which breaks time reversal in $2(2k+1)+1$ space-time dimensions but maintains it in $4k+1$ space-time dimensions.

The transport equation (13.17) is that of transverse transport not of particles but of Landau levels upon the introduction of both a magnetic and an electric field. We would like to now analyze the physical consequences of such a nonlinear transport law. We clearly see that the transport law of equation (13.17) implies the existence of both magnetic and electric fields, and we consider a field configuration $A_x = 0,\ A_y = B_z x,\ A_z = -E_z t,\ A_v = A_t = 0$, where x, y, z, v are the spatial coordinates and t is time. The only nonvanishing components of the external field are $F_{zt} = -E_z$, $F_{xy} = B_z$, which—according to equation (13.17)—generate the current

$$j_v = \frac{C_2}{4\pi^2} B_z E_z.$$

We notice that we have a magnetic field in the z-direction. This magnetic field allows a large degeneracy of the associated Landau level. To capture this degeneracy, equal to the number of fluxes through the planes perpendicular to the magnetic field, we integrate the preceding equation over the x, y-dimensions (with periodic boundary conditions) and with E_z independent of (x, y)) to obtain

$$\int dx\, dy\, j_v = \frac{C_2 N_{xy}}{2\pi} E_z, \tag{13.18}$$

where $N_{xy} = \int dx\, dy\, B_z/2\pi$ is the number of flux quanta through the xy-plane (which is a really a torus because we have periodic boundary conditions), which is quantized to be an integer in the torus geometry (this quantization comes from the solution of the Landau-level problem on the torus and is equivalent to the Dirac quantization of monopole charge). This simple example helps us understand a physical consequence of the second Chern number (and, by generalization, of any higher Chern number): in a (4 + 1)-D insulator with second Chern number C_2, a quantized Hall conductance $C_2 N_{xy}/2\pi$ in the zv-plane is induced by magnetic field with flux $2\pi N_{xy}$ in the perpendicular (xy-) plane. Alternatively, a quantized Hall conductance of C_2 in the zv-plane transports entire Landau levels in the xy-plane under the action of an electric field E_z.

In higher dimensions (such as 6, 8, etc., space dimensions), the equation of motion can be readily generalized, and the physical consequence is clear: we have a Hall current of multiple Landau levels. For example, in 6 dimensions, $x_{1,2,3,4,5,6,}$ we will have, under a magnetic field perpendicular to the 56 directions F_{56} and a magnetic field perpendicular to the 34 directions F_{34} and in the presence of an electric field E_2, a current $j_1 = \frac{C_3}{2\pi}\frac{N_{34}}{2\pi}\frac{N_{56}}{2\pi} E_z$ carrying Landau levels of perpendicular fields.

Similar to the (2 + 1)-D case, the 4-D quantum Hall system will exhibit surface states with open boundary conditions. We can gain further insight into the physical consequences of the second Chern number by studying the surface states of an open-boundary system, which for the (4 + 1)-D case is described by a (3 + 1)-D gapless theory, just as in the edge theory was a (1 + 1)-D gapless theory (2 + 1)-D quantum Hall case. The (3 + 1)-D fermion on the surface of the (4+1)-D topological insulator is stable to disorder and other symmetry-breaking terms, such as TR breaking, etc. Being a single Weyl fermion, it is chiral, with a chirality given by the *first* Chern number of the Fermi surface surrounding the gapless point. On a 3-space-dimensional lattice, Weyl points come in pairs *of opposite chirality*, and they are locally stabilized by translational invariance and the absence of scattering between each other. We cannot have a single Weyl fermion on a purely 3-space-dimensional lattice because of the Nielsen-Nyomiya theorem. On the surface of the (4 + 1)-D topological insulator, there is a single Weyl fermion; hence, even in the absence of translational invariance, the Weyl fermion is stable to the inclusion of perturbation.

In fact, more than a single Weyl fermion at the surface of a 4 + 1 space-time-dimensional topological insulator is stable to perturbations. When the second Chern number is C_2, there are $|C_2|$ branches of (3 + 1)-D gapless surface states with linear dispersion and with identical chirality, so that the low-energy effective theory is described by $|C_2|$ flavors of chiral (Weyl) fermions [49]:

$$H = \text{sign}(C_2) \sum_{\mathbf{k}} \sum_{a=1}^{|C_2|} v_a c_{a\mathbf{k}}^\dagger \sigma \cdot \mathbf{k} c_{a\mathbf{k}}. \tag{13.19}$$

The chirality of the surface states (i.e., the sign of the first Chern number of the Fermi surface surrounding the gapless point) has to be identical to the sign of the second Chern number of the bulk invariant. This gives rise to the multiplicative factor sign(C_2) in equation (13.19). The fact that there are surface states can be understood through several arguments: first, we have a bulk insulator but a nontrivial Hall-type transport law, which implies the existence of electron carriers at the Fermi surface. Because the bulk is gapped, these carriers can come only from the surface states. Second, although the (4 + 1)-D Chern-Simons action is gauge invariant on a closed manifold, it is not so on a manifold with boundaries. Upon performing a gauge transformation of the externally applied field A_μ, the (4 + 1)-D Chern-Simons action acquires a (3 + 1)-D boundary term that cannot be gauged away. This term is an anomaly of 3 + 1 gapless fermions, and it requires their presence. The surface theory gives rise to the same nonlinear equation of motion as the Chern-Simons Lagrangian. We couple the Weyl fermion to an electromagnetic field with the identical configuration as those used in deriving the equation (13.17), Chern-Simons equation of motion ($F_{xy} = B_z$ and $F_{zt} = -E_z$), and we consider the open-boundary condition situation with boundaries at $w = \pm L/2$, $L \to \infty$ large in the thermodynamic limit. The Hamiltonian for a single (3 + 1)-D surface chiral fermion coupled to these fields is

$$h = v\sigma \cdot (\mathbf{k} + \mathbf{A}) = v\sigma_x k_x + v\sigma_y \left(k_y + B_z x\right) + v\sigma_z \left(k_z - E_z t\right).$$

It is now clear that in the weak E_z limit, where the time dependence of $A_z(t) = -E_z t$ can be treated adiabatically, and the instantaneous single-particle energy spectrum at every t corresponds to Landau levels in the xy-plane with dispersion in the z-direction characterized by a momentum k_z, which is adiabatically changed by the presence of A_z. The dispersion has a gapped spectrum of levels. First, we know that a 2-D Dirac Hamiltonian in a magnetic field has dispersion given by $\pm\sqrt{2nB_z}$ with $n = 1, 2, \ldots$, as well as a remaining *single* zero mode at zero energy. However, we are now in three dimensions, but since the momentum along the magnetic field is a good quantum number, we can immediately obtain the form of the dispersion relation as

$$E_{n\pm}(p_z) = \pm v\sqrt{(k_z + A_z)^2 + 2n|B_z|}, \quad n = 1, 2, \ldots,$$

as well as a zero mode:

$$E_0(p_z) = v(k_z + A_z)\text{sign}(B_z). \tag{13.20}$$

This mode crosses the band gap in the z-direction, as in figure 13.2. The problem has now become a gapless chiral (1 + 1)-D problem in which each k_z momentum eigenstate carries a thermodynamically large $N_{xy} = L_x L_y B_z/2\pi$ number of particles (we have considered periodic boundary conditions in the x-, y-directions). We can now perform an identical adiabatic argument to that of Laughlin for quantum Hall edge states [50]. Because we said we will treat the electric field adiabatically, the effect of an infinitesimal electric field E_z is to adiabatically shift the momentum $k_z \to k_z + E_z t$. As shown in figure 13.2, from the time $t = 0$ to $t = T \equiv 2\pi/L_z E_z$, the momentum is shifted as $k_z \to k_z + 2\pi/L_z$. In this time, the electron number of the (upper) 3-D surface increases by N_{xy}, but otherwise the spectrum returns to its original form (of course, the lower 3-D surface has lost a number N_{xy} of electrons). As the surfaces were cut in the v-direction, the fact that the number of surface electrons varies as a function of time implies that a generalized Hall current I_v must be flowing toward the v- direction:

$$I_v = \frac{N_{xy}}{T} = \frac{L_x L_y L_z B_z E_z}{4\pi^2}.$$

This current is taking electrons from the lower surface; it then flows through the bulk and places the electrons on the upper surface.

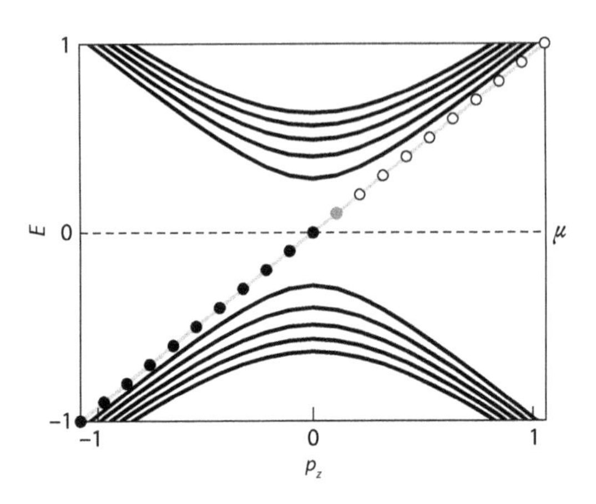

Figure 13.2. Spectrum of the 3-D surface state in a (4 + 1)-D Chern insulator subject to a magnetic and electric field equation (13.20). In the presence of a magnetic field $F_{xy} = B_z$, each level at every momentum p_z in the figure is $N_{xy} = L_x L_y B_z/2\pi$- fold degenerate—the degeneracy of a Landau level. The solid circles are the occupied states of the zeroth Landau level, and they cross the band gap in the z-direction. When the gauge-vector potential A_z is shifted adiabatically from 0 to $2\pi/L_z$, one state (denoted by a red dot) is filled.

In terms of current density, we obtain $j_v = \frac{B_z E_z}{4\pi^2} = \frac{\mathbf{B}\cdot\mathbf{E}}{4\pi^2}$. This is also more compactly shown to be $\frac{\epsilon^{\mu\nu\sigma\tau}F_{\mu\nu}F_{\sigma\tau}}{32\pi^2}$, or equation (13.17), which is the *chiral anomaly* [51, 52] equation of massless (3 + 1)-D Chiral fermions. We have now understood the current density of a (4 + 1)-D topological insulator from a surface-state perspective, almost identical to the understanding of the quantum Hall edge state through the Laughlin adiabatic argument.

13.5 Simple Example of Time-Reversal-Invariant Topological Insulators with Time-Reversal and Inversion Symmetry Based on Lattice Dirac Models

We have been talking in generalities about these higher-dimensional topological insulators, so we now stop to give an example. Unsurprisingly, the example involves a Dirac fermion. Lattice Dirac models hold a special place in the theory of topological insulators. They are the simplest models that exhibit nontrivial topology, based on their surface anomalies. In 2 + 1 dimensions, we showed that a massive Dirac fermion in the continuum (which in that case was a 2 × 2 model) with mass m exhibited a Hall conductance (Chern number) of sign(m)/2. On the lattice, this noninteger Hall conductance does not make sense, but the continuum model does correctly model transitions between states of different Hall conductance, which are to be understood as gap openings and closings of 2-D Dirac fermions. We would like to understand the equivalent statement in 4 + 1 space-time dimensions. To do that, based on our experience, we now look at (4 + 1)-D Dirac models. These are 4 × 4 matrix models containing the 5-D Clifford algebra Γ^μ, $\mu = 1, \ldots, 5$, $\{\Gamma^\mu, \Gamma^\nu\} = 2\delta_{\mu\nu}\mathbb{I}$, with $\mathbb{I}$ the identity matrix.

We note that, unlike the 2 + 1 space-time-dimensional case, where all the $SU(2)$ Pauli matrices were used (per the fact that a Clifford algebra is the same as a Lie algebra in 2 space dimensions), in our case the Dirac Hamiltonian contains only 5 out of the 16 $U(4)$ matrices; as such, it is a fine-tuned point of the generic 4-band Hamiltonian in 4 + 1 space-time dimensions. In fact, the Hamiltonian has doubly-degenerate bands at every point momentum space because it uses only the 5 Clifford-algebra matrices. As such, per our previous discussion, it could be interpreted as the Hamiltonian of a (4 + 1)-D system with TR and inversion symmetry, where the inversion symmetry operator is Γ_5. A (4 + 1)-D Dirac-type Hamiltonian can be written in the compact form as

$$H = \sum_{\mathbf{k}} c_{\mathbf{k}}^\dagger d_a(\mathbf{k}) \Gamma^a c_{\mathbf{k}}. \tag{13.21}$$

This Hamiltonian has two eigenvalues, $E_\pm(\mathbf{k}) = \pm\sqrt{\sum_a d_a^2(\mathbf{k})}$. There is a key difference from the (2 + 1)-D case: both eigenvalues are doubly degenerate. When $\sum_a d_a^2(\mathbf{k}) \equiv d^2(\mathbf{k})$ is nonvanishing in the whole BZ, the system is gapped at half-filling, with the two degenerate bands with $E = E_-(\mathbf{k})$ filled. Because it has only two energy levels (with degenerate bands), the Green's function of the Hamiltonian, equation (13.21), can be determined and used to compute the second Chern number in equation (13.8). Similar to the 2 + 1 space-time-dimensional case, the second Chern number is a Jacobian (winding number) of the map between the 4-torus T^4 of the BZ and the S^4 manifold defined by $\hat{d}_i(\mathbf{k}) = d_i(\mathbf{k})/d(\mathbf{k})$:

$$C_2 = \frac{3}{8\pi^2}\int d^4k\epsilon^{abcde}\hat{d}_a\partial_x\hat{d}_b\partial_y\hat{d}_c\partial_z\hat{d}_d\partial_w\hat{d}_e. \tag{13.22}$$

Let us now compute the second Chern number of the continuum Dirac model. For the continuum Dirac Hamiltonian, we have $d_a(\mathbf{k}) \simeq (k_x, k_y, k_z, k_v, m)+o(|k|^2)$. Taking a momentum space cutoff $\Lambda \ll 2\pi/a$ (where a is a lattice constant), which we then send to infinity, and substituting for the d_i's in the winding-number formula, we obtain, for the second Chern number,

$$C_2 = \frac{3}{8\pi^2}\int_{|\mathbf{k}|\leq\Lambda} d^4k\frac{m}{(m^2+k^2)^{5/2}} = \frac{1}{2}\text{sign}(m). \tag{13.23}$$

Hence, at a phase transition where the gap m goes to zero and changes sign, the change in the Hall conductance is

$$\Delta C_{2\,m=0^-}^{\;m=0^+} = C_{2m=0^+} - C_{2m=0^+} = 1. \tag{13.24}$$

If we regularize the continuum Dirac Hamiltonian by adding a k^2-term to the dispersion (consistent with its symmetries, we can add only terms even in k to terms even in k and terms odd in k to terms odd in k), we can analyze the Hamiltonian $d_a(\mathbf{k}) \simeq (m - k^2, k_x, k_y, k_z, k_w)$, which, in similar fashion to its 2 + 1 space-time-dimensional counterpart, will have (second) Chern number 1 for $m > 0$ and Chern number 0 for $m < 0$. Note that, whereas the first Chern number changed by 1 in the 2 + 1 Dirac Hamiltonian at a crossing between *two* bands, the second Chern number changes by 1 in the 4 + 1 Dirac Hamiltonian at a crossing between *four* bands. The Hamiltonian at the crossing point is just the mass times the inversion operator. The de-generate bands have identical inversion eigenvalues. Hence, a band crossing in the Dirac Hamiltonian corresponds to a switch in *two* inversion eigenvalues at a TR-invariant point. This will become important when providing an alternative proof for the Fu and Kane inversion eigenvalue formula in chapter 14.

The continuum Dirac Hamiltonian has a very simple discretization to the lattice by letting $k_i \to \sin k_i$ and $k_i^2/2 \to 1 - \cos k_i$:

$$d_a(\mathbf{k}) = \left(\sin k_x, \sin k_y, \sin k_z, \sin k_v, \left(m + \sum_i \cos k_i\right)\right).$$

We would like to understand the phases of this model. As usual, we must first look at the points where the system undergoes phase transitions. These transitions are the solutions of equation $\sum_a d_a^2(\mathbf{k}, m) = 0$ (which implies as a necessary but not sufficient condition that k be one of the TR-invariant points), and they lead to five critical values of m and corresponding $\mathbf{k}$ points where the bulk gap closes:

$$m = \begin{cases} -4, \ \mathbf{k} = (0, 0, 0, 0), \\ -2, \ \mathbf{k} \in P\left[(\pi, 0, 0, 0)\right], \\ 0, \ \ \mathbf{k} \in P\left[(\pi, 0, \pi, 0)\right], \\ 2, \ \ \mathbf{k} \in P\left[(\pi, \pi, \pi, 0)\right], \\ 4, \ \ \mathbf{k} = (\pi, \pi, \pi, \pi), \end{cases} \tag{13.25}$$

where $P[\mathbf{k}]$ are all the wavevectors obtained from index permutations of wavevector $\mathbf{k}$; for example, $P\left[(\pi, 0, 0, 0)\right] = \left[(\pi, 0, 0, 0)\right], \left[(0, \pi, 0, 0)\right], \left[(0, 0, \pi, 0)\right], \left[(0, 0, 0, \pi)\right]$ and $P[(\pi, \pi, \pi, 0)]$ consists of $(\pi, \pi, \pi, 0)$, $(0, \pi, \pi, \pi)$, $(\pi, 0, \pi, \pi)$ and $(\pi, \pi, 0, \pi)$. We know that the phases when $m \to \pm\infty$ are atomic limits where $C_2 = 0$. We then must study the change in the second Chern number when the system undergoes a gap-closing-and-opening transition. There are transitions at $m = -4, -2, 0, 2, 4$. Because the first gap closing from the atomic to the nontrivial limit happens at $m = -4$, we wish to study the change of $C_2(m)$ around the critical value $m = -4$. In the limit $\delta m = m + 4 \ll 2$, the system has its minimal gap at $\mathbf{k} = \mathbf{0}$, around which the $d_a(\mathbf{k})$ vector has the approximate form $d_a(\mathbf{k}) \simeq (k_x, k_y, k_z, k_v, \delta m)$. The system then undergoes a phase transition to a Chern insulator of Chern number 1 at $m = -4$. Close to the transition point and for momenta in the vicinity of the momentum at which the transition takes place, the Hamiltonian is hence given by a continuum Dirac Hamiltonian, and the change in Hall conductance is easy to understand as a gap-closing-and-reopening transition of the continuum Dirac Hamiltonian. Because other k points in the BZ remain gapped, it is warranted to analyze the physics only at the point at which the transition takes place: the change of the second Chern number is determined only by the effective continuum model around the level-crossing wavevector(s). Similar analysis can be carried out at the other critical m's: at $m = -2$, the gap closes at 4 points in the BZ, giving rise to a change in Hall conductance of -4 (from 1 to $1 - 4 = -3$). At $m = 0$, the gap closes at 6 points in the BZ, with a change in the Hall conductance $-3 + 6 = 3$. At $m = 2$, the gap closes at another 4 points in the Brillouin zone and the change in Hall conductance is $3 - 4 = -1$. At $m = 4$, there is the last phase transition, where the gap closes at (π, π, π, π) and the Hall conductance changes $-1 + 1 = 0$, which again becomes the atomic limit. The sign of the Dirac Hamiltonian around the points where the phase transitions occurs determines the sign change of the Chern number. This is [13]

$$C_2(m) = \begin{cases} 0, & m < -4 \text{ or } m > 4, \\ 1, & -4 < m < -2, \\ -3, & -2 < m < 0, \\ 3, & 0 < m < 2, \\ -1, & 2 < m < 4 \end{cases} . \tag{13.26}$$

The second Chern number is in one-to-one correspondence with the surface states of the topologically nontrivial phases of this model. We take open boundary conditions for one dimension, say, v, and periodic boundary conditions for all other dimensions, so that $\mathbf{k} = k_x, k_y, k_z$ are still good quantum numbers. The Hamiltonian is a sum of 1-D tight-binding models:

$$H = \sum_{\mathbf{k}, v_1, v_2} c_{\mathbf{k}}^{\dagger}(v_1) \left(\frac{\Gamma^5 - i\Gamma^4}{2} \delta_{v_2, v_1+1} + \left(\sin k_i \Gamma^i + \left(m + \sum_i \cos k_i \right) \Gamma^5 \right) \delta_{v_1, v_2} \right) c_{\mathbf{k}}(v_2), \tag{13.27}$$

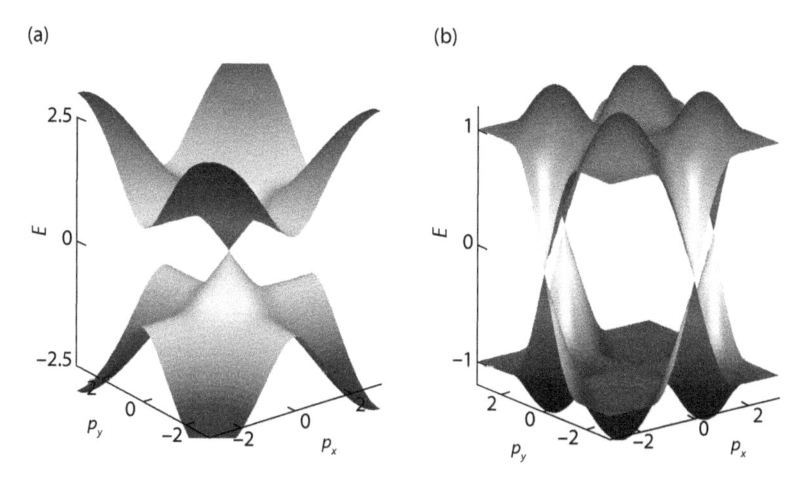

Figure 13.3. The energy spectrum for the surface states of the nontrivial 3-D topological insulator for the following model parameters: (a) $m = -3$, for which the second Chern number is 1, and (b) $m = -1$, for which the Chern number is -3. The surface energy gap vanishes at 1 and 3 (Dirac) points in the BZ respectively. For $m = -3$, the Dirac point is at Γ-point, whereas for $m = -1$, the Dirac point is at the X-points.

where $i = 1, 2, 3$ and $v = 1, 2, \ldots, L$ are the v-coordinates of lattice sites. We immediately realize that the noninteracting problem with open boundary conditions amounts to diagonalizing a 1-D tight-binding problem no matter what dimension we are working in (all the other periodic boundary conditions manifest themselves as good quantum numbers, and per quantum number, we still have to diagonalize a 1-D tight-binding problem). The single-particle energy spectrum is obtained as $E_n(\mathbf{k}),\ j = 1, 2, \ldots, 4L$ (there are four states per site, two orbitals and two spins). We see midgap surface states when $C_2 \neq 0$ (fig. 13.3).

13.6 Problems

1. *2-D Chern Simons Action by Integrating Out Fermions:* Explicitly integrate out the fermions in the 2-D Dirac Hamiltonian $H = \sum_{\mathbf{k}} c^{\dagger}_{\mathbf{k}}[(k_i + A_i)\sigma_i + m\sigma_z]c_{\mathbf{k}}$, where $i = 1, 2$, to obtain the Chern-Simons form $\epsilon_{ij} A_i \partial_0 A_j$ with coefficient $\frac{1}{2}\mathrm{sign}(m)$.

2. *4-D Chern-Simons Action by Integrating Out Fermions:* Explicitly integrate out the fermions in the massive Dirac Hamiltonian in four-dimensions, $H = \sum_{\mathbf{k}} c^{\dagger}_{\mathbf{k}}[(k_a + A_a)\Gamma_a + m\Gamma_5]c_{\mathbf{k}}$, where $a = 1, 2, 3, 4$ and Γ_a are the Clifford-algebra matrices, to obtain the second Chern-Simons form.

3. *Nonlinear Response Diagram:* Show that the fully antisymmetric part of the diagram in figure 13.1 is proportional to

$$\epsilon^{\mu\nu\rho\sigma\tau} \int \frac{d^4k d\omega}{(2\pi)^5} \mathrm{Tr}\left[G\frac{\partial G^{-1}}{\partial q^{\mu}}G\frac{\partial G^{-1}}{\partial q^{\nu}}G\frac{\partial G^{-1}}{\partial q^{\rho}}G\frac{\partial G^{-1}}{\partial q^{\sigma}}G\frac{\partial G^{-1}}{\partial q^{\tau}}\right]. \tag{13.28}$$

4. *First Chern Number and Commutator of Position Operators:* Show that the first Chern number is related to the commutator of the projected position coordinates $[PXP, PYP]$, where P

is the projector into the occupied bands, and $X = \sum_{\mathbf{j}} e^{i2\pi j_x/N_x} |\mathbf{j}, a\rangle \langle \mathbf{j}, a|$, where $\mathbf{j}$ is the position running over all the sites of the lattice, j_x is the x-component of the position, and a is the orbital index.

5. *Second Chern Number and the Extended Commutator of Different Position Operators:* Show that the second Chern number is related to the fully antisymmetric product $\epsilon_{ijlm} P X_i P X_j P X_l P X_m P$, where $X_i, i = 1, 2, 3, 4$, is the ith coordinate in four dimensions.

6. *Streda Formula:* Obtain the Streda-like formula for the (4+1)-D topological insulator relating the density to the magnetic and electric field.

7. *Numerical Implementation of the* (4 + 1)*-D Lattice Dirac Hamiltonian:* Solve for the energy levels of the Hamiltonian $h(\mathbf{k}) = \sum_{i=1}^{4} \sin(k_i)\Gamma_i + (M - \sum_{i=1}^{4} \cos(k_i))\Gamma_5$ with open boundary conditions, and confirm the existence of Weyl fermions on the 3-D surface of this 4-space-dimensional topological insulator. Show the presence of a single Weyl fermion per surface.

8. *Lattice Hamiltonians of Higher Second Chern Number:* Generalize the Hamiltonian of the previous exercise to obtain multiple (2, 3, 4 . . .) Weyl fermions on the surface of the 4-space-dimensional topological insulator.

9. *Stability of the Chiral Fermions on the Surface of a* 4 + 1 *Topological Insulator:* Is the fermion on the surface of the (4 + 1)-space-time-dimensional topological insulator stable to perturbations? What symmetries does it need for its stability?

10. *Fermion Doubling on the Lattice:* Show that in a purely 3-space-dimensional lattice (which is not the surface of a 4-space-dimensional lattice), Weyl fermions have to come in pairs of opposite chirality (chirality = Chern number of the Fermi surface surrounding the Weyl point). This is called Nielsen-Nyomiya theorem.

11. *Equivalence between the Berry Phase of Surface Fermions and the Second Chern Number of a Bulk* (4 + 1)*-D Topological Insulator:* Show by direct calculation that the sign of the first Chern number of the Fermi surface surrounding the gapless Weyl point on the surface of a 4 + 1 space-time-dimensional topological insulator is identical to the sign of the second Chern number of the bulk Hamiltonian.

14

Dimensional Reduction of 4-D Chern Insulators to 3-D Time-Reversal Insulators

In chapter 10, we presented the modern theory of charge polarization. In 1 space dimension, the charge polarization is the integral of the Berry phase over the 1-D BZ defined by $k_x \in [-\pi, \pi]$. We then expressed the 2-D Chern number as the difference between two 1-D charge polarizations when a control parameter, which we called k_y, is varied between 0 and 2π. We could then talk about the charge polarization as being the dimensional reduction of the 2-D Chern insulator. We would like to find a similar construction for the 4-D Chern insulator. As we will see, the dimensional reduction of the Chern insulator in 4 dimensions is the TR invariant insulator in 3 dimensions. Because as time reversal is involved, we must find a way to introduce it into our formalism. Notice that it is crucial that the 4-D Chern insulator can be defined in the presence of TR symmetry for this construction to work. Once we have obtained the dimensional reduction of a 4-D Chern insulator to a 3-D topological insulator, useful and interesting things, such as the effective action of the 3-D topological insulators, can be obtained. The question we want to ask in this chapter is what 3-D polarization difference (depending on a control parameter k_v, which can be thought as the 4th momentum of the topological insulator) gives rise to the 4-D Chern insulator. This will give us an expression of the Z_2 invariant as a winding number, the low-energy field theory of the Z_2 topological insulator, and also a direct relationship between the topological quantum numbers and the physically measurable topological response of the system.

14.1 Low-Energy Effective Action of (3 + 1)-D Insulators and the Magnetoelectric Polarization

The (4 + 1)-space-time-dimensional Dirac model has a Chern-Simons effective low-energy action. To obtain the effective action of a (3 + 1)-D insulator, we can perform a dimensional reduction of this theory. We explicitly perform this for the (4 + 1)-D Dirac model of equation (13.21), and then extend our derivation to the most general insulating models. To be able to consider k_v as a control parameter, we must have k_v as a good quantum number under periodic boundary conditions. The field configuration implementing this is Landau-gauge configuration translationally invariant in the v-direction: $A_{n,n+\hat{i}} = A_{n+\hat{v},n+\hat{v}+\hat{i}}$, where n is any lattice site. We consider the Hamiltonian of the Dirac model eq. (13.21) coupled to the external $U(1)$ gauge field in this Landau gauge:

$$H[A] = \sum_{k_v,\mathbf{x},j} \left[c^{\dagger}_{\mathbf{x},k_v} \left(\frac{c\Gamma^5 - i\Gamma^j}{2} \right) e^{iA_{\mathbf{x},\mathbf{x}+\hat{j}}} c_{\mathbf{x}+\hat{j},k_v} + \text{h.c.} \right] \tag{14.1}$$
$$+ c^{\dagger}_{\mathbf{x},k_v} \left[\sin\left(k_v + A_{\mathbf{x},\mathbf{x}+\hat{v}}\right) \Gamma^4 + \left(m + c\cos\left(k_v + A_{\mathbf{x},\mathbf{x}+\hat{v}}\right)\right) \Gamma^5 \right] c_{\mathbf{x},k_v},$$

where $\mathbf{x}$ runs over the 3-D coordinates and $j = 1, 2, 3$ are the x, y, z-directions. It is now clear why we choose this specific Landau gauge. In this gauge, the (4 + 1)-space-time-dimensional Dirac Hamiltonian in equation (14.2) decouples into a sum of independent

(3 + 1)-D Hamiltonians $H_{3D}[A, \theta]$, dependent on a control parameter $k_v + A_{\mathbf{x},\mathbf{x}+\hat{v}} = \theta$:

$$H_{3D}[A, \theta] = \sum_{\mathbf{x}, j} \left[c_{\mathbf{x}}^{\dagger} \left(\frac{c\Gamma^5 - i\Gamma^j}{2} \right) e^{iA_{\mathbf{x},\mathbf{x}+\hat{j}}} c_{\mathbf{x}+\hat{j}} + \text{h.c.} \right]$$
$$+ c_{\mathbf{x}}^{\dagger} \left[\sin\theta \Gamma^4 + (m + c\cos\theta)\Gamma^5 \right] c_{\mathbf{x}}. \tag{14.2}$$

If the (4 + 1)-D Hamiltonian is gapped, then equation (14.2) also describes a band insulator coupled to an electromagnetic field $A_{\mathbf{x},\mathbf{x}+\hat{s}}$ and an adiabatic parameter field θ.

From now on, everything works out just as in the 4-D Chern insulator case except that the integral over the k_v momentum is not performed—instead, we have the control parameter θ. The response properties of the (3 + 1)-D system is defined from the effective action $S_{3\text{-D}}[A, \theta]$ obtained by integrating out the fermions in the presence of the control parameter θ (remember that "currents" of the control parameter $\frac{\partial H}{\partial \theta}$ are allowed):

$$\exp^{iS_{3D}[A,\theta]} = \int D[\psi]D[\bar{\psi}]e^{i\int dt \left[\sum_{\mathbf{x}} \bar{\psi}_{\mathbf{x}}(i\partial_\tau - A_{\mathbf{x}0})\psi_{\mathbf{x}} - H[A,\theta]\right]}.$$

The nonlinear response term to external fields is now the bubble with three legs. However, the current leg, $\frac{\partial H}{\partial k_v}$, in the (4 + 1)-D theory is now replaced by the "current" of the control parameter θ, $\frac{\partial H}{\partial \theta}$. Also, whereas in the (4+1)-D theory the nonlinear bubble contains an integral over the 4th momentum k_v, in the dimensional reduction, this integral (over θ) is missing. The value of the nonlinear bubble is dependent on θ, and we call it $G_3(\theta)$. It is obviously equal to the integrand of the second Chern number C_2 in equation (13.8) but without the integration over the fourth momentum k_v:

$$G_3(\theta) = \frac{1}{8\pi^2} \int d^3k \epsilon^{ijl} \text{Tr}\left[f_{\theta i} f_{jl} \right], \quad i, j, l = 1, 2, 3, \tag{14.3}$$

where f_{ij}, with $i, j = 1, 2, 3$ is the non-Abelian Berry curvature (field strength) of the Berry-phase gauge field $a_i^{\alpha\beta} = -i \langle \mathbf{k}, \theta; \alpha | \frac{\partial}{\partial k_i} | \mathbf{k}, \theta; \beta \rangle$, whereas $f_{\theta i}$ is the Berry curvature of the θ (and ith) fields, where $a_\theta^{\alpha\beta} = -i \langle \mathbf{k}, \theta; \alpha | \frac{\partial}{\partial \theta} | \mathbf{k}, \theta; \beta \rangle$ is the Berry potential of the θ parameter. The relation of $G_3(\theta)$ to the second Chern number, equation (13.11), is a sum rule,

$$\int_0^{2\pi} G_3(\theta)\, d\theta = C_2. \tag{14.4}$$

The integration assumes that the Hamiltonian stays gapped at all points between $\theta = 0$ and $\theta = 2\pi$. Equation (14.4) is very similar to that of the polarization in 1 dimension, whose winding with respect to an adiabatic parameter (here θ) gives rise to the first Chern number. If we can express $\int_0^{2\pi} G_3(\theta)\, d\theta$ as a winding of some 3-D polarization P_3, then we have found a generalization to 4 dimensions of the well-known relation in 2 dimensions between the winding of a 1-D polarization and the first Chern number. To find the winding of a polarization, we must express $G_3(\theta)$ as a partial derivative $G_3(\theta) = \frac{\partial P_3(\theta)}{\partial \theta}$ of a 3-D polarization $P_3(\theta)$. We will call P_3 the "magnetoelectric" polarization, for reasons that will soon become obvious: it is the coefficient of a current response to applied magnetic *and* electric fields: P_3 couples nonlinearly to $\mathbf{E} \cdot \mathbf{B}$. Hence, if we write,

$$G_3(\theta) = \frac{\partial P_3(\theta)}{\partial \theta}, \tag{14.5}$$

then the second Chern number is the winding of P_3 in a cycle of the control parameter:

$$\int_0^{2\pi} G_3(\theta)\, d\theta = C_2 = P_3(2\pi) - P_3(0). \tag{14.6}$$

To obtain $P_3(\theta)$, we need to find the term whose derivative with respect to the control parameter θ gives the $F \wedge F$ term, where one of the coordinates is θ. In the Landau-gauge configuration translationally invariant in the v direction, $A_{n,n+\hat{i}} = A_{n+\hat{v},n+\hat{v}+\hat{i}}$. By direct calculation, we obtain [13]

$$P_3(\theta_0) = \int d^3k \mathcal{K}^\theta = \frac{1}{16\pi^2} \int d^3k \epsilon^{\theta ijk} \mathrm{Tr}\left[\left(f_{ij} - \frac{i}{3}\left[a_i, a_j\right]\right) \cdot a_k\right], \tag{14.7}$$

where K^θ is one component of the Chern-Simons vector in the 4-D parameter space $(k_x, k_y, k_z, \theta_0)$:

$$\mathcal{K}^A = \frac{1}{16\pi^2} \epsilon^{ABCD} \mathrm{Tr}\left[\left(f_{BC} - \frac{i}{3}\left[a_B, a_C\right]\right) \cdot a_D\right], \tag{14.8}$$

With this definition of the 3-D polarization, we have $G_3(\theta_0) = \int d^3k \partial_\theta \mathcal{K}^\theta \equiv \frac{\partial P_3(\theta_0)}{\partial \theta_0}$.

The more general identity, which is true irrespective of gauge, is that, up to boundary terms

$$\partial_A \mathcal{K}^A = \frac{1}{32\pi^2} \epsilon^{ABCD} \mathrm{Tr}\left[f_{AB} f_{CD}\right], \tag{14.9}$$

where $A, B, C, D = x, y, z, \theta$ and $\mathcal{K}^A$ is the non-Abelian Chern-Simons term.

We have hence determined that the nonlinear response bubble is given (in a certain Landau gauge) by the derivative of the magnetoelectric polarization $P_3(\theta)$ with respect to a control parameter θ. We then found further that $P_3(\theta)$ is given by the integral of the non-Abelian Chern-Simons 3-form over momentum space, in the same way that the charge polarization is defined as the integral of the Berry-connection 1-form over a 1-D path in momentum space. The charge polarization was defined only modulo 1 (times 2π) because a different choice in the position of the Wannier centers (a gauge transformation) would change the charge polarization by an integer. A similar story is true in 3 dimensions: the 3-D integration of the Chern-Simons term is gauge invariant only modulo an integer. Under a gauge transformation $a_i \to u^{-1} a_i u - i u^{-1} \partial_i u$ ($u \in U(N_{\text{occupied}})$ when N_{occupied} bands are occupied), it is trivial to show that the change of P_3 is given by

$$P_3 \to P_3 + \frac{1}{24\pi^2} \int d^3k \epsilon^{\theta ijk} \mathrm{Tr}\left[\left(u^{-1}\partial_i u\right)\left(u^{-1}\partial_j u\right)\left(u^{-1}\partial_k u\right)\right],$$

which is a winding number of the map between the torus T^3 and the non-Abelian gauge transformation manifold $U_{N_{\text{occupied}}}$ and is, therefore, an integer if $N_1 \geq 2$. Thus, $P_3(\theta_0)$, just like $P_1(\theta)$, is defined only modulo 1. However, its change during a full-cycle variation of θ_0 from 0 to 2π is well defined, irrespective of the gauge chosen and gives the second Chern number, C_2.

Now that we have found the expression for the nonlinear response bubble, we would like to obtain the Lagrangian for the 3-D topological insulator, from which we can derive the transport equations. We know that the Lagrangian of the (4 + 1)-D topological insulator was obtained as the Chern-Simons action (eq. (13.7)):

$$S_{4\text{Deff}} = \frac{C_2}{24\pi^2} \int d^4x\, dt \epsilon^{\mu\nu\rho\sigma\tau} A_\mu \partial_\nu A_\rho \partial_\sigma A_\tau \tag{14.10}$$

We remember that the dimensional reduction was obtained by neglecting the integration over the fourth dimension, v. The field in the fourth dimension is the change from an equilibrium value θ_0 of θ, $\delta\theta(\mathbf{x}, t) = \theta(\mathbf{x}, t) - \theta_0$. By replacing the $A_4(\mathbf{x}, t)$ by $\delta\theta(\mathbf{x}, t)$ and not integrating over the fourth dimension, we obtain

$$S_{3\text{Deff}} = \frac{G_3(\theta_0)}{4\pi} \int d^3x\, dt \epsilon^{\mu\nu\sigma\tau} \delta\theta \partial_\mu A_\nu \partial_\sigma A_\tau. \tag{14.11}$$

The coefficient $G_3(\theta_0)$ is determined by the nonlinear bubble Feynman diagram fig. (13.1) but evaluated for the 3-D Hamiltonian (eq. (14.2))—it does not involve an integration over θ. Per our discussion, it can be written as $\frac{\partial P_3(\theta)}{\partial \theta}$, evaluated at the equilibrium field configuration θ_0: $G_3(\theta_0)$.

We now perform an integration by parts to move one of the derivatives from the A-terms onto the $\delta\theta$-term:

$$S_{3\text{Deff}} = \frac{G_3(\theta_0)}{4\pi} \int d^3x\, dt \epsilon^{\mu\nu\sigma\tau} A_\mu \partial_\nu \delta\theta \partial_\sigma A_\tau. \tag{14.12}$$

(Notice that we have swapped the μ, ν labels in order to account for the minus sign coming from the partial integration.) We now assume that the change in the G_3 coefficient in space-time is smooth enough so that it can be introduced inside the integral in terms of the magnetoelectric polarization $G_3 = \frac{\partial P_3}{\partial \theta}$:

$$S_{3\text{D}} = \frac{1}{4\pi} \int d^3x\, dt \epsilon^{\mu\nu\sigma\tau} A_\mu \left(\frac{\partial P_3}{\partial \theta} \right) \partial_\nu \delta\theta \partial_\sigma A_\tau.$$

The term $\frac{\partial P_3}{\partial \theta} \partial_\nu \delta\theta = \frac{\partial P_3}{\partial \theta} \partial_\nu\, \theta$ can now written as $\partial_\nu P_3$, where $P_3(\mathbf{x}, t) = P_3(\theta(\mathbf{x}, t))$ has space-time dependence determined by the θ field (the other momenta are integrated over). Such an expression is meaningful only when the space-time dependence of the θ field is smooth and adiabatic, so that locally θ can still be considered as a parameter—the same reason given for allowing $G_3(\theta)$ inside the integral. Integrating by parts once again, the effective action is finally written as [13]

$$S_{3\text{D}} = \frac{1}{4\pi} \int d^3x\, dt \epsilon^{\mu\nu\sigma\tau} P_3(x, t) \partial_\mu A_\nu \partial_\sigma A_\tau. \tag{14.13}$$

This is the topological part of the field theory of 3-D topological insulators. The magnetoelectric polarization is sometimes called a θ-vacuum or axion background. Notice that up to now, we have not specified whether or not the insulator has TR invariance. Indeed, nonzero magnetoelectric polarization can exist even in trivial insulators. We now look at the conditions imposed on the magnetoelectric polarization P_3 by the requirements of time reversal or inversion.

14.2 Magnetoelectric Polarization for a 3-D Insulator with Time-Reversal Symmetry

In the general case, the magnetoelectric polarization can take any values in the [0, 1] interval (remember P_3 is defined mod 1 in units of 2π). However, in the presence of TR *or* inversion symmetry, the magnetoelectric polarization can take only two values, 0 or $\frac{1}{2}$ (in units of 2π). We will now prove this. With only TR symmetry, we have proved in a previous chapter that the non-Abelian $a(-\mathbf{k})$ is the gauge-transformed $a(\mathbf{k})^*$, the gauge transformation being given by the sewing matrix $B(\mathbf{k})$:

$$a_i(-\mathbf{k}) = B(\mathbf{k})a_i^*(\mathbf{k})B^\dagger(\mathbf{k}) + iB(\mathbf{k})\nabla_i B^\dagger(\mathbf{k}), \tag{14.14}$$

where a is the non-Abelian vector potential and $B(\mathbf{k})$ is the sewing matrix that relates Kramers' pairs:

$$\left|u_\alpha(-\mathbf{k})\right\rangle = B^*_{\alpha\beta}(\mathbf{k})T\left|u_\beta(\mathbf{k})\right\rangle . \tag{14.15}$$

We hence have

$$\begin{aligned} f_{ij}(-\mathbf{k}) &= -\partial_{k_i}a_j(-\mathbf{k}) + \partial_{k_j}a_i(-\mathbf{k}) + i[a_i(-\mathbf{k}), a_j(-\mathbf{k})] \\ &= -B(\partial_i a_j^*(\mathbf{k}) - \partial_j a_i^*(\mathbf{k}) - i[a_i^*(\mathbf{k}), a_j^*(\mathbf{k})])B^\dagger = -B(\mathbf{k})f_{ij}^*(\mathbf{k})B^\dagger(\mathbf{k}); \end{aligned} \tag{14.16}$$

hence, the non-Abelian field strength at $-\mathbf{k}$ is a gauge transform of the complex conjugate of the field strength at $\mathbf{k}$, with the gauge transformation given by the sewing matrix B. From here we can immediately massage the magnetoelectric polarization into a form that unravels its properties under timereversal:

$$\begin{aligned} P_3 &= \frac{1}{16\pi^2}\int d^3k\epsilon_{ijk}\mathrm{Tr}[(f_{ij}(\mathbf{k}) - \frac{2}{3}ia_i(\mathbf{k})a_j(\mathbf{k}))a_k(\mathbf{k})] \\ &= \frac{1}{16\pi^2}\int d^3k\epsilon_{ijk}\mathrm{Tr}[(f_{ij}(-\mathbf{k}) - \frac{2}{3}ia_i(-\mathbf{k})a_j(-\mathbf{k}))a_k(-\mathbf{k})] \\ &= \frac{1}{16\pi^2}\int d^3k\epsilon_{ijk}\mathrm{Tr}[(-B(\mathbf{k})f_{ij}^*(\mathbf{k})B^\dagger - \frac{2}{3}i(B(\mathbf{k})a_i^*(\mathbf{k})a_j^*(\mathbf{k})B^\dagger(\mathbf{k}) \\ &\quad +iB(\mathbf{k})a_i^*(\mathbf{k})\partial_j B^\dagger(\mathbf{k}) - i(\partial_i B(\mathbf{k}))a_j^*(\mathbf{k})B^\dagger(\mathbf{k}) \\ &\quad +\partial_i B\partial_j B^\dagger)(B(\mathbf{k})a_k^*(\mathbf{k})B^\dagger(\mathbf{k}) + iB(\mathbf{k})\partial_k B^\dagger(\mathbf{k}))] \\ &= -\frac{1}{16\pi^2}\int d^3k\epsilon_{ijk}\mathrm{Tr}[(f_{ij}^*(\mathbf{k}) + \frac{2}{3}ia_i^*(\mathbf{k})a_j^*(\mathbf{k}))a_k^*(\mathbf{k})] \\ &\quad -\frac{1}{24\pi^2}\int d^3k\epsilon_{ijk}\mathrm{Tr}[(B(\mathbf{k})\partial_i B^\dagger)(B(\mathbf{k})\partial_j B^\dagger)(B(\mathbf{k})\partial_k B^\dagger)] \\ &\quad -\frac{i}{8\pi^2}\int d^3k\epsilon_{ijk}\partial_i(B(\mathbf{k})a_j^*(\mathbf{k})\partial_k B^\dagger(\mathbf{k})) \\ &= -P_3^* - \frac{1}{24\pi^2}\int d^3k\epsilon_{ijk}\mathrm{Tr}[(B(\mathbf{k})\partial_i B^\dagger(\mathbf{k}))(B(\mathbf{k})\partial_j B^\dagger(\mathbf{k}))(B(\mathbf{k})\partial_k B^\dagger(\mathbf{k}))]. \end{aligned} \tag{14.17}$$

Because P_3 is a real number, we have

$$2P_3 = -\frac{1}{24\pi^2}\int d^3k\epsilon_{ij\mathbf{k}}\mathrm{Tr}[(B(\mathbf{k})\partial_i B^\dagger(\mathbf{k}))(B(\mathbf{k})\partial_j B^\dagger(\mathbf{k}))(B(\mathbf{k})\partial_\mathbf{k} B^\dagger(\mathbf{k}))]. \tag{14.18}$$

The right-hand side of the preceding equation is a winding number of the map from the BZ torus T^3 to the manifold described by the matrix $B(k)$. Because this is a unitary matrix, the manifold is that of one of the unitary groups. Which unitary group it is turns out to be is important in order to obtain a nontrivial winding number. For TR-invariant systems, bands have Kramers' degeneracy at the TR-invariant points. All other degeneracies can be removed because they are accidental. Hence, an insulator with an even number N_{occupied} of occupied bands (and $N_{\mathrm{occupied}}/2$ of Kramers' pairs) can have all degeneracies between all bands removed with the exception of the Kramers' protected ones. As such, the matrix B can be decomposed in diagonal blocks, each being a 2×2 unitary matrix. In other words, the space of the matrix $B(k)$ is the manifold $U(2)^{N_{\mathrm{occupied}}/2} = U(2)\otimes U(2)\otimes\cdots\otimes U(2)$. The homotopy of the map from the 3-torus to the direct product of manifolds reduces to (possibly copies of) the homotopy of the map $T^3 \to U(2)$. The winding number of this map is nontrivial and is given by an integer, because the homotopy group is Z. We have showed that P_3 is defined modulo an integer, so the preceding relation proves that P_3 is either 0 or $\frac{1}{2}$ in units of 2π, depending on whether the winding number of the $B(k)$ map is even or odd. Because P_3 is itself defined mod 1, it means that there are two classes of topological insulators (as we know). The atomic limit obviously has $P_3 = 0$. The topological insulator hence has $P_3 = \pi$. Using a mathematical quantity called the degree of a map, it can be shown that the P_3 formulation of a topological insulator is equivalent to the Pfaffian index description provided by Kane et al. [14].

14.3 Magnetoelectric Polarization for a 3-D Insulator with Inversion Symmetry

We have seen that the magnetoelectric polarization is quantized to have two values in the case when timereversal is present. However, timereversal is not the only symmetry that quantizes the value of P_3 to be either 0 or $\frac{1}{2}$. Inversion does a similar job. The proof follows the TR-invariant case but with some crucial differences related to the fact that $P^2 = 1$ (where P is the inversion matrix), whereas $T^2 = -1$. First, the sewing matrix B is now relating inversion bands:

$$|u_i(-\mathbf{k})\rangle = B^*_{ij}(\mathbf{k})P\,|u_j(\mathbf{k})\rangle\,, \tag{14.19}$$

where i, j run over the occupied bands 1, ..., N_1 (where if the system comes from breaking time reversal on a topological insulator, then N_1 is even, but, in general, N_1 can be both even or odd), is a $U(N_1)$ matrix. For a system with inversion symmetry, the non-Abelian Berry potential at $-k$ is related to the negative of the potential at k by a gauge transformation:

$$\mathbf{a}(-\mathbf{k}) = -B\mathbf{a}(\mathbf{k})B^\dagger + i\mathrm{Tr}[B(\mathbf{k})\nabla B^\dagger(\mathbf{k})], \tag{14.20}$$

where the gauge transformation is the sewing matrix $B(\mathbf{k})$. We can clearly see the differences with the TR-invariant case: a minus sign in front of $BaB^\dagger$ and no complex conjugate. A different gauge covariance of the field strength is manifest:

$$\begin{aligned} f_{ij}(-\mathbf{k}) &= -\partial_{k_i}a_j(-\mathbf{k}) + \partial_{k_j}a_i(-\mathbf{k}) + i[a_i(-\mathbf{k}), a_j(-\mathbf{k})] \\ &= B(\partial_i a_j(\mathbf{k}) - \partial_j a_i(\mathbf{k}) + i[a_i(\mathbf{k}), a_j(\mathbf{k})])B^\dagger = B(\mathbf{k})f_{ij}(\mathbf{k})B^\dagger(\mathbf{k}), \end{aligned} \tag{14.21}$$

where again, compared to the TR-invariant case, there is no minus sign and no complex conjugate. From here the magnetoelectric polarizability is easy to obtain:

$$\begin{aligned}
P_3 &= \frac{1}{16\pi^2}\int d^3k\epsilon_{ijk}\mathrm{Tr}[(f_{ij}(\mathbf{k}) - \frac{2}{3}ia_i(\mathbf{k})a_j(\mathbf{k}))a_k(\mathbf{k})] \\
&= \frac{1}{16\pi^2}\int d^3k\epsilon_{ijk}\mathrm{Tr}[(f_{ij}(-\mathbf{k}) - \frac{2}{3}ia_i(-\mathbf{k})a_j(-\mathbf{k}))a_k(-\mathbf{k})] \\
&= \frac{1}{16\pi^2}\int d^3k\epsilon_{ijk}\mathrm{Tr}[(B(\mathbf{k})f_{ij}(\mathbf{k})B^\dagger - \frac{2}{3}i(B(\mathbf{k})a_i(\mathbf{k})a_j(\mathbf{k})B^\dagger(\mathbf{k}) \\
&\quad -iB(\mathbf{k})a_i(\mathbf{k})\partial_jB^\dagger(\mathbf{k}) + i(\partial_iB(\mathbf{k}))a_j(\mathbf{k})B^\dagger(\mathbf{k}) + \partial_iB\partial_jB^\dagger) \\
&\quad \times(-B(\mathbf{k})a_k(\mathbf{k})B^\dagger(\mathbf{k}) + iB(\mathbf{k})\partial_kB^\dagger(\mathbf{k}))] \\
&= -\frac{1}{16\pi^2}\int d^3k\epsilon_{ijk}\mathrm{Tr}[(f_{ij}(\mathbf{k}) - \frac{2}{3}ia_i(\mathbf{k})a_j(\mathbf{k}))a_k(\mathbf{k})] \\
&\quad -\frac{1}{24\pi^2}\int d^3k\epsilon_{ijk}\mathrm{Tr}[(B(\mathbf{k})\partial_iB^\dagger)(B(\mathbf{k})\partial_jB^\dagger)(B(\mathbf{k})\partial_kB^\dagger)] \\
&\quad +\frac{i}{8\pi^2}\int d^3k\epsilon_{ijk}\partial_i(B(\mathbf{k})a_j(\mathbf{k})\partial_kB^\dagger) \\
&= -P_3 - \frac{1}{24\pi^2}\int d^3k\epsilon_{ijk}\mathrm{Tr}[(B(\mathbf{k})\partial_iB^\dagger)(B(\mathbf{k})\partial_jB^\dagger)(B(\mathbf{k})\partial_kB^\dagger)].
\end{aligned} \tag{14.22}$$

We then obtain a similar relation to that of the TR topological insulators:

$$2P_3 = -\frac{1}{24\pi^2}\int d^3k\epsilon_{ijk}\mathrm{Tr}[(B(\mathbf{k})\partial_iB^\dagger(\mathbf{k}))(B(\mathbf{k})\partial_jB^\dagger(\mathbf{k}))(B(\mathbf{k})\partial_kB^\dagger(\mathbf{k}))]. \tag{14.23}$$

This proves that P_3 is either 0 or $\frac{1}{2}$ in units of 2π, depending on whether the RHS (which is a winding number) is even or odd. Because P_3 is itself defined mod 1, then it means that there are two classes of topological insulators because we know if the winding number of the map $B(k)$ can be nontrivial (different from zero). We would think that the prospect for such a nontrivial winding number is slim: in the absence of time reversal, trivial reasoning would say that the matrix B can now be decomposed into diagonal elements. If you lose TR (generic case for only inversion symmetric insulators), then trivially, it would seem that we can just separate *all* bands and make them have no degeneracies. Only $T^2 = -1$ guarantees the presence of Kramer's doublets—hence, with inversion symmetry only, it would seem that we can split all the degeneracies by breaking time reversal. If this is true, the target space of the map $B(k)$ would, (generically) be $U(1)^{N_1}$. The homotopy group of the maps between T^3 and $U(1)$ is trivial and would indicate that the winding number is always zero.

This trivial reasoning is, fortunately, not true. In fact, it can be proved [53, 54] that, for example, in the case when there are only two occupied bands and the product over the inversion eigenvalues of each of those bands at all inversion-symmetric points is -1, then the two bands *cannot* be separated from each other, and there *must* be degenerate points in the BZ that connect the two bands. These degeneracy points (which come in inversion-symmetric pairs) are protected by a different type of symmetry than in the case of TR invariance. They do not allow the two bands to be separated from one another, and hence the matrix $B(k)$ still belongs to $U(2)$, with nontrivial homotopy group and integer winding number. We have hence proved that either time reversal or inversion can quantize the value of the magnetoelectric polarization to be either 0 or π.

14.4 3-D Hamiltonians with Time-Reversal Symmetry and/or Inversion Symmetry as Dimensional Reductions of 4-D Time-Reversal-Invariant Chern Insulators

Having proven that the magnetoelectric polarization of a system with TR or inversion symmetry can only be 0 or π, we now show that in the case of a TR-invariant 4-D Chern insulator, the Chern number has to be odd if its 3-D reduction is a nontrivial topological insulator. The strategy used is the following: we pick two 3-D Hamiltonians, $h_1(k)$ and $h_2(k)$, satisfying some symmetries (TR and/or inversion). We then pick an interpolation between these Hamiltonians. The interpolation depends on a control parameter $\theta = k_v + A_v$, the canonical momentum in the fourth direction, v. We pick the interpolation to also be TR and/or P invariant in a way shown next. If the interpolation Hamiltonian, which is now a 4-space-dimensional Hamiltonian, has *odd* second Chern number, then $h_1(k)$ and $h_2(k)$ are Hamiltonians that belong to different Z_2 topological classes, so at least one of them is nontrivial.

To be explicit, if we pick two 3-space-dimensional TR-invariant band insulators $h_1(\mathbf{k})$ and $h_2(\mathbf{k})$; an interpolation $h(\mathbf{k}, \theta)$ can be defined between them, satisfying

$$h(\mathbf{k}, 0) = h_1(\mathbf{k}), \; h(\mathbf{k}, \pi) = h_2(\mathbf{k}), \quad T^{-1}h(-\mathbf{k}, 2\pi - \theta)T = h(\mathbf{k}, \theta), \tag{14.24}$$

and $h(\mathbf{k}, \theta)$ is gapped for any $\theta \in [0, 2\pi]$. We interpret the new Hamiltonian $h(k, \theta)$ as a 4-space-dimensional Hamiltonian—if the control parameter is identified with the 4th momentum $\theta = k_w$. The existence of a gapped parametrization is nontrivial and should be explained. It turns out that a correct and sound proof for the existence of a gapped interpolation can be provided only for the case when the system has both TR and inversion symmetry. With just time reversal, the Hamiltonian bands are spin-split, and they are only doubly degenerate at the TR-symmetric points. We now turn the system into an inversion-symmetric system by turning off the inversion-splitting terms. The TR- and inversion-symmetric system has doubly-degenerate bands. Hence, any band crossings involve an effective 4 bands and a Hamiltonian made out of 4×4 Clifford-algebra matrices—they have to be Clifford because the Hamiltonian has two doubly-degenerate bands participating in the band closing and reopening. There are 5 such matrices; hence, in general, the codimension of the effective Hamiltonian is 5 — i.e., we need 5 parameters tuned to zero to close a gap. However, the interpolated Hamiltonian only has 4 variables (the momenta running through the BZ)—and, hence, any crossing is *avoided* in general if we do not tune an extra parameter. As such, a gapped interpolation can always be defined. Breaking inversion symmetry renders the proof for the existence of a gapped interpolation more difficult.

Because the interpolation is periodic in θ, a second Chern number $C_2[h(\mathbf{k}, \theta)]$ of the Berry-phase gauge field can be defined in the $(\mathbf{k}, \theta)$ space. We have already proved that for each of the Hamiltonians in equation (14.25), the magnetoelectric polarization is equal to either 0 or π, because they are TR invariant. In fact, we have proved this irrespective of whether the system is TR invariant, inversion invariant, or both. Besides the TR interpolation in equation (14.25), we can alternatively build an inversion-symmetric interpolation in the case when $h_1(k)$, $h_2(k)$ are inversion-symmetric Hamiltonians:

$$h(\mathbf{k}, 0) = h_1(\mathbf{k}), \; h(\mathbf{k}, \pi) = h_2(\mathbf{k}) \quad P^{\dagger}h(-\mathbf{k}, 2\pi - \theta)P = h(\mathbf{k}, \theta). \tag{14.25}$$

At each value of θ we can define a magnetoelectric polarization $P_3(\theta)$, which is the magnetoelectric polarization of the Hamiltonian $h(k, \theta)$ integrated over the 3-D k-space. Because of time reversal or inversion, with θ involved, the TR properties on the eigenstates of $h(k, \theta)$ give

$$P_3(\theta) = -P_3^*(2\pi - \theta) - \frac{1}{24\pi^2}\int d^3k\epsilon_{ijk}\mathrm{Tr}[(B\partial_i B^\dagger)(B\partial_j B^\dagger)(B\partial_k B^\dagger)] \tag{14.26}$$

with TR symmetry and

$$P_3(\theta) = -P_3(2\pi - \theta) - \frac{1}{24\pi^2}\int d^3k\epsilon_{ijk}\mathrm{Tr}[(B\partial_i B^\dagger)(B\partial_j B^\dagger)(B\partial_k B^\dagger)] \tag{14.27}$$

with inversion symmetry. Because P_3 is real, the two equations give the same constraints. The second Chern number is an integral over the full periodic θ range:

$$C_2[h] = \int_0^{2\pi} d\theta \frac{\partial P(\theta)}{\partial\theta} = 2\int_0^{\pi} d\theta \frac{\partial P(\theta)}{\partial\theta}. \tag{14.28}$$

Hence, for the two Hamiltonians h_1 and h_2, the second Chern number of the interpolation is

$$C_2[h(\mathbf{k}, \theta)] = 2\,(P_3[h_2] - P_3[h_1]) \bmod 2. \tag{14.29}$$

Module 2 comes from the fact that the magnetoelectric polarization is defined only for modulo an integer. The difference of P_3 between the two interpolations determines the relative Chern parity $(-1)^{2(P_3[h_1]-P_3[h_2])}$. Only the Chern parity can be determined because each magnetoelectric polarization is defined mod 1. Because the trivial Hamiltonian h_0 with all hopping elements tuned to zero obviously has $P_3 = 0$, we know that all the Hamiltonians with $P_3 = \frac{1}{2}$ are topologically nontrivial, whereas those with $P_3 = 0$ are trivial. This implies that a 4-D TR- or inversion-invariant Hamiltonian interpolating between a trivial 3-D insulator and a nontrivial 3-D topological insulator must have an odd second Chern number. However, in practice, finding such an interpolation might not always be easy.

14.5 Problems

1. *Derivative of the Chern-Simons Form:* Show by direct computation that $\int d^4k\mathrm{Tr}[F \wedge F] = \int d^4k\mathrm{Tr}[\partial \wedge (F \wedge A - 1/3 A \wedge A \wedge A)]$. (*Hint*: Neglect all the boundary terms and be aware that the identity is true only after taking the trace.)

2. *Explicit Computation of the Magnetoelectric Effect from the Eigenstates of the Massive Dirac Hamiltonian:* Compute the θ-term analytically directly from the eigenstates of the massive Dirac Hamiltonian. Usually this is a nontrivial thing to do because it involves gauge fixing (nobody has yet managed to write the magnetoelectric polarization in terms of projectors solely), but in the case of the Dirac Hamiltonian, the eigenstates are analytically known and derivatives can easily be obtained.

3. *Sewing Matrix and Its Winding Number for the Dirac Hamiltonian:* Use the Dirac Hamiltonian regularized on the lattice to obtain the sewing matrix $B(\mathbf{k})$ and to (try to) compute its winding number directly.

4. *Magnetoelectric Polarization and Point Symmetry Groups:* Prove that under any *improper* rotation the magnetoelectric polarization is quantized to either 0 or π. Proper/improper rotations are defined to have determinant ± 1.

15

Experimental Consequences of the Z_2 Topological Invariant

We now discuss a number of experimental predictions that arise from the physics of 2-D and 3-D topological insulators. Our book focuses on the theoretical aspects of the topological insulator problem—as a result, we do not dwell on the experimental consequences and just mention in passing two of the most important predictions: the topological magneto-electric effect and the presence of a surface Hall conductance of $\frac{1}{2}$ when a ferromagnet gaps the surface of the insulator. Several other important predictions, such as the appearance of Majorana fermions on a vortex core when a superconductor is placed on the surface of a topological insulator, inducing a proximity effect, will be discussed in the topological superconductor part of the book, Chapters 16–18.

Most materials that have been experimentally discovered to be nontrivial topological insulators either have inversion symmetry or are close to an inversion symmetry point in their phase diagram. Once a material has inversion symmetry, it is extremely easy to discover whether it is or is not a topological insulator by using the Fu and Kane inversion eigenvalue formula [9]. In the past years, many people [55, 56, 57, 58, 59, 60, 61] have suggested several candidate materials that could be 3-D topological insulators, including bismuth antimony (BiSb) alloys, strained 3-D HgTe, Bi_2Se_3, Heusler compounds, and others. These 3-D materials are topologically nontrivial because of a band-inversion mechanism identical to that of the HgTe quantum wells [7]. As we saw, the effective model for a 3-D topological insulator is a regularized gapped 3-D Dirac fermion whose mass changes sign in the BZ. If we approximate the materials as having inversion symmetry, every model of a topological insulator close to the minimum gap point is approximated by a massive Dirac Hamiltonian. Its mass changes sign as we move away from the minimum gap point k. The sign changing mass indicates a phase shift of π in the vacuum angle θ (i.e., a shift of P_3 by $\frac{1}{2}$) from its original value in the atomic limit. As an effective model, the interface between the topological insulator and an atomic limit insulator can be modeled as a Lagrangian with a space-varying transition between a $\theta = \pi$ ($P_3 = \frac{1}{2}$) and a $\theta = 0$ ($P_3 = 0$). Notice that although these two values of the angle θ are TR-invariant, *anything* in between these values breaks time reversal. Breaking time reversal on the surface of the topological insulator will gap the edge modes and give rise to a fully insulating system, where quantized responses appear. We now analyze such quantized responses, following the work of [13].

15.1 Quantum Hall Effect on the Surface of a Topological Insulator

The gas of 2-D Dirac Fermions on the surface of a topological insulator has interesting properties under applied electromagnetic fields, which we now try to obtain from the low-energy (3 + 1)-D dimensional Lagrangian in chapter 14, which includes the spatially varying magnetoelectric polarization. The equation of motion is obtained by adding to the action in equation (14.12) a term coupling the current with the vector potential $\int d^3x\, dt j^\mu A_\mu$ and then

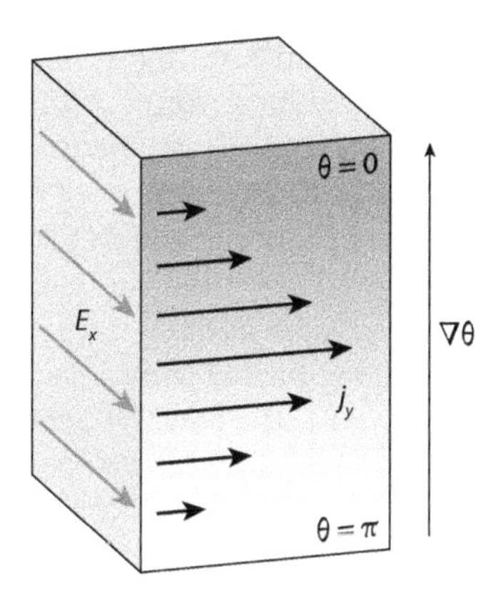

Figure 15.1. A spatial gradient of the magnetoelectric polarization P_3 induces a Hall effect as in the equation of motion (eq (15.1)) in the presence of an electric field E_x on the system. The P_3 changes from π in a topological insulator (bottom of the sample) to 0 in a trivial insulator at the top of the sample.

performing the variation over the gauge field A_μ. We obtain the response equation

$$j^\mu = \frac{1}{2\pi}\epsilon^{\mu\nu\sigma\tau}\partial_\nu P_3 \partial_\sigma A_\tau. \tag{15.1}$$

Notice that as far as the magnetoelectric polarization P_3 is spatially constant, the response equation gives a trivial zero current. However, when the magnetoelectric polarization has a spatial gradient (which happens at the surface between two topologically distinct insulators), the equation of motion (eq. (15.1)) has interesting consequences.

The equation of motion (eq. (15.1)) is that of a quantum Hall effect in a plane perpendicular to $\partial_\nu P_3$. We first consider a topological insulator with open boundary conditions in the z-direction (surfaces perpendicular to the z-direction). In this system, the magnetoelectric polarization $P_3 = P_3(z)$ depends only on z, and the equation of motion (eq. (15.1)) becomes that of a quantum Hall effect in the xy-plane (see fig. 15.1) with the Hall conductivity $\sigma_{xy} = \partial_z P_3/2\pi$:

$$j^\mu = \frac{\partial_z P_3}{2\pi}\epsilon^{\mu\nu\rho}\partial_\nu A_\rho, \tag{15.2}$$

where $\mu, \nu, \rho = t, x, y$ are the time and in-plane coordinates. At the interface between a topological insulator and a trivial insulator, the change in the magnetoelectric polarization P_3 takes place over a region of relatively small width compared to the full size of the topological insulator sample. With a uniform electric field E_x in the x-direction, the total value of the Hall y-current density flowing in the xy-plane is the integral over z in a finite range:

$$J_y^{2D} = \int_{z_1}^{z_2} dz j_y = \frac{1}{2\pi}\left(\int_{z_1}^{z_2} dP_3\right) E_x.$$

The Hall conductance in the xy-plane is

$$\sigma_{xy}^{2D} = \frac{e^2}{\hbar}\int_{z_1}^{z_2}\frac{dP_3}{2\pi}. \tag{15.3}$$

15.2 Physical Properties of Time-Reversal Z_2-Nontrivial Insulators

The Hall conductance of the region $z_1 \leq z \leq z_2$ is a total partial derivative $\sigma_{xy}^{2D} = \frac{e^2}{\hbar}\int_{z_1}^{z_2}\frac{dP_3}{2\pi}$. As such, it depends on the change of P_3 in this region and is not sensitive to any details of the function $P_3(z)$. When time-reversal is present, the values of θ are quantized to either 0 or π; as such, the difference between P_3's of a trivial and topological insulator can be $\pi \mod 2\pi$. The Hall conductance of the region between a topological insulator and a trivial one can hence be $(n + \frac{1}{2})e^2/h$. The $n \in Z$ ambiguity corresponds to the fact that we can add any number of full quantum Hall effects on the surface of a topological insulator. This is similar to the fact that we can add any number of charges at the ends of 1-D wires. The essential factor is $\frac{1}{2}$—it tells us that we have a half-quantized quantum Hall effect at the interface between a trivial insulator and a nontrivial one. This is the equivalent of half-charge on the edge of a 1-D wire. Because the nontrivial insulator has a magnetoelectric polarization $P_3 = \frac{1}{2} \mod 1$, the effective action (eq. (14.13)) of the bulk system is

$$S_{3D} = \frac{2n+1}{8\pi} \int d^3x \, dt \epsilon^{\mu\nu\sigma\tau} \partial_\mu A_\nu \partial_\sigma A_\tau, \tag{15.4}$$

in which $n \in \mathbb{Z}$ is the integer part of P_3.

15.3 Half-Quantized Hall Conductance at the Surface of Topological Insulators with Ferromagnetic Hard Boundary

It is extremely easy to study the surface physics of the 3-D TR-invariant topological insulator when the boundary to the vacuum is very abrupt: $P_3(z) = \frac{1}{2}H(-z)$, where $H(z)$ is the Heaviside step function. We then have

$$\partial_z P_3 = \frac{\delta(z)}{2},$$

up to an integer polarization dependent on the surface details. The effective action is then reduced to a $(2+1)$-D Chern-Simons term on the surface:

$$S_{\text{surf}} = -\frac{2n+1}{8\pi} \int dx \, dy \, dt \epsilon^{3\mu\nu\rho} A_\mu \partial_\nu A_\rho. \tag{15.5}$$

where $n \in \mathbb{Z}$ depends on the nontopological details of the surface. As we saw before, a domain wall of P_3 carries a quantum Hall effect. The Hall conductance of such a surface of a Z_2 nontrivial insulator is, thus, $\sigma_H = (e^2/\hbar)(n + \frac{1}{2})/2\pi$, which is quantized as a half-odd integer times the quanta e^2/h. This Hall conductance comes from the existence of an odd number of Dirac fermions at the boundary between a Z_2 nontrivial strong topological insulator and a trivial insulator. A single Dirac fermion exhibits a half-integer Hall conductance when it is gapped in the presence of a small magnetic field on the surface. This field can be produced by ferromagnetically ordered magnetic moments on a thin layer deposited on the surface. The gapping of a single Dirac fermion gives rise to a *half*-integer quantum Hall effect, as studied earlier. The half quantum Hall effect on the surface is the parity anomaly of massless Dirac fermions [62, 63] upon introducing a mass term. When time reversal is broken, each Dirac cone on the surface generically has a mass. Thereby, the whole system is a gapped insulator, and the Berry-phase curvature is well defined in the occupied bands and the system acquires a Hall conductance. Physically, the TR symmetry-breaking term on the surface can come from magnetic fields or magnetic moments (impurities) localized on the surface coherently coupled with the surface electron liquid. Another alternative is that a gap could open spontaneously by breaking TR symmetry on the surface due to interactions—this is a more-limited effect due to the vanishing density of states of a Dirac fermion. These effects can also be present in topological insulators with inversion symmetry, where the gapping of the surface Dirac fermion can come directly from the bulk magnetism, because inversion is compatible with magnetism.

15.4 Experimental Setup for Indirect Measurement of the Half-Quantized Hall Conductance on the Surface of a Topological Insulator

In a realistic sample, due to the presence of two rather than one boundary surfaces, the half-quantized Hall conductance at the surface of a topological insulator is not directly measurable in a DC current experiment. The Hall conductances of the lower and upper boundary surfaces must be considered together. Depending on the relative orientation of the ferromagnetic moment used to open a gap on the upper or lower surfaces, the Hall conductance of the upper surface is added to or subtracted from the Hall conductance of the lower surface to give either Hall conductance 1 or Hall conductance 0. This is a verifiable prediction, which can be experimentally realized in a ferromagnet-topological insulator heterostructure (fig. 15.2), in which ferromagnets are placed on both surfaces of a topological insulator. An outward-pointing magnetization vector (i.e., toward the direction of the surface normal vector $\hat{n}$) on each surface will give the same Hall effect for the two surfaces. However, even though in this local basis (where the normal on each surface defines a direction), the Hall conductance of the two surfaces is the same; the total Hall conductance in the global basis is the difference of the Hall conductances on the two surfaces because the lower-surface normal vector is the negative of the upper-surface vector. The total Hall conductance is zero. This is similar to the case of a topological insulator sphere where the magnetization always points radially. A gap is always open in the spectrum, and there is no gapless region ("edge mode") that will carry the Hall conductance.

For parallel magnetizations in figure 15.2a, the magnetization vector is outward pointing (with respect to the normal to the surface) on the top surface and inward pointing on the bottom surface. In this local basis, the Hall conductance of the two surfaces is opposite, but in the global x, y, z basis, it is the same. The total Hall conductance adds up to 1. There are two other ways of understanding the nonzero Hall conductances. The Hall conductances of the upper and bottom surfaces are opposite in the *local* basis defined with respect to the normal vector $\hat{n}$. Hence, the Hall conductance of the bottom and top surface differ by 1 in their local basis. The side surface of the topological insulator is then a domain wall between two different quantum Hall regions with Hall conductances that differ by one quantum. Just like a domain wall between $\nu = 0$ and $\nu = 1$, regions in the usual quantum Hall system, such a domain wall will have a chiral edge state, responsible for the net Hall effect. This is similar to the case of a topological insulator sphere in a magnetic field pointing, say, from the south to the north pole: on the equator, the magnetic field is perpendicular to the surface vector $\mathbf{B}\cdot\mathbf{n} = 0$ and it induces no gap into a Dirac Hamiltonian $k_x\sigma_x + k_y\sigma_y$ (but shifts the Dirac point). This is the edge mode that carries Hall conductance. The gapless chiral mode is realized on a line where the normal to the side surface is perpendicular to the magnetization vector. The parallel magnetization gives rise to $\sigma_H = e^2/h$, whereas the antiparallel magnetization then leads to a zero net Hall conductance $\sigma_H = 0$. The experimental observation is the unit difference in Hall conductance between parallel and antiparallel surface magnetization.

15.5 Topological Magnetoelectric Effect

The Dirac fermions on the surface of a topological insulator and their gapping in the presence of a ferromagnetic coating leads to yet another related effect, the topological magnetoelectric effect. This effect is usually presented [13] as a novel effect separate from the existence of the

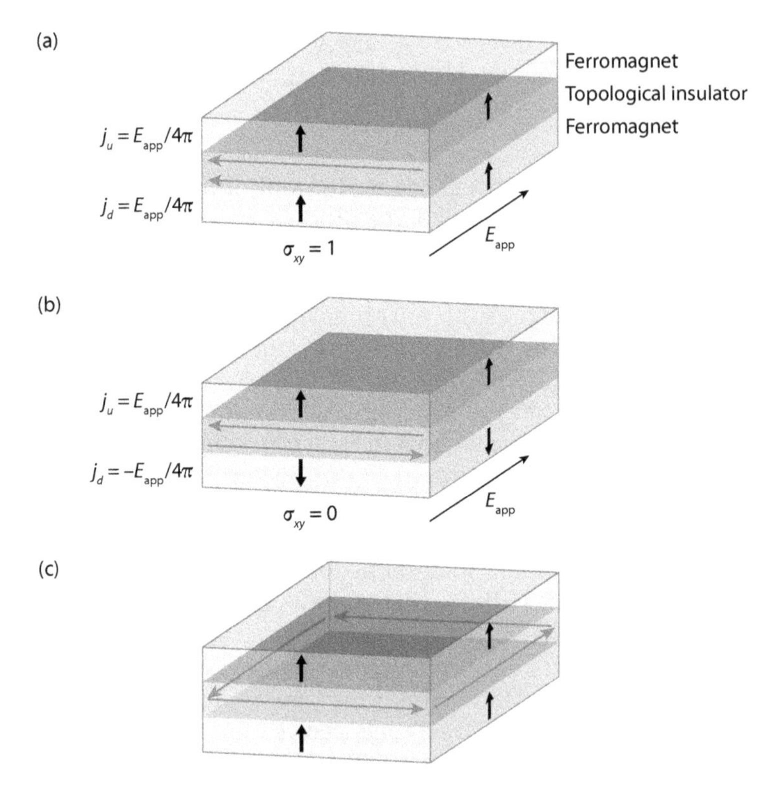

Figure 15.2. The quantum Hall effect in the topological insulator coated with ferromagnets on the surface depends on the direction of the magnetization on the top and bottom surfaces. An electric field E_x (with direction into the paper plane) induces nonzero current for parallel magnetization (a) but induces zero current in the case of antiparallel magnetization (b). In (b), the whole surface of the topological insulator is gapped, and there is no gapless edge state to carry the current—the Hall currents on the two surfaces are opposite and form a circulating current. (c) For the case of parallel magnetization, the chiral edge state exists on the side surfaces of the topological insulator in the regions where $\mathbf{B} \cdot \mathbf{n} = 0$. This edge states carry the quantized Hall current.

half-quantized Hall conductance on the surface of the topological insulator coated with a ferromagnet, but it is not. It is intimately related to and is the direct consequence of the half-integer Hall conductance of the gapped Dirac fermions. Following [13], we consider a sample geometry in which we can gap the Dirac fermions on the full surface of a topological insulator to obtain a full gap in the spectrum. This is possible in the cylindrical geometry shown in figure 15.3a. With a radial magnetization at every point (pointing out of the cylinder's surface), the surface is fully gapped and has a fixed Hall conductance $\sigma_H = \frac{1}{2}e^2/h$ (module a possible integer, but the smallest Hall conductance we can have is $\frac{1}{2}e^2/h$). We now place an electric field parallel to the axis of the cylinder.

The existence of a Hall effect (with any nonzero Hall conductance $\sigma_H \neq 0$) now implies that a tangential current is induced by an electric field parallel to the axis of the cylinder (see fig. 15.3a). A tangential (axial) current along the cylinder generates a magnetic field parallel to the axis of the cylinder (and hence parallel to the electric field), whose value is [13]

$$\mathbf{B}_t = -\frac{4\pi}{c}\sigma_H \mathbf{E} = -(2n+1)\frac{e^2}{\hbar c}\mathbf{E}, \tag{15.6}$$

where n is an integer depending on the details of the surface.

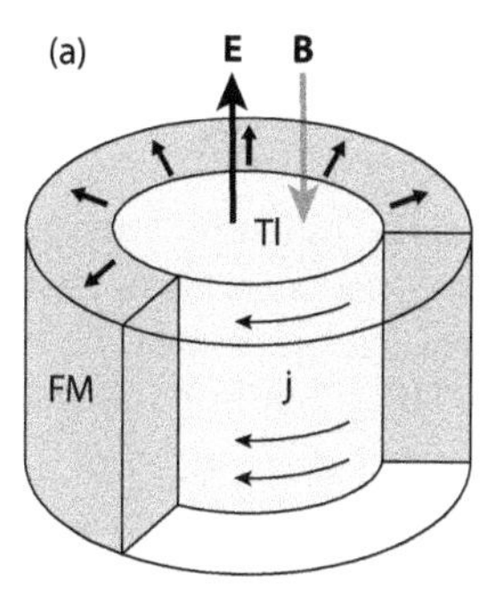

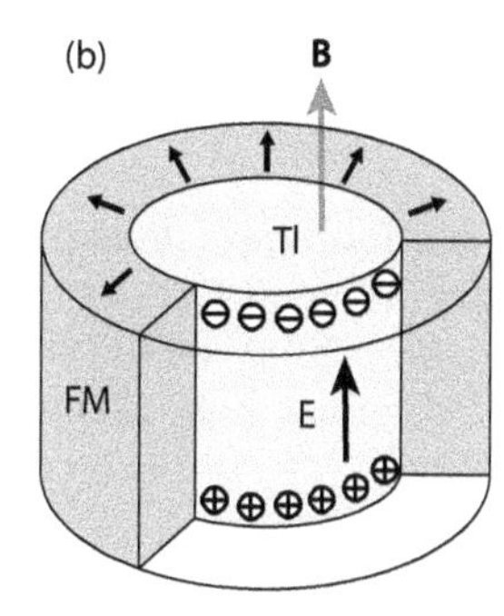

Figure 15.3. (a) A magnetic field is induced by an electric field in a topological insulator coated by a ferromagnet. The ferromagnet opens a gap on the full surface of the topological insulator, thereby giving finite Hall conductance to the Dirac fermions. The Dirac fermions are gapped by the presence of the ferromagnet, and hence the surface of the topological insulator exhibits a quantum Hall effect, with Hall conductance always nonzero. An azimuthal circulating Hall current is induced by the electric field, and this current creates a magnetic field. (b) Electric field induced by a magnetic field parallel to the axis of the cylinder; ⊕ and ⊖ indicate the positive and negative charge accumulated on the top and bottom surfaces, respectively, after the magnetic field is ramped up.

We can also think of the inverse effect: the generation of an electric field by applying a magnetic field. Let us turn on a magnetic field B parallel (see fig. 15.3b) to the axis of the cylinder. In the process of ramping up a magnetic field, a tangential electric field parallel to the side surface and proportional to dB/dt is generated—which induces a Hall current $j\sigma_H dB/dt$. The current causes a charge density proportional to B to be accumulated on the top and bottom surfaces, so that a magnetic field induces an electric field or polarization parallel to it. Such a topological contribution to the electric field is obtained from Maxwell's equations:

$$\mathbf{E}_t = (2n+1)\frac{e^2}{\hbar c}\mathbf{B}. \tag{15.7}$$

In all the preceding, we should emphasize that the value of P_3 in the topological insulator can be determined only when a magnetization is applied to open a gap on the surface—it is quantized only when the full system (including the surface) is gapped. We must also remark that the physics presented here would also take place in a quantum Hall effect sample when placed in the geometry of a cylinder. The only thing that differs between the current setting and the more-conventional quantum Hall physics is the *half*-integer value of the Hall conductance on the surface of a topological insulator.

15.6 Problems

1. *Skin Effect:* Obtain an expression for the skin depth at the surface of a topological insulator when the Fermi level is (a) in the bulk gap or (b) in the bulk valence or conduction band. What is the contribution of the surface state?

2. *Spin-Charge-Coupled Equations:* Find the spin-charge density-coupled diffusion equations for electrons at the surface of a topological insulator.

3. *Trigonal Distortion*: Obtain the trigonal distortion of the Fermi surface of the surface state of a topological insulator assuming the existence of a C_3 symmetry perpendicular to

the surface. (*Hint*: Although we have only a C_3 symmetry axis, the equipotential energy contours will seem to posses a six-fold rotation axis. Why?)

4. *Klein Tunneling in a Topological Insulator:* Obtain the transmission amplitude for an electron at a step boundary on the surface of a topological insulator. Show that for normal incidence, the transmission probability should equal unity, irrespective of the energy of the incoming electron and the potential of the barrier. (*Hint*: This phenomenon is called Klein tunneling.)

16

Topological Superconductors in One and Two Dimensions

BY TAYLOR L. HUGHES

With the explosion of interest in unconventional superconductivity in the past two decades, there have been two primary research foci: (1) the microscopic mechanism that produces the unconventional superconducting pairing potential and (2) new quasiparticle phenomena. In the context of topological superconductors, our presentation will deal only with the quasiparticle physics, and we do not consider any microscopic origin of the unconventional superconductivity. In our discussion we assume that there exists some finite pairing strength, induced by interactions or occasionally through the proximity effect, and that the quasiparticle physics is well described using a mean-field formulation. Thus, we are interested in noninteracting quasiparticles that are coupled to a well-defined background pairing potential, and we ignore the (possibly important) effects that would result from considering a fully self-consistent solution.

16.1 Introducing the Bogoliubov-de-Gennes (BdG) Formalism for *s*-Wave Superconductors

For comparison to more-interesting cases that are discussed later, we begin by introducing the mean-field formulation of the quasiparticle physics for a conventional *s*-wave Bardeen-Cooper-Schrieffer (BCS) superconductor [64, 65]. We start with a simple metal with spin-degeneracy given by the single-particle Hamiltonian

$$H = \left(\frac{p^2}{2m} - \mu\right) I_{2\times 2}, \tag{16.1}$$

where μ is the chemical potential defining the Fermi surface, m is the mass, $I_{2\times 2}$ is the identity matrix in the spin variables, and, assuming isotropy, $p^2 = \sum_{i=1}^{d} p_i^2$ for whatever spatial dimension d we are considering. For the many-body system, the second-quantized Hamiltonian is

$$H = \sum_{\mathbf{p},\sigma} c^\dagger_{\mathbf{p}\sigma} \left(\frac{p^2}{2m} - \mu\right) c_{\mathbf{p}\sigma} \equiv \sum_{\mathbf{p},\sigma} c^\dagger_{\mathbf{p}\sigma} \epsilon(p) c_{\mathbf{p}\sigma}, \tag{16.2}$$

where $c^\dagger_{\mathbf{p}\sigma}$ creates a quasiparticle with momentum $\mathbf{p}$ and spin σ. The many-body ground state of this Hamiltonian is obtained simply by filling in all the levels below the Fermy energy:

$$|\Omega\rangle = \prod_{\mathbf{p}:\, \epsilon(p)<0} \prod_{\sigma} c^\dagger_{\mathbf{p}\sigma} |0\rangle, \tag{16.3}$$

where the vacuum $|0\rangle$ is defined by $c_{\mathbf{p}\sigma}|0\rangle \equiv 0$ for all $\mathbf{p},\ \sigma$.

Formally, we can always write this Hamiltonian as

$$\begin{aligned} H &= \frac{1}{2}\sum_{\mathbf{p}\sigma}\left[c^\dagger_{\mathbf{p}\sigma}\epsilon(p)c_{\mathbf{p}\sigma} - c_{\mathbf{p}\sigma}\epsilon(p)c^\dagger_{\mathbf{p}\sigma}\right] + \frac{1}{2}\sum_{\mathbf{p}}\epsilon(p) \\ &= \frac{1}{2}\sum_{\mathbf{p}\sigma}\left[c^\dagger_{\mathbf{p}\sigma}\epsilon(p)c_{\mathbf{p}\sigma} - c_{-\mathbf{p}\sigma}\epsilon(-p)c^\dagger_{-\mathbf{p}\sigma}\right] + \frac{1}{2}\sum_{\mathbf{p}}\epsilon(p), \end{aligned} \tag{16.4}$$

where in the first equality we have used $\{c^\dagger_{\mathbf{p}\sigma}, c_{\mathbf{p}'\sigma'}\} = \delta_{\sigma\sigma'}\delta_{\mathbf{p}\mathbf{p}'}$ and in the second equality we have relabeled the sum index $\mathbf{p}$ in the second term to $-\mathbf{p}$. If we introduce the spinor $\Psi_{\mathbf{p}} \equiv (c_{\mathbf{p}\uparrow}\ c_{\mathbf{p}\downarrow}\ c^\dagger_{-\mathbf{p}\uparrow}\ c^\dagger_{-\mathbf{p}\downarrow})^T$, we can write our Hamiltonian in a more compact form:

$$H = \sum_{\mathbf{p}} \Psi^\dagger_{\mathbf{p}} H_{\text{BdG}}(\mathbf{p})\Psi_{\mathbf{p}} + \text{constant}, \tag{16.5}$$

$$H_{\text{BdG}}(\mathbf{p}) = \frac{1}{2}\begin{pmatrix} \epsilon(p) & 0 & 0 & 0 \\ 0 & \epsilon(p) & 0 & 0 \\ 0 & 0 & -\epsilon(-p) & 0 \\ 0 & 0 & 0 & -\epsilon(-p) \end{pmatrix}. \tag{16.6}$$

We have introduced the subscript BdG (Bogoliubov-de-Gennes) to label the Hamiltonian written in this redundant formalism; additionally, we will drop the constant from now on. Although the statement is a bit trivial here, we note that the Bloch Hamiltonian $H_{\text{BdG}}(\mathbf{p})$ is invariant under $H_{\text{BdG}}(\mathbf{p}) = -CH^T_{\text{BdG}}(-\mathbf{p})C^{-1}$, where $C = \tau^x \otimes I_{2\times 2}$ and

$$\tau^x = \begin{pmatrix} 0 & 1 \\ 1 & 0 \end{pmatrix}.$$

The full, second-quantized Hamiltonian obeys

$$H = -CH^TC^{-1} = -CH^*C^{-1}, \tag{16.7}$$

where in the second equation we have used the hermiticity of H. This invariance, which will become more important when we consider superconducting pairing, is known as a particle-hole or charge-conjugation "symmetry." We are reserved about calling this a symmetry because what we have really done is to introduce a redundancy into our description of this noninteracting metal. Note that instead of having two degrees of freedom (one band and two spins), the BdG Hamiltonian has four. We now have four energy eigenvalues of H_{BdG}, namely, two copies of $\epsilon(p)$ and two copies of $-\epsilon(-p)$. The important point to note is that only two out of the four bands give *independent* quasiparticle states. Thus, we have created an artificial redundancy by effectively doubling the degrees of freedom. This is complicating our description of what was a simple free-fermion problem.

The point of this formalism is to show that the easiest way to solve for the quasiparticle bands of a mean-field superconductor is to write the Hamiltonian in this BdG form. The pairing potential, which we will now introduce, simply couples the upper and lower blocks of the H_{BdG} we gave for the metal. We begin by studying the conventional s-wave, singlet pairing potential of the form

$$\begin{aligned} H_\Delta &= \Delta c^\dagger_{\mathbf{p}\uparrow}c^\dagger_{-\mathbf{p}\downarrow} + \Delta^* c_{-\mathbf{p}\downarrow}c_{\mathbf{p}\uparrow} \\ &= \frac{1}{2}\left[\Delta\left(c^\dagger_{\mathbf{p}\uparrow}c^\dagger_{-\mathbf{p}\downarrow} - c^\dagger_{-\mathbf{p}\downarrow}c^\dagger_{\mathbf{p}\uparrow}\right) + \Delta^*\left(c_{-\mathbf{p}\downarrow}c_{\mathbf{p}\uparrow} - c_{\mathbf{p}\uparrow}c_{-\mathbf{p}\downarrow}\right)\right], \end{aligned} \tag{16.8}$$

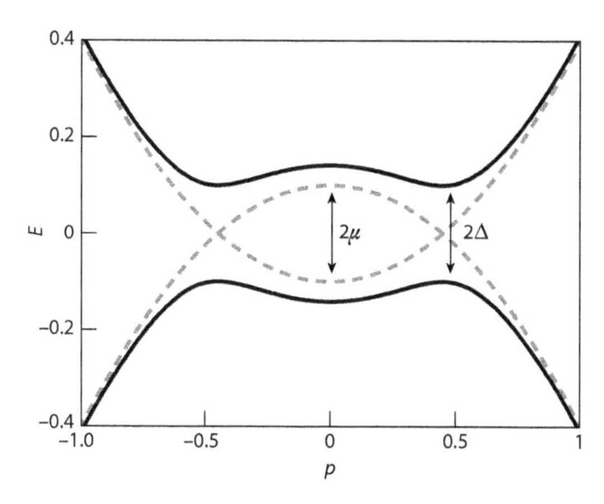

Figure 16.1. Plot of the dispersion relation for an s-wave superconductor. The curves in the figures are plots of the energies $E_{\pm}(p) = \sqrt{\epsilon(p)^2 + |\Delta|^2}$ for (dotted line) $m = 1.0$, $\mu = 0.1$, and $|\Delta| = 0.0$ and (solid line) $m = 1.0$, $\mu = |\Delta| = 0.1$.

where Δ is a complex number representing the superconducting order parameter. This term, at the mean-field level, leads to a nonconservation of charge, i.e., charge is conserved only modulo $2e$. This term captures the physics of two electrons or holes combining to form a Cooper pair or a Cooper pair breaking apart into its constituents. The nonconservation of charge is further manifest in the fact that H_Δ is not invariant under arbitrary gauge transformations ($c_{\mathbf{p}} \to e^{i\phi(\mathbf{p})}c_{\mathbf{p}}$) if we consider Δ to be a conventional, gauge-*invariant* order parameter.

Now let us consider the total Hamiltonian of the metal with a homogeneous pairing potential

$$H + H_\Delta = \sum_{\mathbf{p}} \Psi_{\mathbf{p}}^{\dagger} H_{\mathrm{BdG}}(\mathbf{p}, \Delta)\Psi_{\mathbf{p}} \tag{16.9}$$

$$H_{\mathrm{BdG}}(\mathbf{p}, \Delta) = \frac{1}{2}\begin{pmatrix} \epsilon(p) & 0 & 0 & \Delta \\ 0 & \epsilon(p) & -\Delta & 0 \\ 0 & -\Delta^* & -\epsilon(-p) & 0 \\ \Delta^* & 0 & 0 & -\epsilon(-p) \end{pmatrix}. \tag{16.10}$$

The BdG Bloch Hamiltonian can be decomposed as

$$H_{\mathrm{BdG}}(\mathbf{p}, \Delta) = \epsilon(p)\tau^z \otimes I_{2\times 2} - (\mathrm{Re}\Delta)\tau^y \otimes \sigma^y - (\mathrm{Im}\Delta)\tau^x \otimes \sigma^y, \tag{16.11}$$

where τ^a are Pauli matrices in the particle-hole degrees of freedom and σ^a are spin. Because the three matrices making up $H_{\mathrm{BdG}}(\mathbf{p}, \Delta)$ are mutually anticommuting, we can easily find the energy spectrum because $H_{\mathrm{BdG}}^2(\mathbf{p}, \Delta) = (\epsilon(\mathbf{p})^2 + |\Delta|^2)I_{4\times 4}$. Thus, the energy spectrum is made up of two doubly degenerate bands with energies

$$E_{\pm} = \pm\sqrt{\epsilon(\mathbf{p})^2 + |\Delta|^2}. \tag{16.12}$$

This spectrum has an energy gap whenever $|\Delta| \neq 0$ and is shown in figure 16.1. In fact, the spectrum has similar features to that of a band insulator with a fine-tuned particle-hole symmetry. However, there is an important difference between the fermionic excitations of the gapped insulator state and the gapped superconductor state, namely, the superconductor quasi-particles are combinations of particle *and* hole states. This can be seen by looking at the quasi-particle creation operators that create fermions in the eigenstates of $H_{\text{BdG}}(\mathbf{p}, \Delta)$:

$$\gamma^{\dagger}_{+,\mathbf{p}\uparrow} = e^{i\theta/2} \sin \alpha_{\mathbf{p}}\, c^{\dagger}_{\mathbf{p}\uparrow} + e^{-i\theta/2} \cos \alpha_{\mathbf{p}}\, c_{-\mathbf{p}\downarrow}, \tag{16.13}$$

$$\gamma^{\dagger}_{+,\mathbf{p}\downarrow} = -e^{i\theta/2} \sin \alpha_{\mathbf{p}}\, c^{\dagger}_{\mathbf{p}\downarrow} + e^{-i\theta/2} \cos \alpha_{\mathbf{p}}\, c_{-\mathbf{p}\uparrow}, \tag{16.14}$$

$$\gamma^{\dagger}_{-,\mathbf{p}\uparrow} = e^{i\theta/2} \sin \beta_{\mathbf{p}}\, c^{\dagger}_{\mathbf{p}\uparrow} + e^{-i\theta/2} \cos \beta_{\mathbf{p}}\, c_{-\mathbf{p}\downarrow}, \tag{16.15}$$

$$\gamma^{\dagger}_{-,\mathbf{p}\downarrow} = -e^{i\theta/2} \sin \beta_{\mathbf{p}}\, c^{\dagger}_{\mathbf{p}\downarrow} + e^{-i\theta/2} \cos \beta_{\mathbf{p}}\, c_{-\mathbf{p}\uparrow}, \tag{16.16}$$

where $\gamma^{\dagger}_{\pm\mathbf{p}\sigma}$ creates a quasi-particle in energy band $E_{\pm}$ with momentum $\mathbf{p}$ and spin σ, $\Delta = |\Delta| e^{i\theta}$, and

$$\tan \alpha_{\mathbf{p}} = \frac{\epsilon(p) + \sqrt{\epsilon(p)^2 + |\Delta|^2}}{|\Delta|}, \tag{16.17}$$

$$\tan \beta_{\mathbf{p}} = \frac{\epsilon(p) - \sqrt{\epsilon(p)^2 + |\Delta|^2}}{|\Delta|}. \tag{16.18}$$

Thus, the quasi-particles that are excited in a superconductor are mixtures of creation and annihilation operators of the original quasi-particles in the metal. At low energy ($\epsilon(p) \sim 0$), the quasi-particles are nearly equal-weight superpositions of creation and annihilation operators, whereas at high energy ($\epsilon(p) \gg |\Delta|$), the particles are not influenced by the pairing potential and revert to the metallic quasi-particle form. These operators satisfy $\gamma^{\dagger}_{+,\mathbf{p}\uparrow} = \gamma_{-,-\mathbf{p}\downarrow}$ and $\gamma^{\dagger}_{+,\mathbf{p}\downarrow} = \gamma_{-,-\mathbf{p}\uparrow}$. Thus, it is clear that out of the four creation operators, only two of them create independent excitations. This is due to the artificial redundancy of the BdG formalism.

16.2 p-Wave Superconductors in One Dimension

The simplest models of topological superconductors are mean-field BdG Hamiltonians of spinless fermions in one and two dimensions. Spinless fermions can either be viewed as a toy model for the more complicated spinful case or simply as fermions that are fully spin-polarized due to a source of TR breaking such as a magnetic field. We will first consider a 1-D wire with p-wave superconductivity and then move on to a 2-D chiral p-wave superconductor, both of which exhibit topological superconducting phases.

16.2.1 1-D p-Wave Wire

In this section we discuss an illustrative topological superconductor model first introduced in this context by Kitaev [66]. We begin with a nonsuperconducting 1-D metal of spinless (or spin-polarized) fermions

$$H = \sum_p c^{\dagger}_p \left(\frac{p^2}{2m} - \mu \right) c_p. \tag{16.19}$$

Instead of introducing a momentum-independent s-wave pairing potential, which is not possible for spinless fermions, we will use a momentum-dependent p-wave potential

$$H_{\Delta} = \frac{1}{2} \left(\Delta p c^{\dagger}_p c^{\dagger}_{-p} + \Delta^* p c_{-p} c_p \right). \tag{16.20}$$

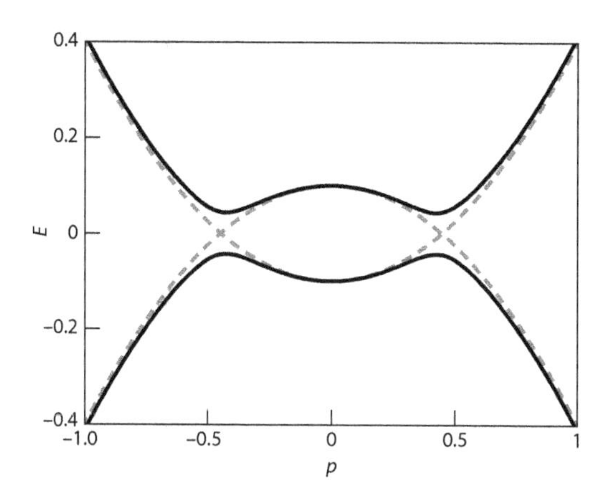

Figure 16.2. Plot of the dispersion relation for a 1-D p-wave superconductor. The curves in the figures are plots of the energies $E_{\pm}(p) = \sqrt{\epsilon(p)^2 + |\Delta|^2 p^2}$ for (dotted line) $m = 1.0$, $\mu = 0.1$, and $|\Delta| = 0.0$ and (solid line) $m = 1.0$, $\mu = |\Delta| = 0.1$.

The BdG Hamiltonian we need to solve is

$$H_{\text{BdG}} = \sum_p \frac{1}{2} \Psi_p^\dagger \begin{pmatrix} \frac{p^2}{2m} - \mu & \Delta p \\ \Delta^* p & -\frac{p^2}{2m} + \mu \end{pmatrix} \Psi_p, \tag{16.21}$$

where $\Psi_p = (c_p \;\; c_{-p}^\dagger)^T$. This Hamiltonian has two energy bands, $E_{\pm} = \pm\sqrt{\epsilon(p)^2 + |\Delta|^2 p^2}$, where $\epsilon(p) = \frac{p^2}{2m} - \mu$. The energy spectrum is shown in figure 16.2 and is gapped as long as $\mu \neq 0$. The critical point $\mu = 0$ separates two physical regimes: the weak-pairing ($\mu > 0$) and strong-pairing ($\mu < 0$) phases. In the weak-pairing phase, the system, without pairing, is metallic, and the resulting superconducting phase is BCS-like, albeit with a momentum-dependent order parameter. In the strong-pairing phase, the system, without pairing, is a gapped insulator and does not fit the usual weak-pairing BCS instability picture because there are no low-energy fermions and no Fermi surface.

One of these two phases is a topological superconductor phase, and we will now derive which one it is. The first step is to notice that in the limit $m \to \infty$, the matrix structure of H_{BdG} is simply a massive 1-D Dirac Hamiltonian: $H_{\text{BdG}}(p) = (\text{Re}\Delta)p\sigma^x - (\text{Im}\Delta)p\sigma^y - \mu\sigma^z$. Because we are considering only a single superconductor, we can use a global phase rotation to make Δ real. Now $H_{\text{BdG}}(p) = |\Delta|p\sigma^x - \mu\sigma^z$, and we easily recognize the form of the Dirac Hamiltonian. This enables us to use all our intuition developed from topological insulators with only slight changes in the physical interpretation. The matrix $H_{\text{BdG}}(p)$ is of the Dirac form and could just as well serve as an insulator Hamiltonian with a strict particle-hole symmetry. However, physically the superconductor and insulator are quite different due to the redundancy mentioned earlier. This is manifest in the form of the Ψ_p operators for the superconductor, which contain both creation and annihilation operators, whereas the analogous quantity for the insulator would contain only annihilation operators; albeit the annihilation operators of *two* independent degrees of freedom.

Immediately, because we have a massive Dirac equation, we can expect to see bound-state zero modes on mass kinks [67] (as we saw for topological insulators in chap. 8). Let us prove this explicitly by assuming that $\mu \to \mu(x)$ and that $\mu(x)$ has a soliton kink profile with $\mu(-\infty) < 0$ and $\mu(+\infty) > 0$. We will look for a single-particle, zero-energy solution—i.e., a $|\psi\rangle$ such that $H_{\text{BdG}}|\psi\rangle = 0|\psi\rangle$. We take the ansatz

$$|\psi(x)\rangle = \exp\left(-\frac{1}{|\Delta|}\int_0^x \mu(x')\,dx'\right)|\phi_0\rangle \tag{16.22}$$

for a constant, normalized spinor $|\phi_0\rangle$. Acting on this ansatz with H_{BdG}, we derive the secular equation

$$\begin{pmatrix} -\mu(x) & i\mu(x) \\ i\mu(x) & \mu(x) \end{pmatrix} |\phi_0\rangle = 0. \tag{16.23}$$

Solving this simple equation leads us to the zero-energy solution

$$|\psi(x)\rangle = \frac{1}{\sqrt{2}} \exp\left(-\frac{1}{|\Delta|}\int_0^x \mu(x')\,dx'\right)\begin{pmatrix} 1 \\ -i \end{pmatrix}. \tag{16.24}$$

Note that although our original Hamiltonian is a 2×2 matrix, there is only one (normalizable) zero-mode solution.

As an aside, we see that the field operator that destroys this zero mode has the form

$$\gamma_0 = \frac{1}{\mathcal{N}}\int dx \exp\left(-\frac{1}{|\Delta|}\int_0^x \mu(x')\,dx'\right)\frac{1}{\sqrt{2}}\left(c(x) - ic^\dagger(x)\right), \tag{16.25}$$

where $c(x)$ destroys a fermion state at the point x and $\mathcal{N}$ is a normalization factor. The anticommutation relations of $c(x)$, $c^\dagger(x)$ imply fermionic anticommutation relations for γ_0, $\gamma_0^\dagger$. Multiplying this by a global $U(1)$ phase $\bar{\gamma}_0 \equiv e^{i\pi/4}\gamma_0$, we find that $\bar{\gamma}_0^\dagger = \bar{\gamma}_0$. Thus, $\bar{\gamma}_0$ is a Majorana fermion-bound state, i.e., it is a fermion state, which is its own antiparticle. In this case it is a 0-D Majorana fermion because it is a localized bound state at a kink and does not propagate. We talk more about Majorana fermions later in this chapter.

Knowing that there is a low-energy fermion on a domain wall between vacua with opposite Dirac masses is tantamount to establishing a topological superconductor phase. The only thing remaining is to determine if it is the weak-pairing or strong-pairing phase that is topological. In the $m \to \infty$ limit, where we are really left with a Dirac Hamiltonian, we cannot distinguish the phases. In this limit, the energy spectra of the superconductors with $\mu < 0$ and $\mu > 0$ are also identical, so there is no distinguishing feature. The key is that we need to keep at least the leading terms in the Hamiltonian that are quadratic in momentum and, thus, need to consider a finite m. Physically, it is easy to understand why this "regularization" is necessary. Let us begin in the limit $m \to \infty$, where the metallic band is completely flat and fixed at $E = 0$. If we first set $\Delta = 0$, we see that there is no physical difference between the phases when $\mu < 0$ and $\mu > 0$ (with $m \to \infty$). Both phases are gapped, even when μ is just slightly tuned away from zero. If we turn on a small Δ, the system will remain gapped and is, thus, adiabatically connected to the trivial gapped limit when $\Delta = 0$. Therefore, neither of the phases in this limit can be considered topological. Heuristically, in this limit the system cannot decide if it is a trivial phase or a topological phase. We must break this degeneracy. To do this, we just pick m to be finite. Now the band disperses either upward or downward in energy, depending on the sign of m. We have been implicitly assuming that $m > 0$, so we will stick with that choice. Now we see there is a clear difference between $\mu > 0$ and $\mu < 0$. If we turn on a finite Δ when $\mu < 0$, the system remains gapped; thus, the trivial insulator phase when $\Delta = 0$ is adiabatically connected with the quasi-particle spectrum of the gapped superconductor phase when $\mu < 0$, $\Delta \neq 0$. This is not the case when $\mu > 0$. At $\Delta = 0$, the system is gapless and becomes gapped only when there is a nonzero Δ. We cannot adiabatically connect the $\mu > 0$, $\Delta \neq 0$ phase to the trivial insulating limit $\mu < 0$, $\Delta = 0$ without passing through a gapless point or region. Besides these physical arguments, we show the energy spectra of these two superconductor phases for a 1-D wire with open boundaries in figure 16.3. For $\Delta \neq 0$, we find zero-mode end states when

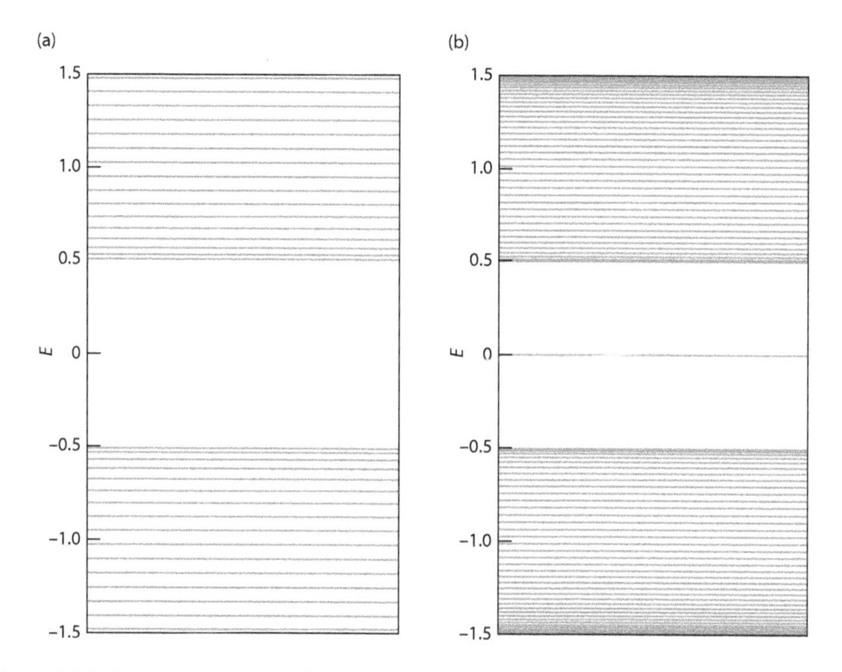

Figure 16.3. Energy spectra for the 1-D *p*-wave wire with open boundary conditions in the (a) trivial phase and (b) nontrivial topological phase with a zero-energy mode on each boundary point.

$\mu > 0$ and no end states when $\mu < 0$. We will give a third argument shortly using a Majorana representation proposed by Kitaev [66].

16.2.2 Lattice *p*-Wave Wire and Majorana Fermions

We now describe the same physics of the 1-D *p*-wave wire using the language of Majorana fermions. We rewrite the BdG Hamiltonian for the *p*-wave wire using Majorana fermion operators, and we reinterpret the topological superconductor phase in terms of these new variables. We continue to choose the gauge where $\Delta = |\Delta|$ is real for simplicity.

We begin with a 1-D lattice with a *complex* fermion c_j on each lattice site j. The lattice BdG Hamiltonian for the 1-D wire with *p*-wave superconductivity is simply

$$H_{\text{BdG}} = \sum_j \left[-t \left(c_j^\dagger c_{j+1} + c_{j+1}^\dagger c_j \right) - \mu c_j^\dagger c_j + |\Delta| \left(c_{j+1}^\dagger c_j^\dagger + c_j c_{j+1} \right) \right]. \tag{16.26}$$

We assume that $t > 0$. With a homogenous $|\Delta|$, we can perform a straight-forward lattice Fourier transform to write the Hamiltonian as

$$H_{\text{BdG}} = \frac{1}{2} \sum_p \Psi_p^\dagger \begin{pmatrix} -2t \cos p - \mu & 2i|\Delta| \sin p \\ -2i|\Delta| \sin p & 2t \cos p + \mu \end{pmatrix} \Psi_p, \tag{16.27}$$

where $\Psi_p = (c_p \;\; c_{-p}^\dagger)^T$. The energy spectrum is $E_\pm(p) = \pm\sqrt{(2t \cos p + \mu)^2 + 4|\Delta|^2 \sin^2 p}$, which, when expanded around $p \sim 0$, recovers the continuum form derived from equation (16.21). As before, for $|\Delta| \neq 0$ this model exhibits critical points when $-2t \cos p - \mu$ vanishes at the same time as $\sin p$. We consider only the critical point when the gap closes at $p = 0$, which implies that the critical $\mu_c = -2t$. This critical point separates two gapped superconducting phases: (1) a trivial phase when $\mu < -2t$ and (2) a topological phase when $\mu > -2t$. There is another critical point where the gap closes at $p = \pi$ for $\mu_c = 2t$, but we do not consider this case.

To see the physics from a different perspective, we will split the complex fermion operators $c_j, c_j^\dagger$ into their Majorana fermion constituents. We replace each complex fermion c_j with *two* Majorana fermions, a_{2j-1}, a_{2j}, via $c_j = \frac{1}{2}(a_{2j-1} + ia_{2j})$. From this splitting we immediately see that $c_j^\dagger = \frac{1}{2}(a_{2j-1} - ia_{2j})$. The Majorana operators are fermionic and satisfy

$$\left\{a_j^\dagger, a_{j'}\right\} = 2\delta_{jj'}, \tag{16.28}$$

but since Majoranas satisfy $a_j = a_j^\dagger$, this means the operators also satisfy the simpler relation

$$\{a_j, a_{j'}\} = 2\delta_{jj'}. \tag{16.29}$$

Given these relations, we can show that

$$\begin{aligned}\left\{c_j^\dagger, c_{j'}\right\} &= \frac{1}{4}\left\{a_{2j-1} - ia_{2j}, a_{2j'-1} + ia_{2j'}\right\} \\ &= \frac{1}{4}\left[\{a_{2j-1}, a_{2j'-1}\} + \{a_{2j}, a_{2j'}\} + i\{a_{2j-1}, a_{2j'}\} - i\{a_{2j}, a_{2j'-1}\}\right] \\ &= \frac{1}{4}\left(4\delta_{jj'}\right) = \delta_{jj'}, \end{aligned} \tag{16.30}$$

which shows that the all the anticommutation relations are consistent with fermionic operators. In fact, we can always break up a complex fermion operator on a lattice site into its real and imaginary Majorana components, although it may not always be a useful representation. As an aside, note that the Majorana anticommutation relation in equation (16.29) is the same as that of the generators of a Clifford algebra, where the generators all square to $+1$. Thus, mathematically we can think of the operators a_i as matrices forming the representation of Clifford-algebra generators.

Using the Majorana representation, the Hamiltonian for the lattice p-wave wire becomes

$$H_{\text{BdG}} = \frac{i}{2}\sum_j \left(-\mu a_{2j-1}a_{2j} + (t + |\Delta|)a_{2j}a_{2j+1} + (-t + |\Delta|)a_{2j-1}a_{2j+2}\right). \tag{16.31}$$

The factor of i in front of the Hamiltonian may seem out of place, but it is required for hermiticity when using the Majorana representation. As a quick example, we can see that an operator like $\left(a_{2j}a_{2j-1}\right)^\dagger = a_{2j-1}^\dagger a_{2j}^\dagger = a_{2j-1}a_{2j} = -a_{2j}a_{2j-1}$ is anti-Hermitian and becomes Hermitian if a factor of i is added, i.e., $ia_{2j}a_{2j-1}$ is Hermitian.

In this representation we can illustrate the key difference between the topological and trivial phases by looking at two special limits.

1. The trivial phase: choose $\mu < 0$ and $|\Delta| = t = 0$. In this case the Hamiltonian reduces to

$$H = -\mu\frac{i}{2}\sum_j \left(a_{2j-1}a_{2j}\right). \tag{16.32}$$

In this phase the Majorana operators on each physical site are coupled, but the Majorana operators between each physical site are decoupled. A representation of this Hamiltonian is shown in figure 16.4a. The Hamiltonian in the physical-site basis is in the atomic limit;

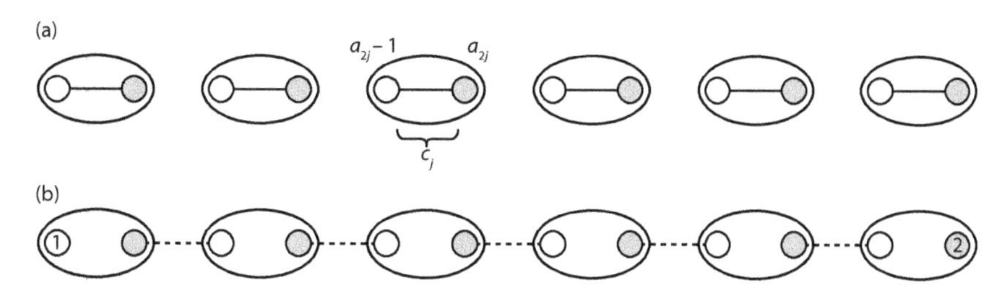

Figure 16.4. Schematic illustration of the lattice *p*-wave superconductor Hamiltonian in the (a) trivial limit and (b) nontrivial limit. The empty and filled circles represent the Majorana fermions making up each physical site (oval). The fermion operator on each physical site (c_j) is split up into two Majorana operators (a_{2j-1} and a_{2j}). In the nontrivial phase, the unpaired Majorana fermion states at the end of the chain are labeled with a 1 and a 2. These are the states that are continuously connected to the zero modes in the nontrivial topological superconductor phase.

thus, the ground state is trivial. If the chain has open boundary conditions, there will be no low-energy states on the end of the chain if the boundaries are cut between *physical* sites. That is, we are not allowed to pick boundary conditions where a physical site is cut in half.

2. The topological phase: $|\Delta| = t > 0$ and $\mu = 0$. For this case the Hamiltonian reduces to

$$H = it \sum_j a_{2j} a_{2j+1}. \tag{16.33}$$

A pictorial representation of this Hamiltonian is shown in figure 16.4b. With open-boundary conditions, it is clear that the Majorana operators a_1 and a_{2L} (where L is the last physical site) are not coupled to the rest of the chain and are "unpaired." In this limit the existence of two Majorana zero modes localized on the ends of the chain is manifest.

These two limits give the simplest representations of the trivial and non-trivial phases. By tuning away from these limits, the Hamiltonian will have some mixture of couplings between Majorana operators on the same physical site and operators between physical sites. However, because the two Majorana modes are localized at different ends of a gapped chain, the coupling between them will be exponentially small in the length of the wire, and they will remain at zero energy. In fact, in the nontrivial phase, the zero modes will not be destroyed until the bulk gap closes at a critical point. Unfortunately, even though this model is simple and clear, there are no confirmed candidates for materials that would realize it.

16.3 2-D Chiral *p*-Wave Superconductor

We now continue our study of topological *p*-wave superconductors, but here we move one dimension higher, to two dimensions. The paradigmatic example is the chiral *p*-wave superconductor whose vortices exhibit anyon excitations that have exotic non-Abelian statistics [68, 69, 70]. For pedagogy we will use both lattice and continuum models of the chiral superconductor. We begin with the lattice Hamiltonian

$$H = \sum_{m,n} \Big\{ -t\left(c^\dagger_{m+1,n}c_{m,n} + \text{h.c.}\right) - t\left(c^\dagger_{m,n+1}c_{m,n} + \text{h.c.}\right) - (\mu - 4t)c^\dagger_{m,n}c_{m,n} + \left(\Delta c^\dagger_{m+1,n}c^\dagger_{m,n} + \Delta^* c_{m,n}c_{m+1,n}\right) + \left(i\Delta c^\dagger_{m,n+1}c^\dagger_{m,n} - i\Delta^* c_{m,n}c_{m,n+1}\right)\Big\}. \qquad (16.34)$$

The fermion operators $c_{m,n}$ annihilate fermions on the lattice site (m, n), and we are considering spinless (or, equivalently, spin-polarized) fermions. We set the lattice constant $a = 1$ for simplicity. The pairing amplitude is anisotropic and has an additional phase of i in the y-direction compared to the pairing in the x-direction. Because the pairing is not on site, just as in the lattice version of the p-wave wire, the pairing terms will have momentum dependence. We can write this Hamiltonian in the BdG form and, assuming that Δ is translationally invariant, can Fourier-transform the lattice model to get

$$H_{\text{BdG}} = \frac{1}{2}\sum_{\mathbf{p}} \Psi^\dagger_{\mathbf{p}} \begin{pmatrix} \epsilon(\mathbf{p}) & 2i\Delta(\sin p_x + i\sin p_y) \\ -2i\Delta^*(\sin p_x - i\sin p_y) & -\epsilon(\mathbf{p}) \end{pmatrix} \Psi_{\mathbf{p}}, \qquad (16.35)$$

where $\epsilon(\mathbf{p}) = -2t(\cos p_x + \cos p_y) - (\mu - 4t)$ and $\Psi_{\mathbf{p}} = (c_{\mathbf{p}}\;\; c^\dagger_{-\mathbf{p}})^T$. For convenience we have shifted the chemical potential by the constant $4t$. As a quick aside, we note that the model takes a simple familiar form in the continuum limit ($\mathbf{p} \to 0$):

$$H^{(\text{cont})}_{\text{BdG}} = \frac{1}{2}\sum_{\mathbf{p}} \Psi^\dagger_{\mathbf{p}} \begin{pmatrix} \frac{p^2}{2m} - \mu & 2i\Delta(p_x + ip_y) \\ -2i\Delta^*(p_x - ip_y) & -\frac{p^2}{2m} + \mu \end{pmatrix} \Psi_{\mathbf{p}}, \qquad (16.36)$$

where $m \equiv \frac{1}{2t}$ and $p^2 = p_x^2 + p_y^2$. We see that the continuum limit has the characteristic $p_x + ip_y$ chiral form for the pairing potential. The quasi-particle spectrum of $H^{(\text{cont})}_{\text{BdG}}$ is $E_\pm = \pm\sqrt{(\frac{p^2}{2m} - \mu)^2 + 4|\Delta|^2 p^2}$, which, with a finite pairing amplitude, is gapped across the entire BZ as long as $\mu \neq 0$. This is unlike some other types of p-wave pairing terms (e.g., $\Delta(p) = \Delta p_x$) which can have gapless *nodal* points or lines in the BZ. In fact, nodal superconductors, having gapless quasi-particle spectra, are not topological superconductors simply by definition (i.e., a bulk excitation gap does not exist).

We immediately recognize the form of $H^{(\text{cont})}_{\text{BdG}}$ as a massive 2-D Dirac Hamiltonian, and indeed equation (16.34) is just a lattice Dirac Hamiltonian, which is what we will consider first. We can use our intuition from topological insulators—in, for example, chapter 8—to understand the different phases of this Hamiltonian. We expect that H_{BdG} will exhibit several phases as a function of Δ and μ for a fixed $t > 0$. For simplicity let us set $t = \frac{1}{2}$ and make a gauge transformation $c_{\mathbf{p}} \to e^{i\theta/2}c_{\mathbf{p}},\; c^\dagger_{\mathbf{p}} \to e^{-i\theta/2}c^\dagger_{\mathbf{p}}$, where $\Delta = |\Delta|e^{i\theta}$. The Bloch Hamiltonian for the lattice superconductor is then

$$H_{\text{BdG}}(\mathbf{p}) = \left(2 - \mu - \cos p_x - \cos p_y\right)\tau^z - 2|\Delta|\sin p_x \tau^y - 2|\Delta|\sin p_y \tau^x, \qquad (16.37)$$

where the τ^a are the Pauli matrices in the particle/hole basis. This has the same matrix structure as a lattice Dirac model with a speed of light $v = 2|\Delta|$ and a Wilson mass term $M(\mathbf{p}) = 2 - \mu - \cos p_x - \cos p_y$. Assuming $|\Delta| \neq 0$, this Hamiltonian has several fully gapped superconducting phases separated by gapless critical points. The quasi-particle spectrum for the lattice model is $E_\pm(\mathbf{p}) = \pm\sqrt{M(\mathbf{p})^2 + 4|\Delta|^2(\sin^2 p_x + \sin^2 p_y)}$ and is gapped (under the assumption that

$|\Delta| \neq 0$) unless $M(p)$, $\sin p_x$, and $\sin p_y$ simultaneously vanish. As a function of (p_x, p_y, μ), we find three critical points. The first critical point occurs at $(p_x, p_y, \mu) = (0, 0, 0)$. The second critical point has two gap closings in the BZ for the same value of μ: $(\pi, 0, 2)$ and $(0, \pi, 2)$. The final critical point, is again, a singly degenerate point at $(\pi, \pi, 4)$. We will show that the phases for $\mu < 0$ and $\mu > 4$ are trivial superconductor phases, whereas the phases $0 < \mu < 2$ and $2 < \mu < 4$ are topological superconductor phases with opposite chirality. In principle we can define a Chern number topological invariant constructed from the projection operator onto the lower quasiparticle band to characterize the phases. However, because electric charge is not conserved, there is no connection between this quantity and the Hall conductance, unlike the case for Chern insulators in chapter 3. We will show this calculation shortly, but first we make some physical arguments about the nature of the phases.

We will first consider the phase transition at $\mu = 0$. The arguments will closely mirror those discussed in chapter 8. The low-energy physics for this transition occurs around $(p_x, p_y) = (0, 0)$, and so we can expand the lattice Hamiltonian around this point; this is nothing but equation (16.36). One way to test the character of the $\mu < 0$ and $\mu > 0$ phases is to make an interface between them. If we can find a continuous interpolation between these two regimes that is always gapped, then they are topologically equivalent phases of matter. If we cannot find such a continuously gapped interpolation, then they are topologically distinct. A simple geometry to study is a domain wall where $\mu = \mu(x)$ such that $\mu(x) = -\mu_0$ for $x < 0$ and $\mu(x) = +\mu_0$ for $x > 0$ for a positive constant μ_0. This is an interface that is translationally invariant along the y-direction, so we can consider the momentum p_y as a good quantum number to simplify the calculation. What we will now show is that there exist gapless, propagating fermions bound to the interface that prevent us from continuously connecting the $\mu < 0$ phase to the $\mu > 0$ phase. This is one indication that the two phases represent topologically distinct classes.

The quasi-1-D, single-particle Hamiltonian in this geometry is

$$H_{\text{BdG}} = \frac{1}{2}\begin{pmatrix} -\mu(x) & 2i|\Delta|\left(-i\dfrac{d}{dx} + ip_y\right) \\ -2i|\Delta|\left(-i\dfrac{d}{dx} - ip_y\right) & \mu(x) \end{pmatrix}, \tag{16.38}$$

where we have ignored the quadratic terms in p, and p_y is a constant parameter, not an operator. This is a quasi-1-D Hamiltonian that can be solved for each value of p_y independently. We propose an ansatz for the gapless interface states:

$$|\psi_{p_y}(x, y)\rangle = e^{ip_y y} \exp\left(-\frac{1}{2|\Delta|}\int_0^x \mu(x')dx'\right)|\phi_0\rangle \tag{16.39}$$

for a constant, normalized spinor $|\phi_0\rangle$. For $p_y = 0$, this Hamiltonian is nearly identical to the one we solved in one dimension for the bound state of the p-wave wire. The secular equation for a zero-energy mode at $p_y = 0$ is

$$H_{\text{BdG}}|\psi_0(x, y)\rangle = 0$$

$$\Longrightarrow \begin{pmatrix} -\mu(x) & \mu(x) \\ -\mu(x) & \mu(x) \end{pmatrix}|\phi_0\rangle = 0. \tag{16.40}$$

The constant spinor that is a solution of this equation is $|\phi_0\rangle = 1/\sqrt{2}\,(1\ \ 1)^T$. This form of the constant spinor immediately simplifies the solution of the problem at finite p_y. We see that the term proportional to p_y in equation (16.38) is $-2|\Delta|p_y\tau^x$. Because $\tau^x|\phi_0\rangle = +|\phi_0\rangle$—i.e., the solution $|\phi_0\rangle$ is an eigenstate of τ^x—then $|\psi_{p_y}(x, y)\rangle$ is an eigenstate of $H_{\rm BdG}$ with energy $E(p_y) = -2|\Delta|p_y$. Thus, we have found a bound-state solution at the interface of two regions with $\mu < 0$ and $\mu > 0$, respectively. This set of bound states, parameterized by the conserved quantum number p_y, is gapless and chiral; i.e., the group velocity of the quasiparticle dispersion is always negative and never changes sign. The chirality is determined by the sign of the "spectral" Chern number mentioned before, which we will calculate next.

These gapless edge states have quite remarkable properties and are not the same chiral fermions that propagate on the edge of integer quantum Hall states but are Majorana fermions as well. They are very special because they must satisfy the condition that the fermions are chiral and that they are simultaneously Majorana. Using Clifford-algebra representation theory, it can be shown that the so-called chiral-Majorana (or Majorana-Weyl) fermions can be found only in space-time dimensions $(8k + 2) - d$, where $k = 0, 1, 2, \ldots$. Thus, we can find chiral-Majorana states only in $1 + 1$ dimensions or in $9 + 1$ dimensions (or higher). In condensed matter, we are stuck with $1 + 1$ dimensions, where we have now seen that they appear as the boundary states of chiral topological superconductors. In fact, the simplest interpretation of such chiral-Majorana fermions is as half of a conventional chiral fermion, i.e., basically, its real or imaginary part. To show this, we will consider the edge state of a $\nu = 1$ quantum Hall system for a single edge:

$$H^{(\rm QH)}_{\rm edge} = \hbar v \sum_p p\eta^\dagger_p \eta_p. \tag{16.41}$$

The fermion operators satisfy $\{\eta^\dagger_p, \eta_{p'}\} = \delta_{pp'}$. Similar to the discussion on the 1-D p-wave wire, we can decompose these operators into their real and imaginary parts:

$$\begin{aligned} \eta_p &= \frac{1}{2}(\gamma_{1,p} + i\gamma_{2,p}), \\ \eta^\dagger_p &= \frac{1}{2}(\gamma_{1,-p} - i\gamma_{2,-p}), \end{aligned} \tag{16.42}$$

where $\gamma_{a,p}$ $(a = 1, 2)$ are Majorana fermion operators satisfying $\gamma^\dagger_{a,p} = \gamma_{a,-p}$ and $\{\gamma_{a,-p}, \gamma_{b,p'}\} = 2\delta_{ab}\delta_{pp'}$. The quantum Hall edge Hamiltonian now becomes

$$\begin{aligned} H^{(\rm QH)}_{\rm edge} &= \hbar v \sum_{p\geq 0} p(\eta^\dagger_p\eta_p - \eta^\dagger_{-p}\eta_{-p}) \\ &= \frac{\hbar v}{4} \sum_{p\geq 0} p\left\{(\gamma_{1,-p} - i\gamma_{2,-p})(\gamma_{1,p} + i\gamma_{2,p}) - (\gamma_{1,p} - i\gamma_{2,p})(\gamma_{1,-p} + i\gamma_{2,-p})\right\} \\ &= \frac{\hbar v}{4} \sum_{p\geq 0} p\,\big(\gamma_{1,-p}\gamma_{1,p} + \gamma_{2,-p}\gamma_{2,p} + i\gamma_{1,-p}\gamma_{2,p} - i\gamma_{2,-p}\gamma_{1,p} \\ &\quad - \gamma_{1,p}\gamma_{1,-p} - \gamma_{2,p}\gamma_{2,-p} + i\gamma_{2,p}\gamma_{1,-p} - i\gamma_{1,p}\gamma_{2,-p}\big) \\ &= \frac{\hbar v}{4} \sum_{p\geq 0} p\,\left(\gamma_{1,-p}\gamma_{1,p} + \gamma_{2,-p}\gamma_{2,p} - \gamma_{1,p}\gamma_{1,-p} - \gamma_{2,p}\gamma_{2,-p}\right) \\ &= \frac{\hbar v}{2} \sum_{p\geq 0} p\,\left(2 + \gamma_{1,-p}\gamma_{1,p} + \gamma_{2,-p}\gamma_{2,p}\right). \end{aligned} \tag{16.43}$$

Thus,

$$H_{\text{edge}}^{(\text{QH})} = \frac{\hbar v}{2} \sum_{p \geq 0} p \left(\gamma_{1,-p}\gamma_{1,p} + \gamma_{2,-p}\gamma_{2,p}\right) \tag{16.44}$$

up to a constant shift of the energy. This Hamiltonian is exactly two copies of a chiral-Majorana Hamiltonian. The edge/domain-wall fermion Hamiltonian of the chiral *p*-wave superconductor will be

$$H_{\text{edge}}^{(p\text{-wave})} = \frac{\hbar v}{2} \sum_{p \geq 0} p\gamma_{-p}\gamma_{p}. \tag{16.45}$$

Finding gapless states on a domain wall of μ is an indicator that the phases with $\mu > 0$ and $\mu < 0$ are distinct. If they were the same phase of matter, we should be able to adiabatically connect these states continuously. However, we have shown a specific case of the more general result that any interface between a region with $\mu > 0$ and a region with $\mu < 0$ will have gapless states that generate a discontinuity in the interpolation between the two regions. This question remains: is $\mu > 0$ or $\mu < 0$ nontrivial? The answer, as we will show now, is that we have a trivial superconductor for $\mu < 0$ and a topological superconductor for $\mu > 0$. Remember that for now we are considering μ only in the neighborhood of 0 and using the continuum model expanded around $(p_x, p_y) = (0, 0)$. We will now define a bulk topological invariant for 2-D superconductors that can distinguish the trivial superconductor state from the chiral topological superconductor state. For the spinless BdG Hamiltonian, which is of the form

$$H_{\text{BdG}} = \frac{1}{2} \sum_{\mathbf{p}} \Psi_{\mathbf{p}}^{\dagger} d_a(\mathbf{p}, \mu)\tau^a \Psi_{\mathbf{p}}, \tag{16.46}$$

$$d_a(\mathbf{p}, \mu) = \left(-2|\Delta|p_y, -2|\Delta|p_x, p^2/2m - \mu\right), \tag{16.47}$$

the topological invariant is the spectral Chern number [71], which simplifies, for this Hamiltonian, to the winding number

$$N_w = \frac{1}{8\pi} \int d^2p\, \epsilon^{ij}\, \hat{d} \cdot \left(\partial_{p_i}\hat{d} \times \partial_{p_j}\hat{d}\right) = \frac{1}{8\pi} \int d^2p\, \frac{\epsilon^{ij}}{|\mathbf{d}|^3}\, \mathbf{d} \cdot \left(\partial_{p_i}\mathbf{d} \times \partial_{p_j}\mathbf{d}\right). \tag{16.48}$$

The unit vector $\hat{d}_a = d_a/|\mathbf{d}|$. This integral has a special form and is equal to the degree of the mapping from momentum space onto the 2-sphere given by $\hat{d}_1^2 + \hat{d}_2^2 + \hat{d}_3^2 = 1$. As it stands, the degree of the mapping $S : (p_x, p_y) \rightarrow (\hat{d}_1, \hat{d}_2, \hat{d}_3)$ is not well defined because the domain is not compact, i.e., (p_x, p_y) is restricted to lie only in the Euclidean plane ($\mathbb{R}^2$). However, for our choice of the map S, we can define the winding number by choosing an equivalent, but compact, domain. To understand the necessary choice of domain, we can simply look at the explicit form of $\hat{d}$:

$$\hat{d} = \frac{\left(-2|\Delta|p_y, -2|\Delta|p_x, p^2/2m - \mu\right)}{\left(4|\Delta|p^2 + (p^2/2m - \mu)^2\right)^{1/2}}. \tag{16.49}$$

We see that $\lim_{|p|\to\infty} \hat{d} = (0, 0, 1)$, and it does not depend on the direction in which we take the limit in the 2-D plane. Because of the uniqueness of this limit, we are free to perform the *one-point compactification* of $\mathbb{R}^2$, which amounts to including the point at infinity in our domain. The topology of $\mathbb{R}^2 \cup \{\infty\}$ is the same as S^2, and so we can consider the degree of our

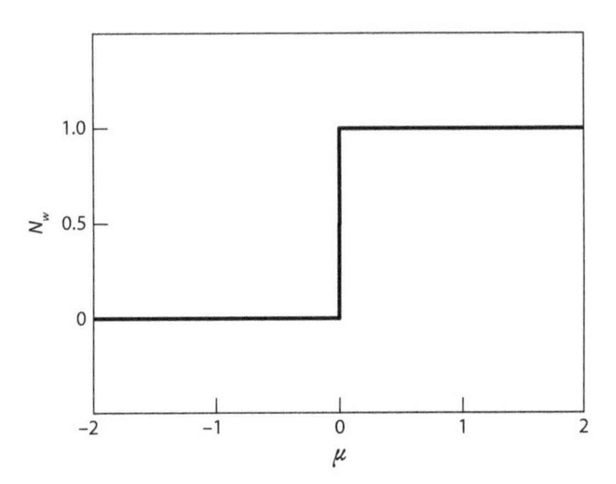

Figure 16.5. The winding number for the continuum chiral p-wave superconductor. The winding index is plotted as a function of the parameter μ, which is tuned through a quantum phase transition at $\mu = 0$.

map from the compactified momentum space (S^2) to the unit d-vector space (S^2). Using the explicit form of the $\mathbf{d}$-vector for this model, we find

$$N_w = \frac{1}{8\pi} \int d^2 p \frac{\left(\frac{p^2}{2m^*} + \mu\right)}{\left(p^2 + \frac{1}{4|\Delta|^2}\left(\frac{p^2}{2m} - \mu\right)^2\right)^{3/2}}. \tag{16.50}$$

The evaluation of this integral can easily be carried out numerically and the result, as a function of μ, is shown in figure 16.5. This figure clearly shows two different phases separated by a quantum critical point at $\mu = 0$. One is topologically nontrivial ($\mu > 0$), and the other is trivial ($\mu < 0$). Thus, we have identified the phase that is in the chiral superconductor state. For the full lattice model there are two other phase transitions: one at $\mu = 2$ and one at $\mu = 4$. The ranges $0 < \mu < 2$ and $2 < \mu < 4$ are topological superconductors and can be distinguished by their edge states when solved with open boundary conditions as shown in figure 16.6.

16.3.1 Bound States on Vortices in 2-D Chiral p-wave Superconductors

We have seen that on domains between regions of chiral superconductors with $\mu < 0$ and $\mu > 0$, there exist chiral Majorana states propagating on the interfaces. For the linear interface, we found an exact zero-mode solution, accompanied by a set of propagating modes. The propagating modes will be separated by an energy gap if there is finite-size quantization due to a finite interface length. For a closed system with periodic boundary conditions, there have to exist an even number of linear domain walls and, thus, an even number of isolated Majorana zero modes. This is important because we are formulating this problem in terms of the redundant BdG Hamiltonian, which strictly enforces a Hilbert space structure with an *even* dimension. That is, there is an even number of fermionic modes. For every state at $E > 0$, there is a partner state at $-E$ related by the particle-hole symmetry, so there must be an even number of states at $E = 0$. If only one (Majorana) mode were present at $E = 0$, there would be an inconsistency.

Now let us imagine a different geometry: consider a disk of radius R, which has $\mu < 0$ surrounded by a region with $\mu > 0$ for $r > R$. We know from our previous discussion that there will be a single branch of chiral Majorana states localized near $r = R$. At first glance, the disk geometry seems to pose a problem because there is no issue with global boundary conditions if we have only one radial interface, and thus we can seemingly get an odd number of Majorana zero modes. However, on any interface in the 2-D plane ($\mathbb{R}^2$), which is a closed curve, there will be no Majorana zero mode and, thus, no inconsistency. This is due to a Berry-phase contribution to the boundary conditions along the closed curve [72]. We can see

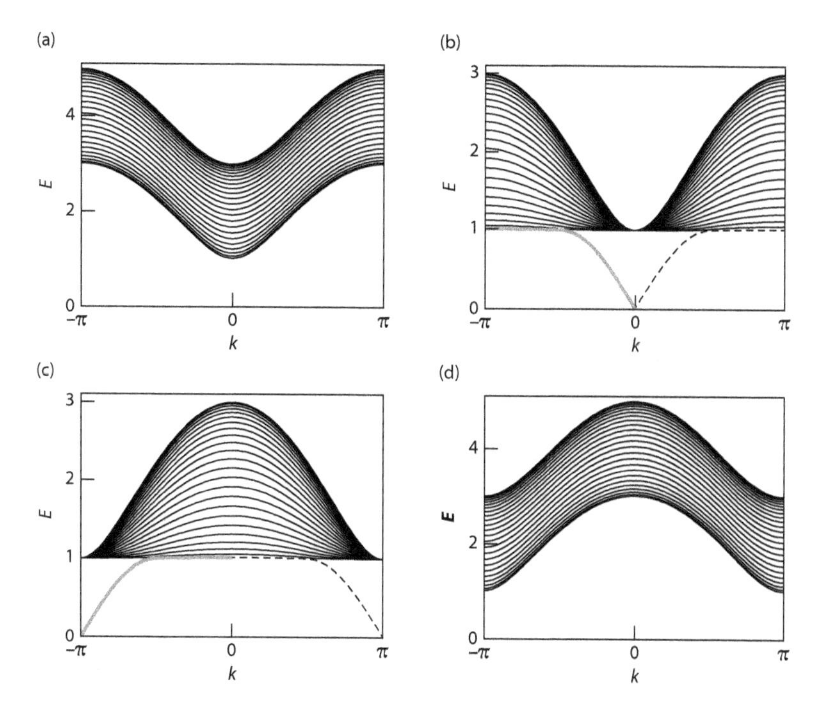

Figure 16.6. Energy spectra for the chiral p-wave lattice model on a cylindrical geometry with one conserved momentum k and one open boundary direction. We plot only the independent energy states, i.e., only the $E \geq 0$ spectrum for the different phases. The states crossing the energy gap (solid and dotted lines) are the edge states where each line style represents a different edge. We show the spectra for $|\Delta| = 0.5$ and different values of μ: (a) $\mu = -1.0$, (b) $\mu = 1.0$, (c) $\mu = 3.0$, and (d) $\mu = 5.0$.

this from the transformation properties of the quasi-particle operator for the spinless chiral superconductor:

$$\gamma^{\dagger}_{\mathbf{p}} = e^{i\theta/2} u(\mathbf{p}) c^{\dagger}_{\mathbf{p}} + e^{-i\theta/2} v(\mathbf{p}) c_{\mathbf{p}}, \tag{16.51}$$

where θ is the phase of of the order parameter and

$$u(\mathbf{p}) = \frac{\epsilon(p) + \sqrt{4|\Delta|^2 p^2 + \epsilon^2(p)}}{(\epsilon(p) + \sqrt{4|\Delta|^2 p^2 + \epsilon^2(p)})^2 + 4|\Delta|^2 p^2},$$

$$v(\mathbf{p}) = \frac{-2i|\Delta|(p_x - ip_y)}{(\epsilon(p) + \sqrt{4|\Delta|^2 p^2 + \epsilon^2(p)})^2 + 4|\Delta|^2 p^2}.$$

A localized quasi-particle operator can be written by Fourier-transforming

$$\gamma^{\dagger}(x) = \int \frac{d^2 p}{(2\pi)^2} e^{i\mathbf{p}\cdot\mathbf{x}} \gamma^{\dagger}_{\mathbf{p}}. \tag{16.52}$$

From equation (16.51) we see that if the superconducting order parameter winds by 2π, then $\gamma^{\dagger}_{\mathbf{p}} \to -\gamma^{\dagger}_{\mathbf{p}}$ and, thus, $\gamma^{\dagger}(x) \to -\gamma^{\dagger}(x)$. This is important because for the chiral superconductor, we have $\Delta(\mathbf{p}) = |\Delta| e^{i\theta} e^{i\phi_{\hat{\mathbf{p}}}} p$, where $\mathbf{p} = p(\cos\phi, \sin\phi)$. From a semiclassical perspective, we see that if we take a quasi-particle around a closed loop, then the phase of $\Delta(\mathbf{p})$, effectively winds by 2π because the direction of the momentum of the quasi-particle is changing by 2π.

Thus, taking any quasi-particle around a closed loop in $\mathbb{R}^2$ sends $\gamma^\dagger(x)$, to $-\gamma^\dagger(x)$, indicating *antiperiodic* boundary conditions on the quasi-particle wavefunctions, i.e., $\gamma^\dagger(r, \theta + 2\pi) = -\gamma^\dagger(r, \theta)$. This also holds for any quasi-particle states bound to the domain wall of μ; thus, states on the circular interface obey antiperiodic boundary conditions. Hence, there is no exact zero mode for the interface states because the lowest-allowed (angular) momentum state is shifted from 0 to $\frac{1}{2}$. This is not true if the interface is a circular boundary of a cylinder because there, $\phi_{\mathbf{p}}$ does not wind by 2π. Thus, solving the chiral superconductor Hamiltonian in a cylindrical geometry can yield exact zero modes, but, of course, there are two zero modes because the cylinder has two boundaries.

We can now ask this simple question: is there a way to get a Majorana zero mode on the boundary of the disk? The answer is yes, and it just requires that we shift the boundary conditions for the states on the disk boundary from antiperiodic to periodic. This is accomplished by inserting magnetic flux into the bulk of the disk so that the fermionic boundary states pick up an Aharonov-Bohm phase of π [72]. So, we can simply insert π-flux into the disk and expect to see a boundary zero mode. This is inconsistent if there is only one Majorana mode, which indicates that there will be a separate Majorana zero mode somewhere else, namely, bound to the π-flux insertion. In a superconductor, a π-flux (i.e., flux $\frac{h}{2e}$) causes the $U(1)$ phase of the order parameter to wind by 2π and creates a vortex. Thus, we intuitively expect Majorana bound states trapped on vortices in the spinless chiral superconductor. If we insert two vortices, assuming they are well separated, the boundary conditions for the edge change back to antiperiodic. Thus, there will not be an edge zero mode; however, there will now be two Majorana zero modes, one bound to each vortex. The degeneracy can be lifted, however, if the vortices are brought too close together, where the zero modes can hybridize.

Let us explicitly show that a vortex in a chiral superconductor will contain a zero mode [68, 69, 73]. This calculation is a variant of our calculation for the existence of a Majorana mode at the interface between topological and trivial superconductors. For this construction, we begin by modeling a vortex as a region of superconductor with $\mu < 0$ for $r < R$ and $\mu > 0$ for $r > R$ [69]. This is the inverse of the geometry we just discussed, i.e., it is a region of trivial superconductor inside a region of topological superconductor. On the interface at $r = R$, there will be a branch of chiral-Majorana modes but with no exact zero mode. If we take the limit $R \to 0$, this represents a vortex, and all the low-energy modes on the interface will be pushed to higher energies. If we put a π-flux inside the trivial region, it will change the boundary conditions such that even in the $R \to 0$ limit there will be a zero mode in the spectrum localized on the vortex.

Now let us take the BdG Hamiltonian in the Dirac limit ($m \to \infty$) and solve the BdG equations in the presence of a vortex located at $r = 0$ in the disk geometry in polar coordinates. Let $\Delta(r, \theta) = |\Delta(r)|e^{ia(r)}$. The profile $|\Delta(r)|$ for a vortex will depend on the details of the model but must vanish inside the vortex core region, e.g., for an infinitely thin core, we just need $|\Delta(0)| = 0$. We take the phase $a(\mathbf{r})$ to be equal to the polar angle at $\mathbf{r}$. This is a completely different way to model the vortex than in the previous paragraph; in fact, this is the more physical, but more complicated, construction. The previous model of a "vortex" is merely suggestive and indicates what we might hope to find on a physical vortex that has a normal metal core, as opposed to a gapped trivial superconductor core. In a real vortex, the core will be a normal metal—but a normal metal with discrete quantized energy levels due to the small core size. Thus, the normal region of a real vortex is gapped for energies much lower than the superconducting gap and is a trivial insulator, which can be adiabatically connected to the $\mu < 0$ superconducting phase, thus justifying the topological equivalence between the two vortex constructions.

The first step in the solution of the bound state for the more-realistic vortex profile is to gauge-transform the phase of $\Delta(r, \theta)$ into the fermion operators via $\Psi(r) \to e^{ia(\mathbf{r})/2}\Psi(r)$.

This has two effects: (1) it simplifies the solution of the BdG differential equations and (2) converts the boundary conditions of $\Psi(r)$ from periodic to antiperiodic around the vortex position $r = 0$. In polar coordinates, the remaining single-particle BdG Hamiltonian is simply

$$H_{\text{BdG}} = \frac{1}{2}\begin{pmatrix} -\mu & 2|\Delta(r)|e^{i\theta}\left(\frac{\partial}{\partial r} + \frac{i}{r}\frac{\partial}{\partial \theta}\right) \\ -2|\Delta(r)|e^{-i\theta}\left(\frac{\partial}{\partial r} - \frac{i}{r}\frac{\partial}{\partial \theta}\right) & \mu \end{pmatrix} \tag{16.53}$$

We want to solve $H\Psi = E\Psi = 0$, which we can do with the ansatz

$$\Psi_0(r,\theta) = \frac{i}{\sqrt{r\mathcal{N}}}\exp\left[-\frac{1}{2}\int_0^r \frac{\mu(r')}{|\Delta(r')|}dr'\right]\begin{pmatrix} -e^{i\theta/2} \\ e^{-i\theta/2} \end{pmatrix} \equiv ig(r)\begin{pmatrix} -e^{i\theta/2} \\ e^{-i\theta/2} \end{pmatrix}, \tag{16.54}$$

where $\mathcal{N}$ is a normalization constant. The function $g(r)$ is localized at the location of the vortex. We see that $\Psi_0(r, \theta + 2\pi) = -\Psi_0(r, \theta)$, as required. From an explicit check, we can see that $H_{\text{BdG}}\Psi_0(r, \theta) = 0$. The field operator that annihilates fermion quanta in this localized state is

$$\gamma = \int r\,dr\,d\theta\; ig(r)\left(-e^{i\theta/2}c(r,\theta) + e^{-i\theta/2}c^{\dagger}(r,\theta)\right), \tag{16.55}$$

from which we can immediately see that $\gamma = \gamma^{\dagger}$. Thus, the vortex traps a single Majorana bound state at zero energy.

16.3.1.1 Non-Abelian Statistics of Vortices in Chiral *p*-Wave Superconductors

We showed in the last section that on each vortex in a spinless chiral superconductor, there exists a single Majorana bound state. If we have a collection of $2N$ vortices that are well separated from each other, a low-energy subspace is generated, which in the thermodynamic limit leads to a ground-state degeneracy of 2^N [74, 75]. For example, two vortices give a degeneracy of 2, which can be understood by combining the two localized Majorana bound states into a single complex fermion state, which can be occupied or unoccupied. From $2N$ vortices, we can form N complex fermion states, giving a degeneracy of 2^N, which can be broken up into the subspace of 2^{N-1} states with even fermion parity and 2^{N-1} states with odd fermion parity. As an aside, because we have operators that mutually anticommute and square to $+1$, we can define a Clifford-algebra structure using the set of $2N\,\gamma_i$.

To illustrate the statistical properties of the vortices under exchanges, we closely follow the work of Ivanov [70]. Let us begin with a single pair of vortices that have localized Majorana operators γ_1, γ_2, respectively, and are assumed to be well separated. We imagine that we adiabatically move the vortices in order to exchange the two Majorana fermions. If we move them slowly enough, then the only outcome of exchanging the vortices is a unitary operator acting on the two degenerate states that make up the ground-state subspace. If we exchange the two vortices, then we have $\gamma_1 \to \gamma_2$ and $\gamma_2 \to \gamma_1$. However, if we look at figure 16.7 we immediately see there is a complication. In this figure we have illustrated the exchange of two vortices, and the dotted lines represent branch cuts, across which the phase of the superconductor order parameter jumps by 2π. Because our solution of the Majorana bound states used the gauge-transformed fermion operators, we see that the bound state on one vortex, which passes through the branch cut of the other vortex, picks up an additional minus

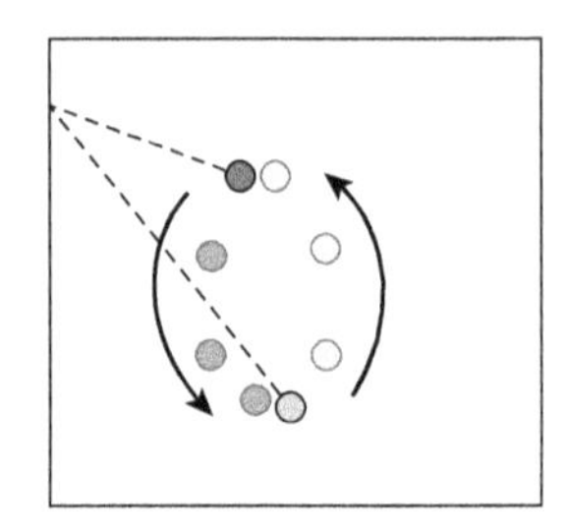

Figure 16.7. Illustration of the exchange of two vortices in a chiral p-wave superconductor. The dotted lines represent branch cuts, across which the phase of the superconducting order parameter jumps by 2π.

sign upon exchange. Thus, the exchange of two vortices is effected by

$$\gamma_1 \to \gamma_2$$
$$\gamma_2 \to -\gamma_1. \tag{16.56}$$

In general, if we have $2N$ vortices, we can think of the different exchange operators $T_{ij}(\gamma_a)$, which, with our choice of conventions, sends $\gamma_i \to \gamma_j$, $\gamma_j \to -\gamma_i$, and $\gamma_k \to \gamma_k$ for all $k \neq i, j$. We can construct a representation of this exchange process on the Hilbert space by finding a $\tau(T_{ij})$ such that $\tau(T_{ij})\gamma_a\tau^{-1}(T_{ij}) = T_{ij}(\gamma_a)$. Such a representation is given by

$$\tau(T_{ij}) = \exp\left(\frac{\pi}{4}\gamma_j\gamma_i\right) = \frac{1}{\sqrt{2}}\left(1 + \gamma_j\gamma_i\right). \tag{16.57}$$

Let us prove this by showing an explicit example for T_{12} that will have the transformation given in equation (16.56):

$$\tau(T_{12})\gamma_1\tau^{-1}(T_{12}) = \frac{1}{2}\left(\gamma_1 - \gamma_1\gamma_2\gamma_1 + \gamma_2 - \gamma_1\right) = \gamma_2, \tag{16.58}$$

$$\tau(T_{12})\gamma_2\tau^{-1}(T_{12}) = \frac{1}{2}\left(\gamma_2 - \gamma_1 + \gamma_2\gamma_1\gamma_2 - \gamma_2\right) = -\gamma_1, \tag{16.59}$$

$$\tau(T_{12})\gamma_3\tau^{-1}(T_{12}) = \gamma_3\tau(T_{12})\tau^{-1}(T_{12}) = \gamma_3. \tag{16.60}$$

Now that we have this representation, we can illustrate the non-Abelian statistics. We start with four vortices with Majorana operators $\gamma_1, \gamma_2, \gamma_3, \gamma_4$. To illustrate the action of the exchange operators on the four-fold degenerate ground-state space, we need to pair these Majorana operators into complex fermions:

$$a = \frac{1}{2}(\gamma_1 + i\gamma_2), \qquad a^\dagger = \frac{1}{2}(\gamma_1 - i\gamma_2),$$
$$b = \frac{1}{2}(\gamma_3 + i\gamma_4), \qquad b^\dagger = \frac{1}{2}(\gamma_3 - i\gamma_4). \tag{16.61}$$

The basis vectors of the ground-state subspace can now be written as $\{|0\rangle_a \otimes |0\rangle_b, |1\rangle_a \otimes |1\rangle_b, |1\rangle_a \otimes |0\rangle_b, |0\rangle_a \otimes |1\rangle_b\}$, where we have ordered the basis so that states of the same fermion parity are together. The notation $|n\rangle_{a,b}$ means $a^\dagger a\,|n\rangle_a = n|n\rangle_a$ and $b^\dagger b\,|n\rangle_b = n|n\rangle_b$. The set of statistical exchanges is generated by T_{12}, T_{23}, T_{34}, and we want to understand how these exchanges act on the ground-state subspace. We can rewrite these three operators as

$$\tau(T_{12}) = \frac{1}{\sqrt{2}}(1 + \gamma_2\gamma_1) = \frac{1}{\sqrt{2}}\left(1 - i(aa^\dagger - a^\dagger a)\right), \tag{16.62}$$

$$\tau(T_{23}) = \frac{1}{\sqrt{2}}\left(1 - i(ba - ba^\dagger + b^\dagger a - b^\dagger a^\dagger)\right), \tag{16.63}$$

$$\tau(T_{34}) = \frac{1}{\sqrt{2}}\left(1 - i(bb^\dagger - b^\dagger b)\right). \tag{16.64}$$

Taking matrix elements in our chosen ground-state basis, we find

$$\tau(T_{12}) = \frac{1}{\sqrt{2}}\begin{pmatrix} (1-i) & 0 & 0 & 0 \\ 0 & (1+i) & 0 & 0 \\ 0 & 0 & (1+i) & 0 \\ 0 & 0 & 0 & (1-i) \end{pmatrix}, \tag{16.65}$$

$$\tau(T_{23}) = \frac{1}{\sqrt{2}}\begin{pmatrix} 1 & -i & 0 & 0 \\ i & 1 & 0 & 0 \\ 0 & 0 & 1 & -i \\ 0 & 0 & i & 1 \end{pmatrix}, \tag{16.66}$$

$$\tau(T_{34}) = \frac{1}{\sqrt{2}}\begin{pmatrix} (1+i) & 0 & 0 & 0 \\ 0 & (1-i) & 0 & 0 \\ 0 & 0 & (1-i) & 0 \\ 0 & 0 & 0 & (1+i) \end{pmatrix}, \tag{16.67}$$

We see that with our basis choice, T_{12}, T_{34} are Abelian phases acting on each state, whereas T_{23} exhibits nontrivial mixing terms between the states with the same fermion parity. Thus, the form of T_{23} represents non-Abelian statistics. Given an initial state $|\psi_i\rangle = |0\rangle_a \otimes |0\rangle_b$, if we take vortex 2 around vortex 3, the final state is $|\psi_f\rangle = (1/\sqrt{2})\left(|0\rangle_a \otimes |0\rangle_b + i|1\rangle_a \otimes |1\rangle_b\right)$. In principle, we must also keep track of the Berry-phase contribution to the statistical phase. Here, we have considered only the wavefunction monodromy; however, it can be proven that the Berry phase does not contribute in this case. The field of topological quantum computation is built on the idea that such exchange or braiding operations will lead to nontrivial quantum evolutions of the ground state, which can be used for quantum computations.

16.4 Problems

1. *Particle-Hole Symmetry:* Given a Bloch Hamiltonian $H(\mathbf{p})$ and a particle-hole/charge-conjugation matrix C, we know that $H(\mathbf{p})$ is particle-hole symmetric if $CH(\mathbf{p})C^{-1} = H^*(-\mathbf{p})$. Suppose that we have

$$H(\mathbf{p}) = \epsilon(\mathbf{p})\mathbb{I}_{2\times 2} + d_a(\mathbf{p})\sigma^a, \tag{16.68}$$

where σ^a are the Pauli matrices and $\epsilon(\mathbf{p})$, $d_a(\mathbf{p})$ are generic functions of $\mathbf{p}$. Find the constraints on $\epsilon(\mathbf{p})$, $d_a(\mathbf{p})$ as functions of $\mathbf{p}$ if $H(\mathbf{p})$ is particle-hole symmetric with (a) $C = \sigma^x$ and (b) $C = \mathbb{I}_{2\times 2}$.

2. *Practice with the BdG Formalism:* Write down the BdG Hamiltonians corresponding to the following Bloch Hamiltonians:

$$\text{(a)}\ \ H_{1\text{D}}(\mathbf{p}) = (p-\mu)\mathbb{I}_{2\times 2}, \tag{16.69}$$

where the two components are spin-$\frac{1}{2}$ degrees of freedom.

$$\text{(b)}\ \ H_{2\text{D}}(\mathbf{p}) = \left(\frac{p^2}{2m}-\mu\right)\mathbb{I}_{2\times 2} + \alpha\mathbf{p}\cdot\boldsymbol{\sigma}, \tag{16.70}$$

where σ^a are 2×2 Pauli matrices representing spin.

$$\text{(c)}\ \ H_{3\text{D}}(\mathbf{p}) = \left(\frac{p^2}{2m}-\mu\right)\mathbb{I}_{4\times 4} + p_a\Gamma^a, \tag{16.71}$$

where $\Gamma^a = \tau^x\otimes\sigma^a$ and where τ^a, σ^a are sets of 2×2 Pauli matrices representing orbital and spin degrees of freedom, respectively. Use the Nambu basis $\Psi = (\psi_{pa}\ \ \psi^{\dagger}_{-pa})^T$, where a represents all the internal spin/orbital degrees of freedom. Be careful to pay attention to the factors of 2 in the construction of the BdG Hamiltonian.

After construction, diagonalize each $H_{\text{BdG}}(\mathbf{p})$ to obtain the "quasi-particle" spectra as functions of $\mathbf{p}$. Does the relation $E(\mathbf{p}) = -E(-\mathbf{p})$ hold for these energy spectra? For the particle-hole symmetry operator $C = \mu^x$, where μ^x is a Pauli matrix in particle/hole space, check that each of the $H_{\text{BdG}}(\mathbf{p})$ satisfies $\mu^x H_{\text{BdG}}(\mathbf{p})\mu^x = -H^*_{\text{BdG}}(-\mathbf{p})$.

3. *Practice with Majorana Fermions:* Suppose we have a two-site lattice model with a single (complex) fermionic degree of freedom per site represented by the annihilation operators c and d, respectively. Enumerate all of the Hermitian terms we can have in a generic Hamiltonian, including all possible on-site terms, a nearest-neighbor-hopping term, and a superconducting pairing term. The total particle-number operator is given by $N = N_c + N_d = c^{\dagger}c + d^{\dagger}d$. Which of the enumerated Hamiltonian terms commute with N? Which don't commute, and why? Now break up each complex fermion into its real and imaginary parts using Majorana fermions. For example, choose $c^{\dagger} = \gamma_1 + i\gamma_2$ and $d^{\dagger} = \gamma_3 + i\gamma_4$, where the γ_i satisfy the properties of Majorana fermions. Construct the corresponding terms of the Hamiltonian in the Majorana basis and explicitly show that the Hamiltonian remains Hermitian in this basis, as it must. Do any terms mix the real part of c with the real part of d? When the on-site energy terms are turned off and the magnitude of the pairing term is tuned to be equal to the hopping, the Majorana Hamiltonian should simplify to just a single term. What is the energy spectrum at this special point?

4. *Stability of Chiral Fermions:* Chiral fermions in $1+1$ dimensions are fermions that can travel in only one direction because there are no states into which one can backscatter. We will consider only systems with translation invariance, so let us begin with the chiral fermion Bloch Hamiltonian

$$H(p) = p - \mu. \tag{16.72}$$

Show that if we add any (translationally invariant) single-particle perturbation (which, in this case, is equivalent to just adjusting μ), the spectrum still remains gapless. Now let us try to open a superconducting gap. To begin, write down the corresponding BdG Hamiltonian H_{BdG} for the chiral fermion. Show that the kinetic energy piece is proportional to the (2×2 in particle-hole space) identity matrix. From this mathematical fact alone, prove that we can never open a gap via (translation-invariant) superconductivity. Now add n copies of chiral fermions with the *same* chirality. Show, by explicitly constructing the BdG Hamiltonian, that we cannot open a gap even with n copies; apply the same proof as before.

Now just add a single extra copy with opposite chirality. Show that we can easily find single-particle, particle-conserving mass terms that will open a gap in the spectrum. Construct the BdG Hamiltonian and find a superconducting pairing term that also opens a gap in the spectrum. Note that all these results still apply to the case with broken translation symmetry; they are just harder to prove and require index theorems in general.

5. *Weak Topological Superconductor in Two Dimensions:* Symmetry class D in $2+1$ dimensions supports weak topological superconductors that are anisotropic states with edge states, which exist only on edges with a particular orientation. We will illustrate such a state in this problem. We begin with an array of decoupled (1+1)-D lattice Hamiltonians for the p-wave wire:

$$H = \sum_{j,n} \left[-t \left(c^{\dagger}_{j,n} c_{j+1,n} + c^{\dagger}_{j+1,n} c_{j,n} \right) - \mu c^{\dagger}_{j,n} c_{j,n} + |\Delta| \left(c^{\dagger}_{j+1,n} c^{\dagger}_{j,n} + c_{j,n} c_{j+1,n} \right) \right], \tag{16.73}$$

where n represents the nth wire in the array. Draw a picture showing the decoupled array of wires. If μ is fixed for each of the wires labeled by n to be in the ($(1+1)$-D) topological phase, which edges will have topological edge states? What is the dispersion relation for these edge states? This system is in a weak topological superconductor state. Now suppose we couple the wires in the transverse direction via the interwire-hopping term

$$H_{\perp} = \sum_{j,n} \left[-t_{\perp} \left(c^{\dagger}_{j,n} c_{j,n+1} + c^{\dagger}_{j,n+1} c_{j,n} \right) \right]. \tag{16.74}$$

Use numerical exact diagonalization on a cylinder (i.e., strip) geometry with one open-boundary condition and one periodic-boundary condition to study the properties of the edge states under this perturbation. Make plots of the energy spectra versus the conserved momentum along the edge to illustrate the edge-state dispersion relation. For fixed μ, t, and $|\Delta|$, over what range of $t_{\perp}$ does the weak topological phase exist?

17

Time-Reversal-Invariant Topological Superconductors

BY TAYLOR L. HUGHES

After using the simplified spinless models of the previous chapter, we now go on to study topological superconductors with spin. We will begin by reviewing the common conventions for singlet- and triplet-pairing terms for single-band metals. We then will discuss TR-invariant topological superconductors in 2 and 3 dimensions, including their classification and edge–surface–state–properties. We will conclude with a brief discussion of a 1-D TR-invariant topological superconductor, which is closely related to the 1-D p-wave wire discussed in the previous chapter.

17.1 Superconducting Pairing with Spin

We are now very familiar with the mean-field description of superconductors, albeit using simplified models. We now discuss a more-general model, which includes the spin of the constituent electrons. The generic form of the pairing term for a translationally invariant model with a single band is

$$H_\Delta = \sum_{\mathbf{p}} \frac{1}{2}\left[c^\dagger_{\mathbf{p}\sigma}\Delta_{\sigma\sigma'}(\mathbf{p})\bar{c}^\dagger_{-\mathbf{p}\sigma'} + \bar{c}_{-\mathbf{p}\sigma}\left(\Delta^\dagger\right)_{\sigma\sigma'}(\mathbf{p})c_{\mathbf{p}\sigma'}\right], \tag{17.1}$$

where $\bar{c}_{\mathbf{p}\sigma} \equiv i\sigma^y_{\sigma\sigma'}c_{\mathbf{p}\sigma'}$. We use this convention of explicitly factoring out the TR matrix so that singlet, s-wave pairing is $\Delta_{\sigma\sigma'} = \mathbb{I}_{\sigma\sigma'}$, which is the convention most commonly found in the literature of unconventional superconductivity. Because $\Delta_{\sigma\sigma'}$ is a 2×2 matrix, we can decompose it into a scalar piece and a vector piece [76]:

$$\Delta_{\sigma\sigma'}(\mathbf{p}) = d_0(\mathbf{p})\mathbb{I}_{\sigma\sigma'} + d_a(\mathbf{p})\sigma^a_{\sigma\sigma'}. \tag{17.2}$$

To understand the meaning of each term, let us explicitly write out the particle pieces of the pairing terms:

$$\begin{aligned}
\mathbb{I}:&\ \frac{1}{2}d_0(\mathbf{p})\left(c^\dagger_{\mathbf{p}\uparrow}c^\dagger_{-\mathbf{p}\downarrow} - c^\dagger_{\mathbf{p}\downarrow}c^\dagger_{-\mathbf{p}\uparrow}\right),\\
\sigma^x:&\ \frac{1}{2}d_1(\mathbf{p})\left(c^\dagger_{\mathbf{p}\downarrow}c^\dagger_{-\mathbf{p}\downarrow} - c^\dagger_{\mathbf{p}\uparrow}c^\dagger_{-\mathbf{p}\uparrow}\right),\\
\sigma^y:&\ \frac{-i}{2}d_2(\mathbf{p})\left(c^\dagger_{\mathbf{p}\uparrow}c^\dagger_{-\mathbf{p}\uparrow} + c^\dagger_{\mathbf{p}\downarrow}c^\dagger_{-\mathbf{p}\downarrow}\right),\\
\sigma^z:&\ \frac{1}{2}d_3(\mathbf{p})\left(c^\dagger_{\mathbf{p}\uparrow}c^\dagger_{-\mathbf{p}\downarrow} + c^\dagger_{\mathbf{p}\downarrow}c^\dagger_{-\mathbf{p}\uparrow}\right).
\end{aligned} \tag{17.3}$$

It is now clear why we chose the matrix convention: the identity term is the singlet-pairing term and the Pauli matrix terms make up the vector components of triplet pairing. It is interesting to note that Fermi statistics puts constraints on the functions $d_0(\mathbf{p})$, $d_a(\mathbf{p})$. As an example, take

$$\frac{1}{2}\sum_{\mathbf{p}} d_0(\mathbf{p})\left(c^\dagger_{\mathbf{p}\uparrow}c^\dagger_{-\mathbf{p}\downarrow} - c^\dagger_{\mathbf{p}\downarrow}c^\dagger_{-\mathbf{p}\uparrow}\right) = \frac{1}{2}\sum_{\mathbf{p}} d_0(-\mathbf{p})\left(-c^\dagger_{\mathbf{p}\downarrow}c^\dagger_{-\mathbf{p}\uparrow} + c^\dagger_{\mathbf{p}\uparrow}c^\dagger_{-\mathbf{p}\downarrow}\right),$$

which vanishes identically if $d_0(\mathbf{p})$ is an odd function of $\mathbf{p}$. Thus, Fermi statistics implies that the only allowed singlet-pairing terms must have

$$d_0(\mathbf{p}) = d_0(-\mathbf{p}). \tag{17.4}$$

Let us now take one of the triplet terms,

$$\frac{1}{2}\sum_{\mathbf{p}} d_1(\mathbf{p})\left(c^\dagger_{\mathbf{p}\downarrow}c^\dagger_{-\mathbf{p}\downarrow} - c^\dagger_{\mathbf{p}\uparrow}c^\dagger_{-\mathbf{p}\uparrow}\right) = \frac{1}{2}\sum_{\mathbf{p}} d_1(-\mathbf{p})\left(-c^\dagger_{\mathbf{p}\downarrow}c^\dagger_{-\mathbf{p}\downarrow} + c^\dagger_{\mathbf{p}\uparrow}c^\dagger_{-\mathbf{p}\uparrow}\right).$$

This implies that the only nonvanishing term is when $d_1(\mathbf{p})$ is an odd function of $\mathbf{p}$. In fact, all the triplet terms transform the same way, such that

$$d_a(\mathbf{p}) = -d_a(-\mathbf{p}) \tag{17.5}$$

is required. Thus, singlet-pairing terms must have even powers of momentum, and triplet-pairing terms must have odd powers. As we saw in the previous chapter, spinless fermions can also have pairing with odd powers of momentum, which is analogous to a spin-polarized system with triplet pairing.

17.2 Time-Reversal-Invariant Superconductors in Two Dimensions

In the previous chapter we discussed the chiral p-wave superconductor, which is the superconductor relative of the quantum Hall effect. Both states have gapless chiral fermions on the boundary (the fermions on the superconductor are also Majorana fermions) and are classified by integer topological invariants. In light of our earlier discussions on the TR-invariant topological insulators, it is natural to ask if there are TR-invariant topological superconductors in two and three dimensions. We begin with the 2-D case and schematically illustrate the relationship between the QAH effect, chiral superconductors, the QSH effect, and helical superconductors in figure 17.1.

Just as the QSH effect can be thought of as two copies of quantum Hall (one copy for each spin and with the opposite sense of TR breaking for each copy), the helical superconductor can be thought of as two copies of the chiral superconductor. The simplest model Hamiltonian we can write is [77, 78]

$$H = \frac{1}{2}\sum_{\mathbf{p}} \Psi^\dagger_{\mathbf{p}} \begin{pmatrix} \frac{p^2}{2m} - \mu & 0 & 0 & -\Delta(p_x + ip_y) \\ 0 & \frac{p^2}{2m} - \mu & \Delta(p_x - ip_y) & 0 \\ 0 & \Delta^*(p_x + ip_y) & -\frac{p^2}{2m} + \mu & 0 \\ -\Delta^*(p_x - ip_y) & 0 & 0 & -\frac{p^2}{2m} + \mu \end{pmatrix} \Psi_{\mathbf{p}}, \tag{17.6}$$

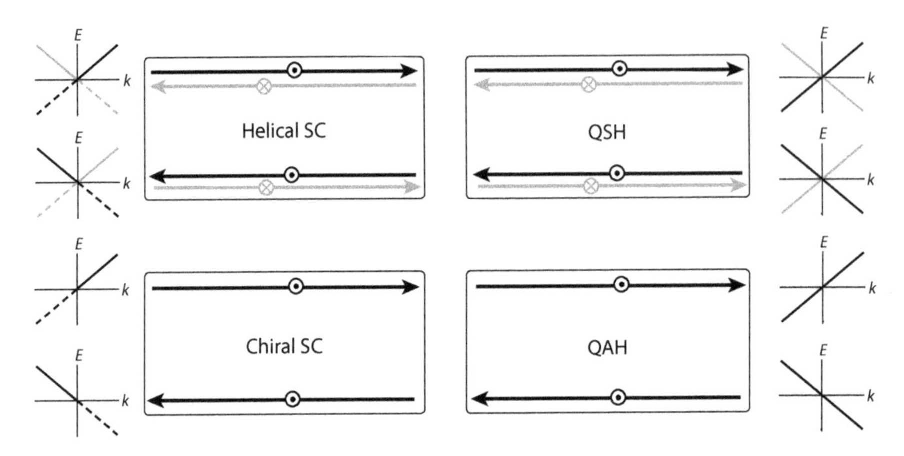

Figure 17.1. Illustration of the relation between 2-D topological superconductors and 2-D topological insulators. In the upper left, we have the TR-invariant topological superconductor with counterpropagating, spin-filtered edge states. The edge states are Majorana fermions. In the upper right, there is the TR-invariant topological insulator with counterpropagating, spin-filtered edge states. The edge states are Dirac fermions. For the lower left, we have the TR-breaking topological superconductor with chiral edge states. The edge states are Majorana-Weyl fermions. Finally, in the lower right, there is the TR-breaking topological insulator with chiral edge states. The edge states are Weyl fermions. In the inset figures, we illustrate the dispersion relations of the various edge states. Dotted lines indicate that these states are not independent of the modes for energies greater than zero.

where $\Psi_{\mathbf{p}} = \begin{pmatrix} c_{\mathbf{p}\uparrow} & c_{\mathbf{p}\downarrow} & -c^{\dagger}_{-\mathbf{p}\downarrow} & c^{\dagger}_{-\mathbf{p}\uparrow} \end{pmatrix}^T$. In the **d**-vector notation, this state has

$$\mathbf{d} = -i\Delta(p_x\hat{y} + p_y\hat{x}). \tag{17.7}$$

The TR operator in this basis is $T = \mathbb{I} \otimes i\sigma^y K$ and the charge-conjugation operator is $C = \tau^y \otimes \sigma^y K$, where τ^a is particle-hole space and σ^a is spin. If each spin had the same chirality, which would break time reversal, then an appropriate d-vector would be $\mathbf{d} = -i\Delta(p_x + ip_y)\hat{y}$, which is a model suggested to be relevant for the pairing state in Sr_2RuO_4. There are no known material candidates for the TR-invariant pairing.

Because this Hamiltonian has the Dirac form, we know immediately from our analysis of the chiral superconductor that when $\mu > 0$, the system will be in a nontrivial topological state with gapless edge modes. On a single edge, there will be a chiral Majorana fermion and an anti-Chiral Majorana fermion with the opposite spin. The edge Hamiltonian is

$$H_{\text{edge}} = \sum_p \Phi_p^{\dagger} \begin{pmatrix} p & 0 \\ 0 & -p \end{pmatrix} \Phi_p = \sum_p \Phi_p^{\dagger} p\sigma^z \Phi_p, \tag{17.8}$$

where $\Phi_p = \begin{pmatrix} \gamma_{p\uparrow} & \gamma_{p\downarrow} \end{pmatrix}$ and the $\gamma_{p\sigma}$ are equal-weight superpositions of particle and hole states with spin σ coming from the solution of equation (17.6) with open boundaries. This Hamiltonian is TR-invariant under $T = i\sigma^y K$. It also satisfies the charge-conjugation symmetry with $C = \mathbb{I}K$. To verify this, one we easily show that

$$TH_{\text{edge}}(p)T^{-1} = p\sigma^y\sigma^z\sigma^y = -p\sigma^z = H_{\text{edge}}(-p), \tag{17.9}$$

where in the last equality we used the fact that $H_{\text{edge}}(p)$ is real. For charge-conjugation,

$$CH_{\text{edge}}(p)C^{-1} = p\sigma^z = -(-p)\sigma^z = -H_{\text{edge}}(-p). \tag{17.10}$$

Time-reversal symmetry is a required symmetry, and the charge-conjugation "symmetry" is an invariance of our use of the BdG formalism in equation (17.6). We can now test to see if we can open a gap in the edge-state spectrum while preserving T (we must preserve C as well to maintain consistency). The possible mass terms we can add to the Hamiltonian are

$$H'_{\text{edge}} = H_{\text{edge}} + m_0\mathbb{I} + m_a\sigma^a \tag{17.11}$$

for constants m_0, m_x, m_y, m_z. Adding the identity term breaks C and is not allowed, even though this term would not open a gap anyway. The m_x, m_y, m_z terms all break T and are thus forbidden. For completeness, we note that the m_x, m_z terms also break C. If we did not require T symmetry, we could just turn on a finite $m_y\sigma^y$ and open a gap. Thus, the robustness of the edge states relies on TR symmetry.

From the edge states, we should also be able to see if the classification should be $\mathbb{Z}$ or $\mathbb{Z}_2$. The following example is simply testing whether $1+1=2$ or $1+1=0$, i.e., if having two identical sets of edge states on a single edge is still stable or if a gap can be opened while preserving T (and C). The Hamiltonian for two sets of edge states is simply

$$H^{(2)}_{\text{edge}}(p) = p(\mathbb{I}\otimes\sigma^z). \tag{17.12}$$

This Hamiltonian satisfies the same symmetries as the single copy. To find a possible mass term, we need a matrix $\mathcal{M}$ that satisfies T and C symmetries as well as $\{\mathcal{M}, \mathbb{I}\otimes\sigma^z\} = 0$. The last condition is specified because a Hamiltonian $H^{(2)}_{\text{edge}}(p) + m\mathcal{M}$ will have the gapped energy spectrum $\pm\sqrt{p^2+m^2}$ if the anti-commutation property holds. We can simply guess the form of $\mathcal{M}$. Let's take

$$\mathcal{M} = \tau^y \otimes \sigma^x, \tag{17.13}$$

where τ^y is a Pauli matrix in the edge-state flavor index. Under T we have

$$Tm\mathcal{M}T^{-1} = m(-\tau^y)\otimes(-\sigma^x) = m\mathcal{M}, \tag{17.14}$$

which is, thus, invariant. Under C we have

$$Cm\mathcal{M}C^{-1} = m\mathcal{M} = -m\mathcal{M}^*. \tag{17.15}$$

Hence, we have found a mass term that is invariant under both symmetries. This indicates that the classification should be $\mathbb{Z}_2$ because two identical copies can cancel. This matches the classification of the QSH insulator; however, we will see that 3-D topological superconductors with T symmetry are classified by integers, unlike their insulator counterparts.

In the limit where the spin-$\uparrow$ and spin-$\downarrow$ blocks remain decoupled, we can define two winding numbers, $N_\uparrow$ and $N_\downarrow$. Because time reversal is preserved, we always find that $N_\uparrow + N_\downarrow = 0$. However, in this decoupled limit, we can define a $\mathbb{Z}_2$ index by

$$\nu_{Z_2} = \frac{1}{2}(N_\uparrow - N\downarrow) \mod 2. \tag{17.16}$$

The winding number, when defined, indicates the number of branches of edge modes. If $\nu_{Z_2} = 0$, there is an even number of *pairs* of edge states, and we showed before that this means there is no generic robustness of the edge modes. When $\nu_{Z_2} = 1$, there is always one pair of edge modes that is stable. We can couple the different spin blocks via s-wave singlet pairing—i.e., a nonzero d_0. As long as the pairing is weak enough to leave the bulk gap open, the topological character of the state is unchanged. More-generic bulk topological invariants that do not require decoupled spin blocks, have been defined in the literature, but we will not discuss these quantities here.

17.2.1 Vortices in 2-D Time-Reversal-Invariant Superconductors

The form of the order parameter we have been using is

$$\Delta = \begin{pmatrix} 0 & -\Delta(p_x + ip_y) \\ \Delta(p_x - ip_y) & 0 \end{pmatrix}. \tag{17.17}$$

We want to consider two different types of vortex defects, both of which can be written in the form

$$\Delta_{\text{vortex}} = \begin{pmatrix} 0 & -\Delta e^{i\phi_\uparrow}(p_x + ip_y) \\ \Delta e^{i\phi_\downarrow}(p_x - ip_y) & 0 \end{pmatrix}. \tag{17.18}$$

Vortices with the form $\phi_\uparrow = \theta(r)$, $\phi_\downarrow = 0$ (and vice versa) are felt by only one spin component [79]. From our study of the spinless chiral superconductor, we know that vortices of this form will contain a single, spin-polarized Majorana bound state, which leads to non-Abelian statistics.

Another interesting vortex is the TR-invariant vortex with $\phi_\uparrow = -\phi_\downarrow = \theta(r)$ [78, 79]. Note that we are still considering a simplified model with conserved S_z. This type of vortex will contain a Kramers' pair of Majorana bound states, $\gamma_\uparrow, \gamma_\downarrow$. For the entire system to preserve TR symmetry, there must exist another Kramers' pair of modes $(a_\uparrow, a_\downarrow)$ localized on another vortex or on the boundary. Thus, there will be four Majorana modes in total, which give rise to a single Kramers' pair of complex fermions. We can make the combinations

$$\begin{aligned} \psi_\uparrow &= \frac{1}{2}(\gamma_\uparrow + ia_\uparrow), & \psi_\uparrow^\dagger &= \frac{1}{2}(\gamma_\uparrow - ia_\uparrow), \\ \psi_\downarrow &= \frac{1}{2}(\gamma_\downarrow + ia_\downarrow), & \psi_\downarrow^\dagger &= \frac{1}{2}(\gamma_\downarrow - ia_\downarrow). \end{aligned} \tag{17.19}$$

Under time reversal, we know that $\psi_\uparrow \to -\psi_\downarrow$ and $\psi_\downarrow \to \psi_\uparrow$, which implies that

$$\begin{aligned} \gamma_\uparrow &\to -\gamma_\downarrow, & \gamma_\downarrow &\to \gamma_\uparrow, \\ a_\uparrow &\to -a_\downarrow, & a_\downarrow &\to a_\uparrow. \end{aligned} \tag{17.20}$$

On a single vortex there is an interesting consequence of the TR symmetry [78]. Instead of the ψ fermions described in equation (17.19), let us define

$$\begin{aligned} \phi_1 &= \frac{1}{2}(\gamma_\uparrow + i\gamma_\downarrow), \\ \phi_1^\dagger &= \frac{1}{2}(\gamma_\uparrow - i\gamma_\downarrow), \\ \phi_2 &= \frac{1}{2}(a_\uparrow + ia_\downarrow), \\ \phi_2^\dagger &= \frac{1}{2}(a_\uparrow - ia_\downarrow), \end{aligned} \tag{17.21}$$

which is a local fermion on each vortex. Now consider the following operators

$$
\begin{aligned}
i\gamma_\uparrow\gamma_\downarrow &= 2\phi_1^\dagger\phi_1 - 1, \\
ia_\uparrow a_\downarrow &= 2\phi_2^\dagger\phi_2 - 1.
\end{aligned} \tag{17.22}
$$

We see that these two operators are related to the local fermion occupation number on each vortex. Acting with time reversal, we find, for example,

$$
T(i\gamma_\uparrow\gamma_\downarrow)T^{-1} = (-i)(-\gamma_\downarrow)(\gamma_\uparrow) = -i\gamma_\uparrow\gamma_\downarrow; \tag{17.23}
$$

thus, time reversal changes the local fermion occupation from one to zero, or vice versa. It thus acts to locally change fermion parity. Of course, time reversal acting on the entire system will flip the parities of both vortices, leaving the whole system with the same fermion parity with which it started. Unfortunately, we cannot use the TR-invariant vortices to establish a topological quantum computation architecture because there are two Majorana modes on each vortex; thus, we can locally couple to a single vortex.

17.3 Time-Reversal-Invariant Superconductors in Three Dimensions

With our understanding of 3-D TR-invariant topological insulators and our recent considerations for topological superconductors, we can immediately guess the form of a BdG Hamiltonian for a 3-D topological superconductor:

$$
H_{\text{BdG}} = \frac{1}{2}\sum_{\mathbf{p}} \Psi_{\mathbf{p}}^\dagger \begin{pmatrix} \frac{p^2}{2m} - \mu & 0 & |\Delta|p_z & |\Delta|p_- \\ 0 & \frac{p^2}{2m} - \mu & |\Delta|p_+ & -|\Delta|p_z \\ |\Delta|p_z & |\Delta|p_- & -\frac{p^2}{2m} + \mu & 0 \\ |\Delta|p_+ & -|\Delta|p_z & 0 & -\frac{p^2}{2m} + \mu \end{pmatrix} \Psi_{\mathbf{p}}, \tag{17.24}
$$

where $p_\pm = p_x \pm ip_y$, $\Psi_{\mathbf{p}} = (e^{-i\theta/2}c_{\mathbf{p}\uparrow} \quad e^{-i\theta/2}c_{\mathbf{p}\downarrow} \quad -e^{i\theta/2}c_{-\mathbf{p}\downarrow}^\dagger \quad e^{i\theta/2}c_{-\mathbf{p}\uparrow}^\dagger)^T$, and we have gauged away the phase θ of the order parameter from the Hamiltonian matrix into the fermion operators. Although no known realistic superconductors have this type of Hamiltonian, the B-phase of helium-3 exhibits such a *superfluid* state [11, 76, 78, 79, 80]. As a function of the chemical potential μ, this system exhibits a phase transition between the strong-pairing, trivial superfluid state ($\mu < 0$) and the weak-pairing, topological superfluid state ($\mu > 0$). If we compare the form of the matrix $H_{\text{BdG}}(\mathbf{p})$ to the Bloch Hamiltonian for a 3-D TR-invariant topological insulator, we immediately see the analogy. The order parameter $|\Delta|$ serves the same purpose as the Fermi velocity or speed of light, for the insulator case, and the chemical potential μ in the superfluid represents the mass parameter in the insulator case.

Using the analogy with the topological insulator, we know that in the topological phase there will be gapless fermionic modes on the boundary that are protected from disorder. Like all topological superconductors and superfluids, the boundary states will be neutral and have a Majorana character. The surface-state-Hamiltonian for the 3-D topological superconductor, with a surface normal to the z-axis, is

$$H_{\rm surf} = \sum_{\mathbf{p}} \Phi_{\mathbf{p}}^{\dagger} \left(p_x \sigma^x + p_y \sigma^y \right) \Phi_{\mathbf{p}}, \tag{17.25}$$

where $\Phi_{\mathbf{p}} = \left(\gamma_{\mathbf{p}\uparrow} \; \gamma_{\mathbf{p}\downarrow} \right)^T$ and $\gamma_{\mathbf{p}\sigma}$ satisfy $\gamma_{\mathbf{p}\sigma}^{\dagger} = \gamma_{-\mathbf{p}\sigma}$. The surface states are thus propagating (nonchiral) Majorana fermions. The surface Hamiltonian has two symmetries, time reversal, with $T = i\sigma^y K$, and charge conjugation, with $C = -i\sigma^x K$. We have already shown that $p_x\sigma^x$ and $p_y\sigma^y$ are T invariant, so let us quickly show they are C invariant. The Hamiltonian must satisfy $\sigma^x \left(p_x\sigma^x + p_y\sigma^y \right) \sigma^x = -(-p_x\sigma^{x*} - p_y\sigma^{y*})$, which it clearly does. Because the Hamiltonian satisfies both T and C, we can define a chiral-symmetry operator $\chi = CT = i\sigma^z$. A Hamiltonian has a chiral symmetry if

$$\chi H(\mathbf{p}) \chi^{-1} = -H(\mathbf{p}). \tag{17.26}$$

Clearly, $H_{\rm surf}$ has this property and thus is chiral symmetric.

If we were to naively classify the topological states coming from the Bloch matrix part of equation (17.24), we might say that there should be a Z_2 classification of the topological superconductor states, based purely on the analogy with the topological insulator case. After all, the primary difference in the matrix Hamiltonians for the superconductor and insulator case is just the addition of a strict charge-conjugation symmetry. Interestingly, in this case it turns out that the additional C symmetry strengthens the classification from Z_2 to $\mathbb{Z}$ [11, 80]. The reason the classification changes can be seen two ways: (1) from an understanding of the surface-state stability and (2) from a bulk topological invariant. We will cover both these cases and begin with the surface-state picture because we have already introduced the Hamiltonian. The test of whether the surface states are Z_2 or $\mathbb{Z}$ stable was mentioned in the section on 2-D TR-invariant superconductors: namely, we want to see if two copies of the surface states add up together or cancel out. Before we perform this test, let us comment on the stability of a single copy. For a single copy, we could try to destroy the surface states by opening a gap with the perturbation $H' = m_z\sigma^z$. However, we can easily see that this term breaks, for example, χ, and is not allowed:

$$\chi m_z \sigma^z \chi^{-1} = m_z\sigma^z \neq -m_z\sigma^z = -H(\mathbf{p}). \tag{17.27}$$

Thus a single copy of the surface states is stable to perturbations that preserve the required symmetries T, C and the induced symmetry χ.

Now let us consider two identical copies of the surface-state Hamiltonian

$$H_{\rm surf}^{(2)}(\mathbf{p}) = p_x \mathbb{I} \otimes \sigma^x + p_y \mathbb{I} \otimes \sigma^y. \tag{17.28}$$

To find a mass term, we must find a matrix $\mathcal{M}$ that anticommutes with $H_{\rm surf}^{(2)}$ and preserves all the symmetries. To anticommute, we must have $\mathcal{M} = X \otimes \sigma^z$ for a 2×2 matrix X. We can easily see that any matrix of this form breaks the chiral symmetry $\chi = \mathbb{I} \otimes i\sigma^z$ and thus is not allowed. So, we cannot find any allowable mass terms to open a gap, and two such surface-state copies are stable. The only way to allow for a gap opening is to add a surface state with the opposite chirality, where by chirality we mean the Fermi-surface Berry phase for the massless fermion surface states. The Berry phases are quantized to be $\pm\pi$, and the sign determines the chirality. If we began instead with the Hamiltonian

$$H_{\rm surf}^{(2')}(\mathbf{p}) = p_x \tau^z \otimes \sigma^x + p_y \mathbb{I} \otimes \sigma^y, \tag{17.29}$$

we could look for mass terms of the form $\mathcal{M} = \tau^y \otimes \sigma^x$. This mass term satisfies $T = \mathbb{I} \otimes i\sigma^y K$, $C = -i\mathbb{I} \otimes \sigma^x K$, and $\chi = \mathbb{I} \otimes i\sigma^z$ symmetries and opens a gap. Thus an antichiral state can cancel the chiral state.

These surface-state arguments indicate that the classification should be characterized by a topological integer, and Schnyder, Ryu, Furusaki, and Ludwig introduced just such a quantity [11]:

$$N_w = \frac{1}{24\pi^2} \int d^3 p \epsilon^{\mu\nu\rho} \mathrm{Tr}\left[\left(q^{-1}(\mathbf{p})\partial_\mu q(\mathbf{p})\right) \left(q^{-1}(\mathbf{p})\partial_\nu q(\mathbf{p})\right) \left(q^{-1}(\mathbf{p})\partial_\rho q(\mathbf{p})\right) \right] \tag{17.30}$$

where $q(\mathbf{p})$ is a special matrix projection operator that will be defined shortly. Calculating this invariant involves several steps. Given a Bloch Hamiltonian $H_{\mathrm{BdG}}(\mathbf{p})$, we first need to calculate the occupied Bloch wavefunctions. As an explicit example, we will use the Hamiltonian given in equation (17.24). The occupied quasi-particle Bloch functions for this Hamiltonian are

$$|u_1(\mathbf{p})\rangle = \frac{1}{\sqrt{2E(p)(E(p)+M(p))}} \begin{pmatrix} -p_- \\ p_z \\ 0 \\ E(p)+M(p) \end{pmatrix}, \tag{17.31}$$

$$|u_2(\mathbf{p})\rangle = \frac{1}{\sqrt{2E(p)(E(p)+M(p))}} \begin{pmatrix} -p_z \\ -p_+ \\ E(p)+M(p) \\ 0 \end{pmatrix}, \tag{17.32}$$

where $M(p) = \frac{p^2}{2m} - \mu$ and $E(p) = \sqrt{|\Delta|^2 p^2 + M^2(p)}$. To find $q(\mathbf{p})$, we will need the projection operator onto the occupied states:

$$P(\mathbf{p}) = |u_1(\mathbf{p})\rangle\langle u_1(\mathbf{p})| + |u_2(\mathbf{p})\rangle\langle u_2(\mathbf{p})| \tag{17.33}$$

$$= \frac{1}{2E(p)} \begin{pmatrix} \frac{|\Delta|^2 p^2}{E(p)+M(p)}\mathbb{I} & -|\Delta|\mathbf{p}\cdot\sigma \\ -|\Delta|\mathbf{p}\cdot\sigma & (E(p)+M(p))\mathbb{I} \end{pmatrix}. \tag{17.34}$$

We can now form the combination

$$Q(\mathbf{p}) \equiv 2P(\mathbf{p}) - 1 = \frac{1}{E(p)} \begin{pmatrix} -M(p)\mathbb{I} & -|\Delta|\mathbf{p}\cdot\sigma \\ -|\Delta|\mathbf{p}\cdot\sigma & M(p)\mathbb{I} \end{pmatrix}$$

$$= -\frac{1}{E(p)} \left(|\Delta| p_i \left(\tau^x \otimes \sigma^i\right) + M(p)\left(\tau^z \otimes \mathbb{I}\right) \right). \tag{17.35}$$

By definition, to extract the necessary $q(\mathbf{p})$ matrix, we need to transform $Q(\mathbf{p})$ into block off-diagonal form [11], which we do by performing a basis change, essentially rotating $Q(\mathbf{p})$ around the τ^x-axis to send $\tau^x \to \tau^x$ and $\tau^z \to \tau^y$, leaving us with

$$Q(\mathbf{p}) = -\frac{1}{E(p)} \begin{pmatrix} 0 & |\Delta|\mathbf{p}\cdot\sigma - iM(p)\mathbb{I} \\ |\Delta|\mathbf{p}\cdot\sigma + iM(p)\mathbb{I} & 0 \end{pmatrix}$$

$$\equiv \begin{pmatrix} 0 & q(\mathbf{p}) \\ q^\dagger(\mathbf{p}) & 0 \end{pmatrix}. \tag{17.36}$$

Thus,

$$q(\mathbf{p}) = -\frac{1}{E(p)} (|\Delta|\mathbf{p} \cdot \sigma - iM(p)\mathbb{I}) . \tag{17.37}$$

We know that $Q(\mathbf{p})^2 = 4P(\mathbf{p})^2 - 4P(\mathbf{p}) + 1 = 4P(\mathbf{p}) - 4P(\mathbf{p}) + 1 = 1$; thus, $q^\dagger(\mathbf{p})q(\mathbf{p}) = q(\mathbf{p})q^\dagger(\mathbf{p}) = 1$, so $q(\mathbf{p})$ is unitary. The topological information of $H_{\text{BdG}}(\mathbf{p})$ is stored in $q(\mathbf{p})$, which is a map from $(p_x, p_y, p_z) \to U(2)$ in this case. If we take the limit as $\mathbf{p} \to \infty$ in any direction, $q(\mathbf{p})$ has a unique limit: $\lim_{\mathbf{p}\to\infty} q(\mathbf{p}) = -i\mathbb{I}$. Thus, we can one-point compactify momentum space to get $\mathbb{R}^3 \cup \{\infty\} \equiv S^3$. Then, all the topological information is contained in the map from compactified momentum space to $U(2)$, i.e., $S^3 \to U(2)$. This set of maps is classified by the homotopy group $\pi_3(U(2)) = \mathbb{Z}$. The topological invariant that distinguishes these classes was already listed in equation (17.30). Although the existence of an integer-valued index is necessary for an integer classification, it is not sufficient. In principle, we should also carry out an analysis of the properties of equation (17.30) under symmetry-preserving gauge transformations to show that the class cannot be modified. For example, as we saw in chapter 14, the $\mathbb{Z}_2$ topological invariant for 3-D topological insulators can be cast in the form of an integer-valued winding number. However, the number can be modified by gauge transformations and is well defined only as a $\mathbb{Z}_2$ quantity. We leave the analysis for the 3-D superconductor case as an exercise for the interested reader.

Plugging the explicit $q(\mathbf{p})$ matrix into equation (17.30), we find, after tedious algebra,

$$\begin{aligned} N_w &= \frac{1}{24\pi^2} \int d^3p \frac{12\left(\mu + \frac{p^2}{2m}\right)}{E^4(p)} \\ &= \frac{2}{\pi} \int_0^\infty dp \frac{p^2\left(\mu + \frac{p^2}{2m}\right)}{E^4(p)} . \end{aligned} \tag{17.38}$$

This integral is quantized and always yields an integer. It can be carried out numerically, and we find that for $\mu < 0$, $N_w = 0$, and for $\mu > 0$, $N_w = -1$, as we expected. The winding number N_w gives a bulk integer classification of 3-D TR-invariant superconductors and superfluids.

17.4 Finishing the Classification of Time-Reversal-Invariant Superconductors

The 2-D and 3-D examples of TR-invariant topological superconductors (with $T^2 = -1$) were illustrated via the analogy with the corresponding topological insulator in those dimensions. We have seen one clear distinction between insulators and superconductors, which is that, in three dimensions, the topological classification is integer instead of Z_2. The two TR-invariant topological superconductors we discussed in this chapter belong to the same symmetry family, denoted DIII [11, 81], whereas the TR-invariant insulators are in the family AII. The two families have the following classification table:

Dim	AII	DIII
1	0	Z_2
2	Z_2	Z_2
3	Z_2	$\mathbb{Z}$

Although we have seen examples of four of these classes, we have not discussed the 1-D class in the DIII family. This superconductor will be a 1-D chain with TR symmetry.

The simplest nontrivial example we can give is two copies of the 1-D p-wave chain, one for each spin:

$$H_{\text{BdG}} = \sum_{j,\sigma} \left[-t \left(c^\dagger_{j\sigma} c_{j+1\sigma} + c^\dagger_{j+1\sigma} c_{j\sigma} \right) - \mu c^\dagger_{j\sigma} c_{j\sigma} + |\Delta| \left(c^\dagger_{j+1\sigma} c^\dagger_{j\sigma} + c_{j\sigma} c_{j+1\sigma} \right) \right]. \quad (17.39)$$

From our study of the 1-D p-wave chain, we know that this Hamiltonian will have two nontrivial phases as a function of μ. When $\mu > 0$, the chain will be in a topological phase with a Kramers' pair of Majorana modes, on each end of an open chain. Let $a_\uparrow$, $a_\downarrow$, $b_\uparrow$, $b_\downarrow$ be the four Majorana modes, with the a modes on one end and the b modes on the other end. We want to see how time reversal acts on the Majorana modes. To do this, we will see how time reversal acts on the complex fermion states formed from these Majorana modes:

$$\psi_\uparrow = \frac{1}{2} \left(a_\uparrow + i b_\uparrow \right), \quad (17.40)$$

$$\psi_\downarrow = \frac{1}{2} \left(a_\downarrow + i b_\downarrow \right). \quad (17.41)$$

We know that under T, $\psi_\uparrow \to -\psi_\downarrow$ and $\psi_\downarrow \to \psi_\uparrow$. From this transformation, we can see that under T,

$$a_\uparrow \to -a_\downarrow,$$
$$a_\downarrow \to a_\uparrow,$$
$$b_\uparrow \to -b_\downarrow,$$
$$b_\downarrow \to b_\uparrow,$$

which are the same transformations as the T-invariant modes trapped on the T-invariant defects of the 2-D TR-invariant topological superconductor (see equation (17.20)). Now, unlike the single p-wave chain, because we have two Majorana modes on each end, we can add a local perturbation to each end of the chain to lift the degeneracy and destroy the stability of the end states:

$$H_{\text{pert}} = \lambda \left(i a_\uparrow a_\downarrow + i b_\uparrow b_\downarrow \right). \quad (17.42)$$

But, TR symmetry saves us because

$$\begin{aligned} T H_{\text{pert}} T^{-1} &= \lambda \left(-i(-a_\downarrow) a_\uparrow - i(-b_\downarrow) b_\uparrow \right) \\ &= -H_{\text{pert}}, \end{aligned} \quad (17.43)$$

and thus H_{pert} is odd under time reversal and is forbidden.

To see the Z_2 nature of the classification, we can add an extra Kramers' pair of Majorana modes on each end and see if we can lift the end-state degeneracy while preserving T. Let us focus on the end with the a-modes, which will contain an extra set of modes: $\bar{a}_\uparrow$, $\bar{a}_\downarrow$. Now we can add the perturbation

$$H_{\text{pert}} = \lambda \left(i a_\uparrow \bar{a}_\uparrow - i a_\downarrow \bar{a}_\downarrow \right). \quad (17.44)$$

Acting with T, we find

$$TH_{\text{pert}}T^{-1} = \lambda\left(-i(-a_{\downarrow})(-\bar{a}_{\downarrow}) + ia_{\uparrow}(\bar{a}_{\uparrow})\right)$$
$$= +H_{\text{pert}}. \tag{17.45}$$

For this case we see that this Hamiltonian perturbation is T-invariant, and it pushes the energies to $\pm|\lambda|$. Thus, the classification is Z_2. If this were an insulator Hamiltonian instead of a superconductor Hamiltonian, then we could relax the charge-conjugation symmetry. In this case, even without symmetrically splitting the modes, we could push a single pair of modes away from zero energy while preserving T by adding a potential to the end of the wire. This is why the classification for the *insulator* (family AII) does not have a nontrivial state according to the standard classification table.

17.5 Problems

1. *TR Symmetry and Superconductivity:* Consider a lattice with a single site that has a single orbital with a spin-$\frac{1}{2}$ degree of freedom. We have the fermion annihilation operators: $c_{\uparrow}$, $c_{\downarrow}$. For this pathological case, we can write a spin-singlet "pairing-term"

$$H_{SC} = \Delta\left(c_{\uparrow}^{\dagger}c_{\downarrow}^{\dagger} - c_{\downarrow}^{\dagger}c_{\uparrow}^{\dagger}\right) + \text{h.c.} \tag{17.46}$$

Under time reversal with $T = i\sigma^{y}K$, where K is complex conjugation and σ^{y} is a Pauli spin-$\frac{1}{2}$ matrix, the annihilation operators transform as $Tc_{\uparrow}T^{-1} = c_{\downarrow}$ and $Tc_{\downarrow}T^{-1} = -c_{\uparrow}$. First, show that $TH_{\text{SC}}T^{-1} = H_{\text{SC}}$ and is thus TR-invariant. Now perform a $U(1)$ gauge transformation $c_{\sigma} \to e^{i\theta}c_{\sigma}$ and calculate how H_{SC} transforms. Show that unless $2\theta = n\pi$, where n is an integer, that $TH_{\text{SC}}(\theta)T^{-1}$ does not appear to be TR-invariant. Show that a new T operator, $T = U(\theta)K$, can be constructed that satisfies $T^2 = -1$ such that $T(\theta)H_{\text{SC}}(\theta)T^{-1}(\theta) = H_{\text{SC}}(\theta)$ and is, thus, invariant.

2. *Particle Hole and TR Symmetry:* Suppose we also have a TR operator $T = UK$, where U is a unitary matrix and K is complex conjugation. If $H(\mathbf{p})$ is to be TR-invariant, it must satisfy $TH(\mathbf{p})T^{-1} = H(-\mathbf{p})$ or, equivalently, $UH^{*}(\mathbf{p})U^{T} = H(-\mathbf{p})$. Suppose that we have

$$H(\mathbf{p}) = \epsilon(\mathbf{p})\mathbb{I}_{2\times 2} + d_a(\mathbf{p})\sigma^{a}, \tag{17.47}$$

where σ^{a} are the Pauli matrices representing spin and $\epsilon(\mathbf{p})$, $d_a(\mathbf{p})$ are generic functions of $\mathbf{p}$. Let $T = i\sigma^{y}K$. Find the constraints on $\epsilon(\mathbf{p})$ and $d_a(\mathbf{p})$ as functions of $\mathbf{p}$ if $H(\mathbf{p})$ is TR and particle-hole symmetric with (a) $C = -i\sigma^{x}$ and (b) $C = \mathbb{I}_{2\times 2}$. For (a), prove that the highest dimension in which we can write a Dirac-like kinetic term of the form $H(\mathbf{p}) = p_a\sigma^{a}$ is $2+1$. This is an indicator that no TR-invariant topological superconductor in class DIII can exist in $4+1$ dimensions.

3. *Symmetry Protection of Majorana End States:* In $(1+1)$-D class D superconductors, which are protected by no symmetries, exhibit a Z_2 classification. Let us illustrate this stability by focusing on the Majorana fermion bound states at the end of the p-wave wire. If we begin with a single Majorana mode γ_1, then the state is absolutely stable because we cannot eliminate it via any local perturbation. If we add a second exact copy, γ_2, show that the term

$H = i\Delta\gamma_1\gamma_2$, where Δ is a real number, has an energy gap and no states at zero energy if $\Delta \neq 0$. Argue that if we add a third Majorana, γ_3, at least one mode must remain at zero energy.

We would now like to add an additional symmetry so that the classification is increased from Z_2 to Z. Show that if we require TR symmetry T with $T^2 = +1$, i.e., $T = UK$, where $UU^* = +1$, then all terms of the form $i\gamma_i\gamma_j$ are forbidden, and thus the spectrum remains gapless on the end of the wire no matter how many copies we add. The addition of the TR symmetry with $T^2 = +1$ converts the symmetry class to BDI, which indeed has an integer classification.

18

Superconductivity and Magnetism in Proximity to Topological Insulator Surfaces

BY TAYLOR L. HUGHES

The basic models for all topological insulators and superconductors are massive Dirac Hamiltonians. Using the basic principle that mass-domain walls in a Dirac Hamiltonian lead to gapless fermion states, we are immediately led to the reason for the existence of the topological boundary states on nontrivial insulators and superconductors. One of the primary motivations in the field is to find real material examples of such states; so far, the success in finding nontrivial topological insulators has outpaced the search for nontrivial superconductors. There are two paradigms in the search for topological superconductors.

1. Take a material with a simple band structure, e.g. spinless fermions or helium-3 (^{3}He), which have the energy spectrum $\frac{p^2}{2m}$, and add momentum dependent pairing terms that are topological in nature. Up to now these are the only types of examples we have considered. The chiral p-wave superconductor and the ^{3}He-B phase are in this category.
2. Take a material with a "topological" band structure and add simple superconducting pairing terms. This is an approach that has risen in recent years.

It is the second method that is the focus of this chapter. We will show novel ways to generate states that mimic topological insulators and superconductors by using the Dirac bandstructure on the *boundary* of a TR-invariant topological insulator.

18.1 Generating 1-D Topological Insulators and Superconductors on the Edge of the Quantum-Spin Hall Effect

We want to consider the 2-D topological insulator in the QSH state. The bulk is gapped and completely inert for our purposes. Our focus is on the edge which has the Hamiltonian

$$H_{\text{edge}} = \sum_p p c^\dagger_{p\sigma} \sigma^z_{\sigma\sigma'} c_{p\sigma'}. \tag{18.1}$$

This is a 1-D Dirac Hamiltonian, and as long as TR symmetry is preserved, the edge will remain metallic and gapless. But in fact, we know that to create topological states, we just need to open a gap in the Dirac Hamiltonian the right way. One way we can open a gap is to apply a T-breaking field that will induce perturbations

$$H_T = m_a \sigma^a, \tag{18.2}$$

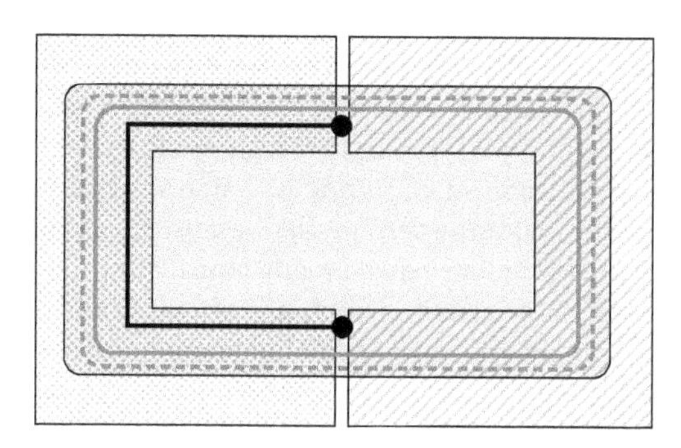

Figure 18.1. A top-down view of a QSH effect (rounded rectangle) covered by two different magnets (U-shapes). Places where there is a proximity coupling between the two different magnets and the QSH effect are shaded with slashes and dots respectively. The solid and dashed gray lines represent the spin-up and spin-down counter-propagating helical edge modes. The black dots show the points at which low-energy fermion-bound states are trapped between the two magnets at the domain wall. The black line connecting the black dots suggests that one can think about this geometry as mimicking a 1-D topological insulator or superconductor with topological end-states located at the black dots.

$a = x, y, z$, on the edge. There is one additional way to open a gap in the edge states, by breaking a symmetry that we have been implicitly assuming the whole time in our discussion of topological *insulators*: charge-*conservation* symmetry. Namely, we can also induce a superconducting gap in the edge by adding an s-wave pairing potential. The generic Hamiltonian for dealing with both types of mass terms is a BdG Hamiltonian:

$$H_{\text{edge}}^{(\text{BdG})} = \frac{1}{2} \sum_p \Psi_p^\dagger \begin{pmatrix} p\sigma^z + m_x\sigma^x - \mu & |\Delta| i\sigma^y \\ -|\Delta| i\sigma^y & p\sigma^z - m_x\sigma^x + \mu \end{pmatrix} \Psi_p, \tag{18.3}$$

where $\Psi_p = \begin{pmatrix} c_{p\uparrow} & c_{p\downarrow} & c_{-p\uparrow}^\dagger & c_{-p\downarrow}^\dagger \end{pmatrix}$, the c , $c^\dagger$ operators have been gauge-transformed to absorb the phase of Δ, and, for simplicity, we have kept only one magnetic perturbation, $m_x\sigma^x$.

To clearly understand the physics, we will study this model in several limits and choose $\mu = 0$ to simplify the picture. The spectrum of $H_{\text{edge}}^{\text{BdG}}$ is

$$\begin{aligned} E_1 &= \sqrt{p^2 + (|\Delta| + m_x)^2}, \\ E_2 &= \sqrt{p^2 + (|\Delta| - m_x)^2}, \\ E_3 &= -\sqrt{p^2 + (|\Delta| + m_x)^2}, \\ E_4 &= -\sqrt{p^2 + (|\Delta| - m_x)^2}, \end{aligned} \tag{18.4}$$

which is just two copies of the massive Dirac spectrum with masses given by $|\Delta| \pm m_x$. Now that we have been given this type of spectrum, we would like to do something interesting with it. The basic idea of which we will take advantage is that mass domain walls in the 1-D Dirac Hamiltonian trap fermion-bound states. By creating mass domain walls on the edge of QSH, we can mimic the properties of different 1-D topological states. Suppose we have the geometry shown in figure 18.1. The edge of the system is a closed curve, and we have set up two regions with different materials lying on top of the edge region. Now we will engineer three different 1-D topological states by creating domain walls.

First, take each material to be magnetic and have a T-breaking proximity coupling into the edge. In this limit $|\Delta| \equiv 0$, and there can be an induced magnetic mass m_1 and m_2 in the two regions. If we find a way to engineer the magnets such that in one region the induced mass is m_1 and in the other region it is $-\text{sign}(m_1)|m_2|$, then on the two domain walls there will be trapped fermionic modes (note that only the sign needs to be opposite; the magnitudes do not need to be exactly the same). This system has some of the same features as the 1-D topological insulator in class D, which has a Z_2 classification linked to the number of complex fermion zero modes located on the ends of the wire. Here, the charge-conjugation symmetry C required for this class would be fine-tuned although we can still predict measurable consequences of this state [82].

The next limit we can take is when $m_x \equiv 0$ and the edge is covered by two different s-wave superconductors. By the proximity effect, an s-wave pairing potential can be induced in the edge states, producing superconducting gaps $|\Delta_1|$ and $|\Delta_2|$ in the two regions. If these can be adjusted so that there is a π phase shift between them—i.e., $\text{sign}\Delta_1 = -\text{sign}\Delta_2$—then on each domain wall there will exist a Kramers' pair of Majorana zero modes because the system remains TR invariant as the order parameter is real [83]. We have already considered such a 1-D topological superconductor state: the doubled p-wave chain, which is in class DIII and has a Z_2 classification. Thus, π-shifted superconductors on the edge of QSH produce a state that mimics the (yet unobserved) TR invariant 1-D topological superconductor.

The most interesting case is the case where one material is a magnet and the other is a superconductor. In this case, we have two separate massive fermions, one with mass $|\Delta| - m_x$ and the other with $|\Delta| + m_x$. Thus, there is a mass domain wall when, for example, $|\Delta| - m_x$ jumps from positive to negative. This occurs when the magnitude of m_x becomes equal to $|\Delta|$. If we simply have a region of magnet next to a region of superconductor, then somewhere between them, as the magnitudes of $|\Delta|$ and m_x drop toward zero, there will be a domain wall and a domain-wall bound state [83].

As mentioned earlier, the number of degrees of freedom of the entire superconducting edge Hamiltonian is equivalent to that of the doubled p-wave wire, which is a superconducting wire with two spin components. However, no matter what the initial sign of m_x is, only half of the degrees of freedom see the mass domain wall. As an explicit case, let $m_x > 0$ for $x > 0$ and $|\Delta| > 0$ for $x < 0$. In this case, the only bands that see the domain wall are the E_2 and E_4 bands. This means that only the fermion masses for the E_2 and E_4 bands switch sign. This generates one fermionic bound state at the interface of the superconductor and the magnet. Because we are in the BdG formalism, only one of E_2 and E_4 should be considered as an independent state, and thus the fermion bound state is Majorana in nature. Thus, this geometry mimics the 1-D p-wave wire with unpaired Majorana bound states. If we can construct complicated magneto-superconductor junction structures, it may be possible to realize a topological quantum computation architecture in 2-D topological insulators because these Majorana bound states will have non-Abelian statistics.

18.2 Constructing Topological States from Interfaces on the Boundary of Topological Insulators

The previous section illustrated how to mimic 1-D topological insulators and superconductors from the edge of the QSH insulator. In this section, we will show how to construct 2-D topological insulators and superconductors from the surface of a 3-D strong topological insulator. For simplicity, we will focus on a single surface with a surface normal $\hat{n} = \hat{z}$ and assume that the nontrivial TR-invariant topological insulator has one set of gapless surface modes located near $\mathbf{p} = 0$ in the surface BZ. The low-energy surface-state Hamiltonian is

$$H_{\text{surf}} = v_F (\mathbf{p} \times \sigma) \cdot \hat{n} = v_F \left(p_x \sigma^y - p_y \sigma^x\right), \tag{18.5}$$

where σ^a represents spin. This Hamiltonian is a massless Dirac fermion with TR symmetry.

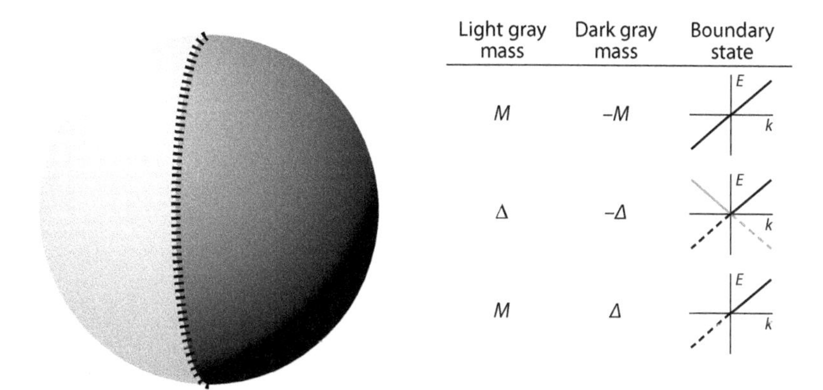

Figure 18.2. An illustration of a spherical ball of a 3-D topological insulator with two materials (light and dark) covering the eastern and western hemispheres of the surface. These materials can be ferromagnetic or *s*-wave superconducting layers. The dashed longitudinal line represents a domain wall on the surface between the two regions. The table on the right summarizes the types of gapless domain-wall states that are present when two different materials are placed adjacent to one another. For the magnetic case, a $\pm M$ domain wall yields chiral modes localized on the domain wall. A π-phase junction between two s-wave pairing potentials with order parameters $\pm|\Delta|$ yields counterpropagating Majorana modes. Finally, a domain between a magnetic region and an s-wave superconductor region gives a chiral-Majorana mode. The dotted lines in the inset dispersion plots indicate the Majorana nature, i.e., that the particle and antiparticle states are not independent. These three different cases mimic an integer quantum Hall state, a TR-invariant topological superconductor, and a chiral topological superconductor, respectively.

With the gapless Dirac spectrum on the edge of the QSH effect, we have just seen that we can use gap-opening perturbations to create topological states. Here we will use the same procedure. Let us begin with applying a TR-breaking field to the surface, e.g., a proximity coupling to a ferromagnetic layer. Only one direction of the proximity-induced magnetization is effective at opening a gap, namely, the direction parallel to $\hat{n}$. In our specific case, we need $M_z \neq 0$, which will yield the Hamiltonian

$$H_{\text{surf}} = v_F \left(p_x\sigma^y - p_y\sigma^x\right) + M_z\sigma^z, \tag{18.6}$$

which has the energy spectrum of a massive Dirac fermion: $E_\pm = \pm\sqrt{p_x^2 + p_y^2 + M_z^2}$ where we have set $v_F = 1$. A 2-D massive Dirac fermion is known to produce a half-quantum Hall effect with the Hall conductance $\sigma_{xy} = \frac{e^2}{2h}\text{sign}\, m$ (see chap. 15). If we open a consistent magnetization gap across the entire surface, then we should have a *topological* insulator on the surface with a half-Hall conductance. Unfortunately, we cannot measure this Hall conductance in transport because, with a uniform mass gap over the entire surface there is no place to attach metallic leads. However, there is a way around this: we can simply have a surface region coated with a magnetization parallel to $\hat{n}$ neighboring a region with magnetization antiparallel to $\hat{n}$ (see fig. 18.2). This will create an effective mass-domain wall in H_{surf} and lead to a set of gapless states propagating along the domain wall. In fact, we already know that a mass-domain wall for 2-D Dirac fermions leads to a set of chiral modes on the wall. Thus, we have a metallic region of the surface to which we can attach leads, but clearly this will simply measure $\sigma_{xy} = \pm e^2/h$ and not a half-integer value because we just have the same type of gapless state that lies on the edge of a $\nu = \pm 1$ integer quantum Hall state. What we have hence found is that a magnetic domain wall on the surface of a 3-D topological insulator mimics an integer quantum Hall state, which is a 2-D TR-breaking topological insulator [13].

Just as on the edge of two dimensions, we can also induce a gap in the surface states by breaking charge conservation symmetry in the presence of a proximity-coupled s-wave superconductor. If we coat the whole surface with a superconducting layer, then the entire surface will become gapped. In this case, the BdG surface Hamiltonian is

$$H_{\text{surf}}^{(\text{BdG})} = \frac{1}{2} \sum_{\mathbf{p}} \Psi_{\mathbf{p}}^{\dagger} \begin{pmatrix} p_x \sigma^y - p_y \sigma^x - \mu & \Delta i \sigma^y \\ -\Delta i \sigma^y & -p_x \sigma^y - p_y \sigma^x + \mu \end{pmatrix} \Psi_{\mathbf{p}}, \tag{18.7}$$

where $\Psi_{\mathbf{p}} = (c_{\mathbf{p}\uparrow}\ c_{\mathbf{p}\downarrow}\ c_{-\mathbf{p}\uparrow}^{\dagger}\ c_{-\mathbf{p}\downarrow}^{\dagger})$. This Hamiltonian has an energy spectrum

$$\pm E_{\pm} = \pm\sqrt{(p \pm \mu)^2 + |\Delta|^2}, \tag{18.8}$$

where $p = \sqrt{p_x^2 + p_y^2}$. If we can coat the surface with two separate superconductors that have a relative π phase shift (see fig. 18.2) then we will again have a type of mass domain wall. In this case, the domain is TR invariant, and there will be a Kramers' pair of counter-propagating fermionic states. Let us work this out explicitly. Assume that $\mu = 0$, $\Delta = \Delta(y)$ and that $\Delta(y > 0) = |\Delta_0| = -\Delta(y < 0)$. We can use the standard ansatz at $p_x = 0$ (remember p_x is a good quantum number, so we can solve for the bound states for each p_x independently):

$$\psi_0 = \exp\left[-\int_0^y \Delta(y)\right] \phi_0, \tag{18.9}$$

where ϕ_0 is a constant, four-component spinor. We act on this ansatz with $H_{\text{surf}}^{(\text{BdG})}$ to get the secular equation

$$\begin{aligned} H_{\text{surf}}^{(\text{BdG})} \psi_0 &= E\psi_0 = 0 \\ &\Longrightarrow \begin{pmatrix} 0 & -i\Delta(y) & 0 & \Delta(y) \\ -i\Delta(y) & 0 & -\Delta(y) & 0 \\ 0 & -\Delta(y) & 0 & -i\Delta(y) \\ \Delta(y) & 0 & -i\Delta(y) & 0 \end{pmatrix} \phi_0 = 0. \end{aligned}$$

This yields two possible solutions

$$\begin{aligned} \phi_0^{(1)} &= \frac{1}{\sqrt{2}} (0\ \ 1\ \ 0\ \ i)^T, \\ \phi_0^{(2)} &= \frac{1}{\sqrt{2}} (1\ \ 0\ \ -i\ \ 0)^T. \end{aligned} \tag{18.10}$$

For finite p_x we need to choose the combinations

$$\begin{aligned} \phi_0^{(+)} &= e^{i\pi/4} \left(i\phi_0^{(1)} + \phi_0^{(2)}\right), \\ \phi_0^{(-)} &= e^{i\pi/4} \left(-i\phi_0^{(1)} + \phi_0^{(2)}\right). \end{aligned} \tag{18.11}$$

By acting with the full $H^{(BdG)}_{surf}$ on $\psi^{(\pm)}_{p_x} = e^{ip_x x}\psi^{(\pm)}_0$, we see that these states have energies $E_\pm = \pm p_x$. We can also show that that $\psi^{(+)}_{p_x}$ is the TR partner of $\psi^{(-)}_{-p_x}$ and that $\psi^{(\pm)\dagger}_{p_x} = \psi^{(\pm)}_{-p_x}$, as required by a Majorana fermion. Thus we have a Kramers' pair of propagating Majorana fermions. We can compare this to the 2-D TR-invariant topological superconductor discussed earlier, and we see that the edge states of the 2-D superconductor match the domain-wall states for the π-junction on the topological insulator surface. So far, this is the only way to realize such a 2-D topological superconducting state.

The final such domain wall state is the novel Fu-Kane [83] proposal for realizing Majorana fermions on the surface of a 3-D topological insulator. The basic structure is to have a domain wall between a magnetic region and an s-wave superconductor region (see fig. 18.2. The resultant BdG Hamiltonian with magnetic and superconducting proximity effects is

$$H^{(BdG)}_{surf} = \frac{1}{2}\sum_{\mathbf{p}} \Psi^\dagger_{\mathbf{p}} \begin{pmatrix} p_x\sigma^y - p_y\sigma^x + M\sigma^z - \mu & |\Delta| i\sigma^y \\ -|\Delta| i\sigma^y & -p_x\sigma^y - p_y\sigma^x - M\sigma^z + \mu \end{pmatrix} \Psi_{\mathbf{p}}, \tag{18.12}$$

where $\Psi_{\mathbf{p}} = (c_{\mathbf{p}\uparrow} \;\; c_{\mathbf{p}\downarrow} \;\; c^\dagger_{-\mathbf{p}\uparrow} \;\; c^\dagger_{-\mathbf{p}\downarrow})$ and the global phase of Δ has been transformed into the fermion operators. To be explicit, let us assume that the magnetic proximity layer has a magnetization parallel to the surface normal so that $M > 0$. The energy spectrum for $\mu = 0$ is

$$\begin{aligned} E_1 &= \sqrt{p^2 + (|\Delta| + M)^2}, \\ E_2 &= \sqrt{p^2 + (|\Delta| - M)^2}, \\ E_3 &= -\sqrt{p^2 + (|\Delta| + M)^2}, \\ E_4 &= -\sqrt{p^2 + (|\Delta| - M)^2}. \end{aligned} \tag{18.13}$$

It is instructive to analyze this system coming from the limit of opposing magnetization domains. The magnetic case, which exhibits a chiral domain-wall state between the two magnetic regions. We want to begin turning on a finite pairing potential in one of the magnetic regions, for example, the one with $M > 0$. We can see from the dispersions in equation (18.13) that for small $|\Delta|$, the system remains gapped. However, we do see some interesting behavior. We know that the chiral domain-wall state is equivalent to two chiral-Majorana fermions. When a finite Δ is turned on, these two states become independent instead of being constrained to be the real and imaginary parts of a charge-conserving chiral fermion. One of these states is exponentially localized on the domain wall (in the transverse direction) within a length $\xi_1 = \frac{\hbar v_F}{||\Delta|+M|}$, whereas the other is bound within $\xi_2 = \frac{\hbar v_F}{||\Delta|-M|}$. Because $M > 0$, we see that as $|\Delta|$ increases from zero, ξ_1 decreases and ξ_2 increases. This means that one of the chiral-Majorana fermions is beginning to spread out in the bulk away from the domain. Eventually, when $|\Delta| = M$, ξ_2 diverges and there is a phase transition at a gap-closing point. Once $|\Delta| > M$, the system is again gapped, and there will be only one chiral-Majorana fermion remaining on the domain wall. We can then adiabatically tune M to zero in this region. This indicates that on a domain wall between a magnetic region and an s-wave superconductor, there will be a single chiral-Majorana fermion. Thus, this system mimics the 2-D TR breaking $p_x + ip_y$ topological superconductor.

We have seen that a combination of TR-breaking fields and superconductivity can be used to mimic the properties of a 2-D chiral p-wave superconductor but with TR symmetry preserved. We will now show that in a certain limit, the low-energy model of a topological insulator surface coupled to an s-wave superconductor essentially matches the model for

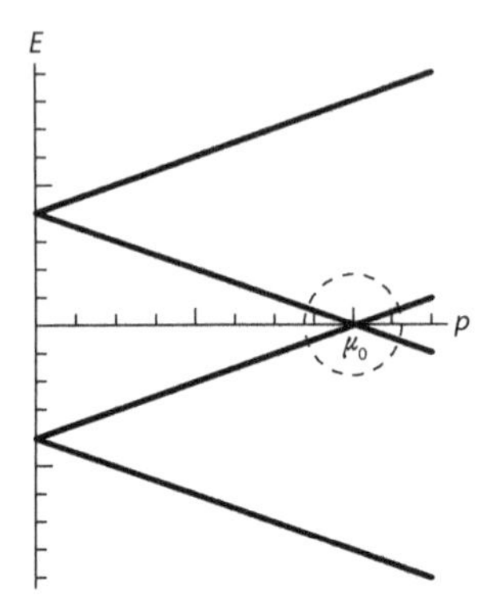

Figure 18.3. Energy spectra of E_1, E_2, E_3, E_4 from equation (18.14). The low-energy states used in the derivation of the effective Hamiltonian are encircled by the dotted line.

the spinless chiral p-wave superconductor that we studied in chapter 16. This argument is due to Fu and Kane [83]. Consider the Hamiltonian in equation (18.12) in the limit with $M = 0$ and $\mu \gg |\Delta|$. In fact, as an approximation, we will first consider the case $|\Delta| = 0$ and then use perturbation theory in Δ to look at the effective low-energy model. The low-energy quasiparticle spectrum in this limit has the eigenstates

$$\begin{aligned} E_1 &= p - \mu_0, \\ E_2 &= -p - \mu_0, \\ E_3 &= p + \mu_0, \\ E_4 &= -p + \mu_0, \end{aligned} \tag{18.14}$$

which are plotted in figure 18.3. We are going to focus on the low-energy physics at $E = 0$ and $p = \mu_0$. We will consider polar coordinates in momentum space $(p, \theta_{\mathbf{p}})$, and expand around the (ring-shaped) Fermi surface with $p = \mu_0$ to get the $E = 0$ wavefunctions

$$\begin{aligned} \chi^+ &= \frac{1}{\sqrt{2}} \left(-ie^{-i\theta_{\mathbf{p}}} \; 1 \; 0 \; 0\right)^T, \\ \chi^- &= \frac{1}{\sqrt{2}} \left(0 \; 0 \; -ie^{i\theta_{\mathbf{p}}} \; 1\right)^T. \end{aligned} \tag{18.15}$$

Now, we want to perturb these degenerate states by adding an s-wave-pairing potential, a finite momentum, i.e., $p \to \mu_0 + \delta p$, and a chemical potential shift, $\mu \to \mu_0 + \delta\mu$. The effective 2×2 Hamiltonian is

$$\begin{aligned} H^{(\text{eff})}_{\alpha\beta}(p, \theta_{\mathbf{p}}) &= \langle \chi^\alpha | \delta H^{(\text{Bdg})}_{\text{surf}}(p, \theta_{\mathbf{p}}) | \chi^\beta \rangle, \\ \delta H^{(\text{BdG})}_{\text{surf}}(p, \theta_{\mathbf{p}}) &= \frac{1}{2} \begin{pmatrix} -\delta\mu & -ie^{-i\theta_{\mathbf{p}}}\,\delta p & 0 & |\Delta| \\ ie^{i\theta_{\mathbf{p}}}\,\delta p & -\delta\mu & -|\Delta| & 0 \\ 0 & -|\Delta| & \delta\mu & ie^{i\theta_{\mathbf{p}}}\,\delta p \\ |\Delta| & 0 & -ie^{-i\theta_{\mathbf{p}}}\,\delta p & \delta\mu \end{pmatrix}. \end{aligned} \tag{18.16}$$

We find the result that

$$H^{(\text{eff})}(p, \theta_{\mathbf{p}}) = \begin{pmatrix} \delta p - \delta\mu & -i|\Delta|e^{-i\theta_{\mathbf{p}}} \\ i|\Delta|e^{i\theta_{\mathbf{p}}} & -\delta p + \delta\mu \end{pmatrix}, \tag{18.17}$$

which is essentially the Hamiltonian for a spinless chiral p-wave superconductor expanded around the Fermi surface at finite μ. One key difference between this Hamiltonian and the chiral superconductor studied earlier is that this Hamiltonian actually preserves TR symmetry. There seems to be some inconsistency because we know that if there are chiral edge states, then we can easily pick out a sense of TR breaking. However, if we go back and look at how we derived this effective model, we can see that we assumed that the s-wave superconductor covers the whole system, so in fact there are no chiral domain-wall states. Of course, there cannot be any chiral *edge* states in any sense because this is a surface Hamiltonian, and a surface has no edge. We already know how to generate the interesting domain-wall chiral-Majorana fermions by placing a superconducting region next to a magnetic region, but then we have explicitly broken time-reversal with the magnet, and everything is consistent. The state with only the superconductor is just waiting to be driven into either the chiral state by applying some TR breaking field or the helical state by engineering a π-phase junction in the surface superconductor.

Perhaps the most interesting property of the 2-D chiral superconductor is the existence of non-Abelian anyons (Majorana zero modes) bound to vortices. We may ask, Do such bound states exist in this hybrid version? The answer is yes, as we will now show. Assume that we have a vortex at $r = 0$ given by $\Delta = |\Delta(r)|e^{i\theta(r)}$ and a zero-mode ansatz:

$$\psi_0 = \exp\left[-\int_0^r |\Delta(r')|dr'\right]\begin{pmatrix} u \\ v \end{pmatrix}, \tag{18.18}$$

where u, v are constant two-component spinors. Note that the vortex configuration *does* break time-reversal, and thus we expect it should have similar features to vortices in a chiral superconductor. The BdG Hamiltonian in polar coordinates with $\mu = 0$, acting on ψ_0, leads to two two-component equations:

$$\begin{aligned} -i\,(\cos\theta\sigma^y - \sin\theta\sigma^x)\,(-|\Delta(r)|)u + i\sigma^y|\Delta(r)|e^{i\theta}v = 0, \\ -i\sigma^y|\Delta(r)|e^{-i\theta}u - i\,(-\cos\theta\sigma^y - \sin\theta\sigma^x)\,(-|\Delta(r)|)v = 0. \end{aligned} \tag{18.19}$$

We can divide by $|\Delta(r)|$ and multiply both equations by $-i\sigma^y$ to get

$$\begin{aligned} (\cos\theta + i\sin\theta\sigma^z)\,u + e^{i\theta}v = 0 \\ e^{-i\theta}u + (\cos\theta - i\sin\theta\sigma^z)\,v = 0. \end{aligned} \tag{18.20}$$

These equations are easy to solve, and we find that

$$\psi_0 = \frac{i}{\sqrt{2}}\exp\left[-\int_0^r |\Delta(r')|dr'\right]\begin{pmatrix} 1 \\ 0 \\ -1 \\ 0 \end{pmatrix}. \tag{18.21}$$

The field operator that creates a mode in this state is

$$\gamma^\dagger = \int d^2x \frac{i}{\sqrt{2}}\exp\left[-\int_0^r |\Delta(r')|dr'\right]\left(c_\uparrow^\dagger(r,\theta) - c_\uparrow(r,\theta)\right), \tag{18.22}$$

from which we can see that $\gamma^\dagger = \gamma$. Thus, we find that we can create Majorana bound states on the surface of a proximity-coupled 3-D topological insulator.

As an aside, it is worth briefly mentioning what happens if we go up one higher dimension to consider the (4+1)-D TR-invariant topological insulator. This system, which is described by a massive Dirac equation with two spins and two orbitals, has an integer classification and thus supports chiral fermions on the boundary. The boundary is $3+1$ dimensions, and the chiral fermions are simply the conventional Weyl fermions in $3+1$ dimensions. The boundary Hamiltonian for a boundary perpendicular to the fourth spatial direction is simply

$$H_{\mathrm{bdry}}(\mathbf{p}) = p_x\sigma^x + p_y\sigma^y + p_z\sigma^z, \tag{18.23}$$

where σ^a is spin. We see that we again have gapless fermions, and we can attempt to use the same tricks from lower dimensions to mimic $(3+1)$-D topological insulators and superconductors on the surface of the (4+1)-D parent insulator. That is, we want to open a gap on the boundary and create domain walls. Because we have chiral fermions, this is not as easy to do. If we add any magnetic coupling $m_a\sigma^a$, we see that no gap is induced. This means that we cannot get a quantum Hall–type state, nor a chiral superconductor–type state because we cannot induce a TR breaking mass. It turns out this is consistent because there are no isotropic TR-breaking topological insulators and superconductors in $(3+1)$ dimensions. However, one trick still works, and we can induce an s-wave superconducting gap, which couples like

$$H_{\mathrm{bdry}}^{(\mathrm{BdG})}(\mathbf{p}) = \frac{1}{2}\begin{pmatrix} p_a\sigma^a - \mu & i\sigma^y|\Delta| \\ -i\sigma^y|\Delta| & p_a(\sigma^a)^T + \mu \end{pmatrix}, \tag{18.24}$$

where $a = x, y, z$. If we have a π-junction on the surface, we will now generate a domain-wall theory with (2+1)-D Majorana fermions propagating on the domain. This mimics the (3+1)-D topological superconductor in the DIII class, i.e., the same class to which ^{3}He-B belongs.

18.3 Problems

1. *Competing and Compatible Mass Terms in the Dirac Hamiltonian:* Consider a generic massless Dirac Hamiltonian

$$H_D = \sum_{a=1}^{d} p_a\Gamma^a, \tag{18.25}$$

where $d+1$ is the space-time dimension and Γ^a are Dirac matrices with the property that $\{\Gamma^a, \Gamma^b\} = 2\delta^{ab}$ (which also implies that $(\Gamma^a)^2 = 1$). The energy spectrum of this Hamiltonian is $E_\pm = \pm|p|$, which is gapless. Now, suppose we want to add a mass-perturbation term to the Hamiltonian $H_m = m_1\Lambda^1$, where $\{\Lambda^1, \Gamma^a\} = 0$ for all a and $(\Lambda^1)^2 = 1$. Calculate the energy spectrum and show that the energy spectrum is gapped. (*Hint*: Consider $(H_D + H_m)^2$). Next, suppose that we turn on a mass-domain wall in m_1 in the x^1-direction, so that $m_1 = m_1(x_1)$. Write and simplify the general matrix equation that must be solved for zero modes bound to this domain wall at $p_2 = p_3 \ldots p_d = 0$).

Now, let us consider adding a second mass perturbation so that $H_m = m_1\Lambda^1 + m_2\Lambda^2$, where $\{\Lambda^2, \Gamma^a\} = 0$ for all a and $(\Lambda^2)^2 = 1$. Let us focus on the two possibilities: (1) $\{\Lambda^1, \Lambda^2\} = 0$ and (2) $[\Lambda^1, \Lambda^2] = 0$. Case 1 is called *compatibile* mass terms, and case 2 is called *competing* mass terms. Calculate the energy spectra of $H_D + H_m$ as a function of m_1 and m_2 for cases 1 and 2. Draw a "phase" diagram as a function of m_1 and m_2 that shows the places in the m_1, m_2-plane where the spectrum is gapless for cases 1 and 2. For case 2, argue that if at

$x_1 = -\infty$, $H_m = \Lambda^1$, and at $x_1 = +\infty$, $H_m = \Lambda^2$ (for all values of the other x_i), then there must be a zero-mode bound state somewhere in between. Write and simplify the matrix equation for the domain-wall zero mode(s) at $p_2 = p_3 \ldots p_d = 0$. For case 1 argue that there will be no zero-mode state.

2. *Classification of Mass Terms for Coupled QSH Edge States:* Suppose we bring the edges of two different systems exhibiting the QSH effect into close proximity so that we permit tunneling and interactions. The Hamiltonian for such a system, where the upper edge of one QSH sample is coupled to the lower edge of a separate sample, in the decoupled limit is just

$$H_{\text{edge}} = p\,(\tau^z \otimes \sigma^z), \tag{18.26}$$

where σ^a indicates spin and τ^z indicates upper and lower edges. Why must we use τ^z and not just $\mathbb{I}_{2\times 2}$ in the tensor product? Show that this Hamiltonian is TR invariant for $T = \mathbb{I}_{2\times 2} \otimes i\sigma^y K$, where K is complex conjugation. Write down mass terms corresponding to TR-invariant interedge tunneling (i.e., real tunneling matrices without spin flips) and TR-breaking intraedge Zeeman terms. Draw schematic pictures indicating the physical meaning of the T-breaking magnetic mass terms. Show that the tunneling term *competes*, i.e., commutes in the matrix sense with at least one magnetic term. Add both a homogeneous tunneling term and a homogenous competing magnetic term and solve for the energy spectrum. When is the spectrum gapless? Finally, find the eigenstate for the domain-wall zero-mode bound state between a region where tunneling dominates and a region where the competing magnetic mass dominates.

APPENDIX

3-D Topological Insulator in a Magnetic Field

As in the case of the Chern insulator in a magnetic field, we would like to analyze the problem of a 3-D topological insulator in a magnetic field. The question we would like to have answered is the following: once we break time reversal in a 3-D topological insulator by applying a strong magnetic field, is there still a way of telling whether the insulator was nontrivial or trivial in zero field? Although the lattice expansion of a topological insulator with Peierls substitution is impossible to solve analytically in the presence of a magnetic field, the small k-expansion around the point of minimum gap is manageable. Our calculation should be then valid for topological insulators very close to the transition point, with small gap. We perform a small k-expansion of a topological insulator and place a magnetic field in the z-direction. We quantize the k_x, k_y momenta perpendicular to the field in terms of creation and annihilation operators. We use the shorthand notation, $M_0(k_z) = M - B_1 k_z^2$, where M and B_1 are constants in the same notation and with the same values in [84] and where we also introduced a Zeeman coupling g, which acts on the basis states described in [84]. Define constants $\epsilon_\pm = \pm M_0(k_z)$ and $B_\pm = 2(D_2 \pm B_2)/l^2$, whose values are given in [84]. l is the magnetic length. For an integer $n > 1$ ($|n\rangle$ being the Harmonic oscillator wavefunction), we can again choose the wavefunction $(\phi_1, \phi_2, \phi_3, \phi_3) = (\theta_1|n-1\rangle, \theta_2|n-1\rangle, \theta_3|n\rangle, \theta_4|n\rangle)$, and the energies

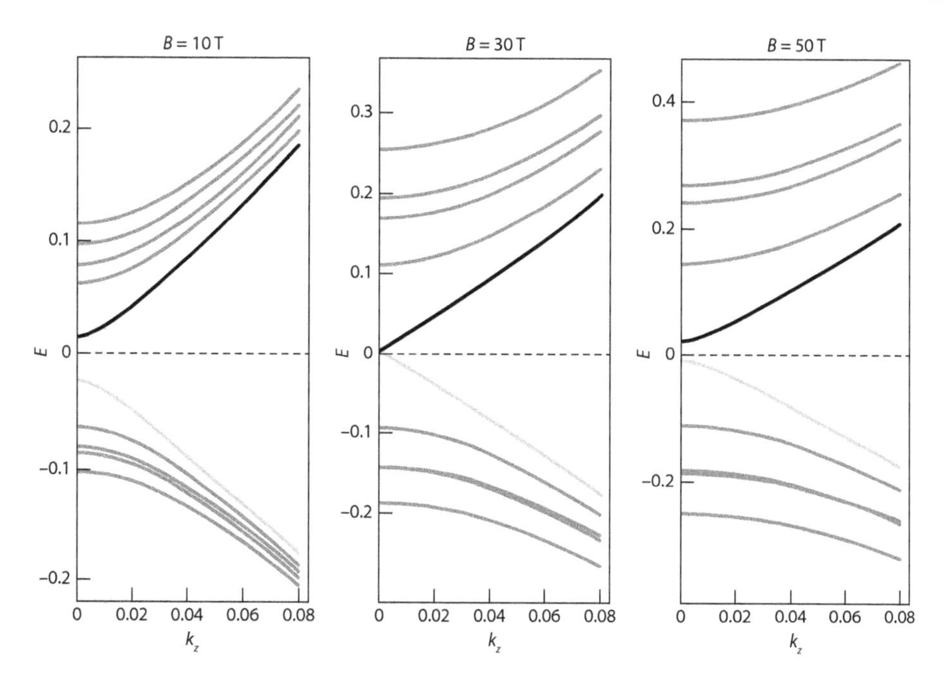

Figure A.1. Evolution of the Landau levels with increasing magnetic field for the topologically nontrivial case. The g-factor was chosen to be zero, and the mass gap was chosen to be $\frac{1}{10}$ of that in [84], all other constants being identical to those in the reference. The gap is chosen at $\frac{1}{10}$ of the real value to allow for the level crossing at experimentally reachable magnetic fields.

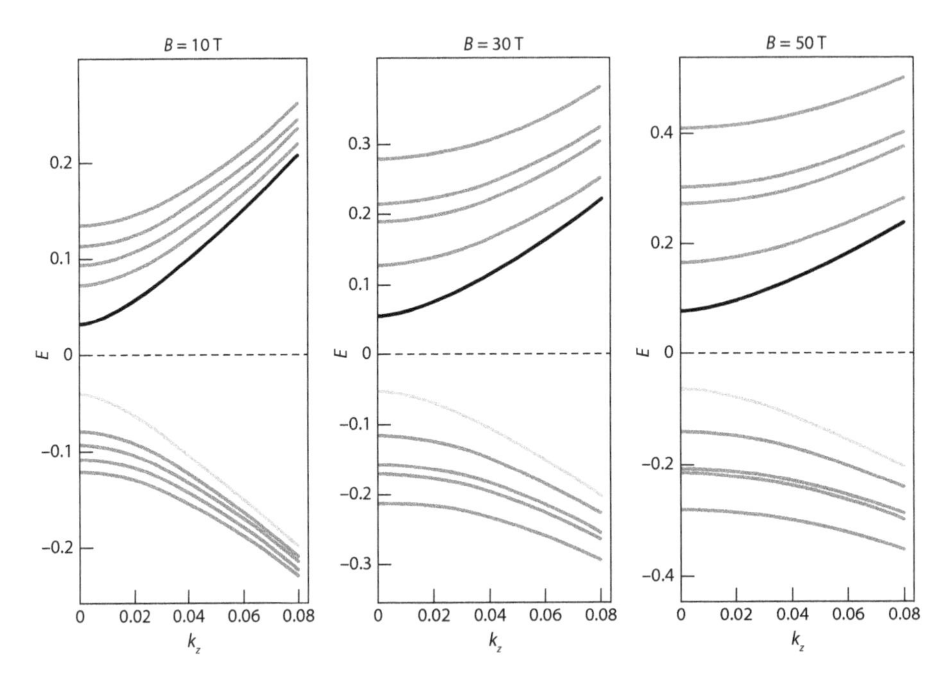

Figure A.2. Evolution of the Landau levels with increasing magnetic field for the topologically trivial case. The g-factor was chosen to be zero, and the mass gap was chosen to be $-1/10$ of that of [84], all other constants being identical.

are obtained by diagonalizing the matrix:

$$\begin{pmatrix} \epsilon_+ + B_-\left(n-\frac{1}{2}\right)+ & A_1k_z & 0 & A_2\frac{\sqrt{2}}{l}\sqrt{n} \\ A_1k_z & \epsilon_- + B_+\left(n-\frac{1}{2}\right) & A_2\frac{\sqrt{2}}{l}\sqrt{n} & 0 \\ 0 & A_2\frac{\sqrt{2}}{l}\sqrt{n} & \epsilon_+ + B_-\left(n+\frac{1}{2}\right) & -A_1k_z \\ A_2\frac{\sqrt{2}}{l}\sqrt{n} & 0 & -A_1k_z & \epsilon_- + B_+\left(n+\frac{1}{2}\right) \end{pmatrix} \tag{A.1}$$

A Zeeman term $gB\sigma$ can easily be added. The matrix can be diagonalized, but unfortunately, the full expressions for the energies are too long to be written down—they are solutions of a fourth-order equation. Interesting things again take place in the $n = 0$ level. If we take the Hamiltonian equation (A.1) and diagonalize it for $n = 0$, we get four energies, which is wrong—-the system has only two $n = 0$ states, similar to the previous case. The reason is that in this case we have $(\phi_1, \phi_2, \phi_3, \phi_3) = (0, 0, \theta_3|n\rangle, \theta_4|n\rangle)$, so we need only to diagonalize a 2-by-2 Hamiltonian: we find the characteristic equation to be

$$\left(\epsilon_+ + \frac{B_-}{2}\right)\theta_3 - A_1k_z\theta_4 = E\theta_3, \qquad \left(\epsilon_- + \frac{B_+}{2}\right)\theta_4 - A_1k_z\theta_3 = E\theta_4, \tag{A.2}$$

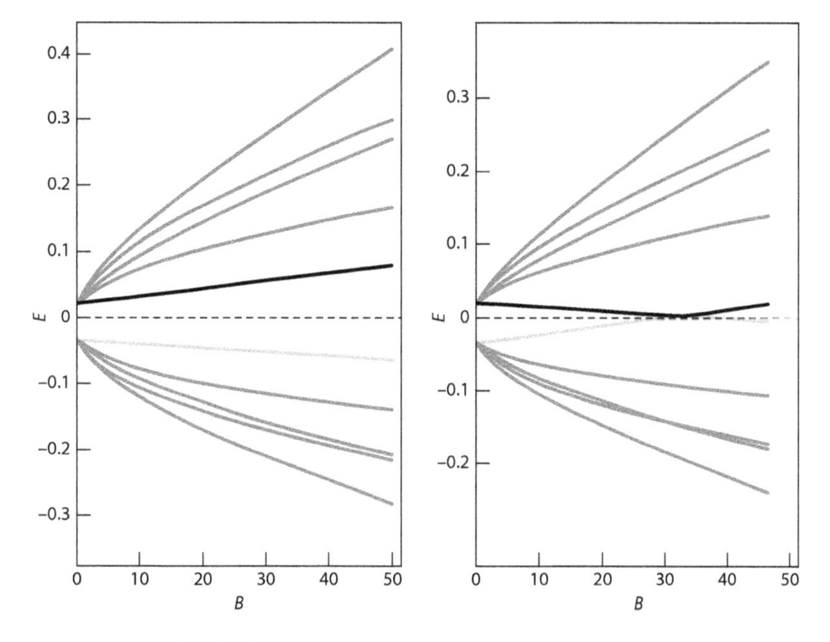

Figure A.3. Landau-level behavior as a function of magnetic field B (only a few Landau levels are plotted) for a 3-D topological insulator at $k_z = 0$. Left: topologically trivial insulator. Right: topologically nontrivial insulator.

which gives the two energies

$$E_{\pm} = \frac{(\epsilon_+ + \frac{B_-}{2}) + (\epsilon_- + \frac{B_+}{2}) \pm \sqrt{\left((\epsilon_+ + \frac{B_-}{2}) - (\epsilon_- + \frac{B_+}{2})\right)^2 + 4A_1^2 k_z^2}}{2} \tag{A.3}$$

The dependence of these energies on k_z and B is given in figure A.1, A.2 and A.3, where the topological side is easily distinguishable from the trivial one.

There is a level crossing at $k_z = 0$ and $\epsilon_+ + B_-/2 - \epsilon_- + B_+/2 = 0$, which is identical to the condition: $M = B_2/l^2$. This condition obviously requires $M/B_2 > 0$; this is the same condition that guarantees the existence of a nontrivial topological insulator. At $M = B_2/l^2$, the closing point is a 1-D Dirac point.

References

[1] C. Nayak, S. H. Simon, A. Stern, M. Freedman, and S. D. Sarma, "Non-abelian anyons and topological quantum computation," *Rev. Mod. Phys.*, vol. 80, p. 1083, 2008.

[2] A. Y. Kitaev, "Fault-tolerant quantum computation by anyons," *Annals Phys.*, vol. 303, pp. 2–30, 2003.

[3] F. D. M. Haldane, "Model for a quantum Hall effect without Landau levels: Condensed-matter realization of the "parity anomaly"," *Phys. Rev. Lett.*, vol. 61, p. 2015, 1988.

[4] C. L. Kane and E. J. Mele, "Quantum spin Hall effect in graphene," *Phys. Rev. Lett.*, vol. 95, p. 226801, 2005.

[5] C. L. Kane and E. J. Mele, "z_2 topological order and the quantum spin Hall effect," *Phys. Rev. Lett.*, vol. 95, p. 146802, 2005.

[6] B. Bernevig and S. C. Zhang, "Quantum spin Hall effect," *Phys. Rev. Lett.*, vol. 96, p. 106802, 2006.

[7] B. A. Bernevig, T. L. Hughes, and S. C. Zhang, "Quantum spin Hall effect and topological phase transition in HgTe quantum wells," *Science*, vol. 314, p. 1757, 2006.

[8] M. König, S. Wiedmann, C. Brüne, A. Roth, H. Buhmann, L. Molenkamp, X.-L. Qi, and S.-C. Zhang, "Quantum spin Hall insulator state in HgTe quantum wells," *Science*, vol. 318, pp. 766–770, 2007.

[9] L. Fu and C. L. Kane, "Topological insulators with inversion symmetry," *Phys. Rev. B*, vol. 76, no. 4, p. 045302, 2007.

[10] D. Hsieh, D. Qian, L. Wray, Y. Xia, Y. S. Hor, R. J. Cava, and M. Z. Hasan, "A topological Dirac insulator in a quantum spin Hall phase," *Nature*, vol. 452, p. 970, 2008.

[11] A. P. Schnyder, S. Ryu, A. Furusaki, and A. W. W. Ludwig, "Classification of topological insulators and superconductors in three spatial dimensions," *Phys. Rev. B*, vol. 78, p. 195125, 2008.

[12] A. Y. Kitaev, "Periodic table for topological insulators and superconductors," *Adv. in Theo. Phys.: AIP Conf. Proc.*, vol. 1134, p. 22, 2009.

[13] X.-L. Qi, T. L. Hughes, and S.-C. Zhang, "Topological field theory of time-reversal invariant insulators," *Phys. Rev. B*, vol. 78, p. 195424, 2008.

[14] L. Fu and C. L. Kane, "Time reversal polarization and a z_2 adiabatic spin pump," *Phys. Rev. B*, vol. 74, no. 19, p. 195312, 2006.

[15] J. E. Moore and L. Balents, "Topological invariants of time-reversal-invariant band structures," *Phys. Rev. B*, vol. 75, no. 12, p. 121306, 2007.

[16] R. Roy, "Topological phases and the quantum spin Hall effect in three dimensions," *Phys. Rev. B*, vol. 79, p. 195322, 2009.

[17] E. Prodan, "Robustness of the spin-Chern number," *Phys. Rev. B*, vol. 80, p. 125327, 2009.

[18] E. Prodan, "Noncommutative tools for topological insulators," *New J. Phys*, vol. 12, p. 065003, 2010.

[19] A. M. Essin, J. E. Moore, and D. Vanderbilt, "Magnetoelectric polarizability and axion electrodynamics in crystalline insulators," *Phys. Rev. Lett.*, vol. 102, p. 146805, 2009.

[20] M. Freedman, C. Nayak, K. Shtengel, K. Walker, and Z. Wang, "A class of p, t-invariant topological phases of interacting electrons," *Ann. Phys.*, vol. 310, p. 428, 2004.

[21] A. Kitaev and J. Preskill, "Topological entanglement entropy," *Phys. Rev. Lett.*, vol. 96, p. 110404, 2006.

[22] M. Levin and X.-G. Wen, "Detecting topological order in a ground state wave function," *Phys. Rev. Lett.*, vol. 96, p. 110405, 2006.

[23] H. Li and F. D. M. Haldane, "Entanglement spectrum as a generalization of entanglement entropy: Identification of topological order in non-Abelian fractional quantum Hall effect states," *Phys. Rev. Lett.*, vol. 101, p. 010504, 2008.

[24] S. Ryu and Y. Hatsugai, "Entanglement entropy and the Berry phase in the solid state," *Phys. Rev. B*, vol. 93, p. 245115, 2006.

[25] N. Bray-Ali, L. Ding, and S. Haas, "Topological order in paired states of fermions in two dimensions with breaking of parity and time-reversal symmetries," *Phys. Rev. B*, vol. 80, p. 180504(R), 2009.

[26] S. T. Flammia, A. Hamma, T. L. Hughes, and X.-G. Wen, "Topological entanglement Renyi entropy and reduced density matrix structure," *Phys. Rev. Lett.*, vol. 103, p. 261601, 2009.

[27] R. Thomale, A. Sterdyniak, N. Regnault, and B. A. Bernevig, "The entanglement gap and a new principle of adiabatic continuity," *Phys. Rev. Lett.*, vol. 104, p. 180502, 2010.

[28] R. Thomale, D. P. Arovas, and B. A. Bernevig, "Non-local order in gapless systems: Entanglement spectrum in spin chains," *e-print arxiv*, p. 0912.0028, 2009.
[29] F. Pollmann, A. M. Turner, E. Berg, and M. Oshikawa, "Entanglement spectrum of a topological phase in one dimension," *Phys. Rev. B*, vol. 81, p. 064439, 2010.
[30] A. M. Turner, Y. Zhang, and A. Vishwanath. arxiv: 0909.3119.
[31] L. Fidkowski, "Entanglement spectrum of topological insulators and superconductors," *Phys. Rev. Lett.*, vol. 104, p. 130502, 2010.
[32] M. Kargarian and G. A. Fiete. arxiv: 1005.3815.
[33] E. Prodan, T. L. Hughes, and B. A. Bernevig, "Entanglement spectrum of a disordered topological Chern insulator," *Phys. Rev. Lett.*, vol. 105, p. 115501, 2010.
[34] A. Sterdyniak1, N. Regnault, and B. A. Bernevig, "Extracting excitations from model state entanglement," *Phys. Rev. Lett.*, vol. 106, p. 100405, 2011.
[35] M. V. Berry, "Quantal phase factors accompanying adiabatic changes," *Proc. R. Soc. Lond.*, vol. 392, p. 45, 1984.
[36] X.-G. Wen and A. Zee, "Winding number, family index theorem, and electron hopping in magnetic fields," *Nucl. Phys. B(FS)*, vol. B136, p. 641, 1989.
[37] D. J. Thouless, M. Kohmoto, P. Nightingale, and M. den Nijs, "Quantized Hall conductance in a two-dimensional periodic potential," *Phys. Rev. Lett.*, vol. 49, p. 405, 1982.
[38] Y. Hatsugai, "Chern number and edge states in the integer quantum Hall effect," *Phys. Rev. Lett.*, vol. 71, p. 3697, 1993.
[39] A. J. Niemi and G. W. Semenoff, "Axial-anomaly-induced fermion fractionization and effective gauge-theory actions in odd-dimensional space-times," *Phys. Rev. Lett.*, vol. 51, p. 2077, 1983.
[40] Xiao-Liang Qi, Yong-Shi Wu, and Shou-Cheng Zhang, "Topological quantization of the spin Hall effect," *Phys. Rev.*, vol. B74, p. 085308, 2006.
[41] Y. A. Bychkov and E. I. Rashba, "Oscillatory effects and the magnetic susceptibility of carriers in inversion layers," *J. Phys. C*, vol. 17, p. 6039, 1984.
[42] Y. Yao, F. Ye, X. L. Qi, S. C. Zhang, and Z. Fang. cond-mat/0606350.
[43] H. Min et al., "Intrinsic and Rashba spin-orbit interactions in graphene sheets," *Phys. Rev. B*, vol. 74, p. 165310, 2006.
[44] A. Pfeuffer-Jeschke. Ph.D. Thesis, University of Wurzburg, Germany, 2000.
[45] C. Wu, B. A. Bernevig, and S. C. Zhang, "Helical liquid and the edge of quantum spin Hall systems," *Phys. Rev. Lett.*, vol. 96, p. 106401, 2006.
[46] D. N. Sheng, L. Sheng, Z. Y. Weng, and F. D. M. Haldane, "Spin Hall effect and spin transfer in disordered Rashba model," *Phys. Rev. B*, vol. 72, p. 153307, 2005.
[47] R. Resta and David Vanderbilt, "Theory of polarization: A modern approach" in *Physics of Ferroelectrics: A modern perspective*, ed. K. M. Rabe, C. H. Ahns, and J. M. Triscone, 31 1/N 68. Berlin: Springer Verlag, 2007.
[48] S. Murakami, "Gap closing and universal phase diagrams in topological insulators," *Physica E-Low-Dimensional Systems & Nanostructures*, vol. 43, pp. 748–754, January 2011. International Symposium on Nanoscience and Quantum Physics, Tokyo, Japan, February 23–25, 2009.
[49] M. Creutz, "Aspects of chiral symmetry and the lattice," *Rev. Mod. Phys.*, vol. 73, p. 119, 2001.
[50] R. B. Laughlin, "Anomalous quantum Hall effect: An incompressible quantum fluid with fractionally charged excitations," *Phys. Rev. Lett.*, vol. 50, pp. 1395–1398, May 1983.
[51] S. Adler, "Axial-vector vertex in spinor electrodynamics," *Phys. Rev.*, vol. 177, p. 2426, 1969.
[52] J. S. Bell and R. Jackiw, "A pcac puzzle: $\pi_0 \to \gamma\gamma$ in the σ-model," *Nuovo Cimento A*, vol. 60, p. 47, 1969.
[53] T. L. Hughes, E. Prodan, and B. A. Bernevig, "Inversion-symmetric topological insulators," *Phys. Rev. B*, vol. 83, p. 245132, 2011.
[54] A. M. Turner, Y. Zhang, R. S. K. Mong, and A. Vishwanath, "Quantized response and topology of insulators with inversion symmetry." arXiv:1010.4335.
[55] D. Hsieh, Y. Xia, R. J. Cava, and M. Z. Hasan, "Observation of time-reversal-protected Single-Dirac-cone topological-insulator states in Bi(2)Te(3) and Sb(2)Te(3)," *Phys. Rev. Lett.*, vol. 103, October 2, 2009.
[56] Y. Xia, D. Qian, R. J. Cava, and M. Z. Hasan, "Observation of a large-gap topological-insulator class with a single Dirac cone on the surface," *Nature Physics*, vol. 5, pp. 398–402, June 2009.
[57] Y. S. Hor, A. Yazdani, M. Z. Hasan, N. P. Ong, and R. J. Cava, "*p*-type Bi(2)Se(3) for topological insulator and low-temperature thermoelectric applications," *Physical Review B*, vol. 79, May 2009.
[58] D. Hsieh, R. J. Cava, and M. Z. Hasan, "Observation of unconventional quantum spin textures in topological insulators," *Science*, vol. 323, pp. 919–922, February 13, 2009.

[59] D. Hsieh, R. J. Cava, and M. Z. Hasan, "A topological Dirac insulator in a quantum spin Hall phase," *Nature*, vol. 452, pp. 970–U5, April 24, 2008.
[60] H. Zhang, X. Dai, Z. Fang, S.-C. Zhang, et al., "Topological insulators in Bi(2)Se(3), Bi(2)Te(3) and Sb(2)Te(3) with a single Dirac cone on the surface," *Nature Physics*, vol. 5, pp. 438–442, June 2009.
[61] Y. L. Chen, J. G. Analytis, X. Dai, Z. Fang, S. C. Zhang, I. R. Fisher, Z. Hussain, Z.-X. Shen, et al., "Experimental realization of a three-dimensional topological insulator, Bi(2)Te(3)," *Science*, vol. 325, pp. 178–181, July 10, 2009.
[62] A. N. Redlich, "Parity violation and gauge noninvariance of the effective gauge field action in three dimensions," *Phys. Rev. D*, vol. 29, p. 2366, 1984.
[63] E. Fradkin, E. Dagotto, and D. Boyanovsky, "Physical realization of the parity anomaly in condensed matter physics," *Phys. Rev. Lett.*, vol. 57, pp. 2967–2970, 1986.
[64] J. Bardeen, L. N. Cooper, and J. R. Schrieffer, "Theory of superconductivity," *Phys. Rev.*, vol. 108, p. 1175, 1957.
[65] M. Tinkham, *Introduction to Superconductivity: Second Edition*. Dover Publications, 2004.
[66] A. Y. Kitaev, "Unpaired Majorana fermions in quantum wires," *Phys.-Usp.*, vol. 44, p. 131, 2001.
[67] R. Jackiw and C. Rebbi, "Solitons with fermion number 1/2," *Phys. Rev. D*, vol. 13, p. 3398, 1976.
[68] G. E. Volovik, "Fermion zero modes on vortices in chiral superconductors," *JETP Lett.*, vol. 70, p. 609, 1999.
[69] N. Read and D. Green, "Paired states of fermions in two dimensions with breaking of parity and time-reversal symmetries and the fractional quantum Hall effect," *Phys. Rev. B*, vol. 61, p. 10267, 2000.
[70] D. A. Ivanov, "Non-Abelian statistics of half-quantum vortices in *p*-wave superconductors," *Phys. Rev. Lett.*, vol. 86, p. 268, 2001.
[71] G. E. Volovik, *The Universe in a Helium Droplet*. Oxford University Press, USA, 2003.
[72] M. Stone and R. Roy, "Edge modes, edge currents, and gauge invariance in px + ipy superfluids and superconductors," *Phys. Rev. B*, vol. 69, p. 184511, 2004.
[73] N. B. Kopnin and M. M. Salomaa, "Mutual friction in superfluid He-3: Effects of bound states in the vortex core," *Phys. Rev. B*, vol. 44, p. 9667, 1991.
[74] G. Moore and N. Read, "Non-Abelions in the fractional quantum Hall effect," *Nucl. Phys. B*, vol. 360, p. 362, 1991.
[75] C. Nayak and F. Wilczek, "$2n$ quasihole states realize $2(n-1)$ dimensional spinor braiding statistics in paired quantum Hall states," *Nucl. Phys. B*, vol. 479, p. 529, 1996.
[76] A. J. Leggett, "A theoretical description of the new phases of liquid He-3," *Rev. Mod. Phys.*, vol. 47, p. 331, 1975.
[77] R. Roy, "Topological invariants of time reversal invariant superconductors." arxiv:condmat/0608064 (unpublished), 2006.
[78] X.-L. Qi, T. L. Hughes, S. Raghu, and S.-C. Zhang, "Time-reversal-invariant topological superconductors and superfluids in two and three dimensions," *Phys. Rev. Lett.*, vol. 102, p. 187001, 2009.
[79] R. Roy, "Topological superfluids with time reversal symmetry." arxiv:0803.2868 (unpublished), 2008.
[80] Alexei Kitaev. Unpublished.
[81] A. Altland and M. R. Zirnbauer, "Nonstandard symmetry classes in mesoscopic normal-superconducting hybrid structures," *Phys. Rev. B*, vol. 55, p. 1142, 1997.
[82] X.-L. Qi, T. L. Hughes, and S.-C. Zhang, "Fractional charge and quantized current in the quantum spin Hall state," *Nat. Phys.*, vol. 4, p. 273, 2008.
[83] L. Fu and C. L. Kane, "Superconducting proximity effect and Majorana fermions at the surface of a topological insulator," *Phys. Rev. Lett.*, vol. 100, p. 096407, 2008.
[84] Haijun Zhang et al., "Topological insulators at room temperature." arXiv: 08121622.

Index

www.ingramcontent.com/pod-product-compliance
Ingram Content Group UK Ltd.
Pitfield, Milton Keynes, MK11 3LW, UK
UKHW010047010325
455718UK00005B/43